D1271419

Map Projections

Map Projections

A Reference Manual

LEV M. BUGAYEVSKIY

JOHN P. SNYDER

UK Taylor & Francis Ltd, 4 John St, London WC1N 2ET
USA Taylor & Francis Inc., 1900 Frost Road, Suite 101, Bristol PA19007

British Library Cataloguing in Publication Data

A catalogue record for this book is available from the British Library.

ISBN 0 7484 0303 5 (cased)
0 7484 0304 3 (paper)

Library of Congress Cataloging in Publication Data are available

Cover design by Youngs.

Typeset by Santype International Ltd., Salisbury, Wilts.

Printed in Great Britain by Burgess Science Press on paper which has a specified pH value on final manufacture of not less than 7.5 and is therefore 'acid free'.

Contents

Symbols

The following symbols are used frequently enough to merit general listing. Occasionally, some are used for other purposes. Symbols used temporarily are not listed here.

a	as linear value: (1) equatorial radius or semimajor axis of the ellipsoid of revolution, or (2) maximum scale factor at point on map. As angle: azimuth in spherical (or spheroidal) polar coordinate system.
b	(1) for ellipsoid of revolution, the polar radius or semiminor axis, or (2) minimum scale factor at point on map. For triaxial ellipsoid, the equatorial radius at a right angle to the semimajor axis a.
c	for triaxial ellipsoid, the polar radius or semiaxis.
e	(1) for ellipsoid of revolution, the (first) eccentricity of the ellipse defined by dimensions a and b, where $e = [(a^2 - b^2)/a^2]^{1/2}$; for triaxial ellipsoid, the eccentricity of the ellipse forming the equator; or (2) a Gaussian coefficient (see section 1.3.1).
e'	for ellipsoid of revolution, the second eccentricity, where $e' = e/(1 - e^2)^{1/2}$.
f	(1) function of the following parenthetical argument such as ϕ, or (2) a Gaussian coefficient (see section 1.3.1).
h	(1) elevation of a given point above the surface of the reference ellipsoid, or (2) a Gaussian coefficient (see section 1.3.2).
ln	natural logarithm, or logarithm to base e, where e = 2.718 28 ....
M	radius of curvature of meridian at a given point on the ellipsoid (see section 1.1.1 and Appendix 2).
m	linear scale factor along meridian (see section 1.4.1).
N	radius of curvature of ellipsoid surface in a plane orthogonal to the meridian (see section 1.1.1 and Appendix 2).
n	linear scale factor along parallel of latitude (see section 1.4.1).
n'	an auxiliary function of ellipsoid semiaxes a and b, namely $n' = (a - b)/(a + b)$.
p	area scale factor.
q	isometric latitude.
r	radius of parallel of latitude ϕ, where $r = N \cos \phi$.
S	area of portion of surface of ellipsoid or sphere.
s	linear distance along surface of sphere or ellipsoid.
u, v	coordinate system for intermediate use.
v	distortion variable (see sections 1.6.4 and 1.6.5).

X, Y, Z — three-dimensional axes with center at a given point on the surface of the sphere or ellipsoid, the Y-axis in the meridian plane pointing to the North Pole, the Z-axis coinciding with the normal to the ellipsoid surface, and the X-axis positive east from this point.

X_G, Y_G, Z_G — three-dimensional axes for the sphere or ellipsoid, with the center at the center of the ellipsoid, and the X_G-, Y_G-, and Z_G-axes increasing in the direction of the Greenwich meridian in the equatorial plane, the meridian 90°E in the equatorial plane, and the North Pole, respectively.

x — rectangular coordinate: distance to the right of the vertical line (Y-axis) passing through the origin or center of a projection (if negative, it is distance to the left).

x_ϕ — partial derivative $\partial x/\partial \phi$; similarly for y_λ, etc.

$x_{\phi\phi}$ — second derivative $\partial^2 x/\partial \phi^2$, etc.

y — rectangular coordinate: distance above the horizontal line (X-axis) passing through the origin or center of a projection (if negative, it is distance below).

z — (1) spherical angle from selected point on surface of sphere or ellipsoid to some other point on surface, as viewed from center, or (2) rectangular coordinate in direction of Z-axis.

γ — convergence of meridian (deviation from direction of Y-axis on map).

δ — azimuth in plane polar coordinates.

ε — deviation of graticule intersection from right angle on map.

λ — longitude east of Greenwich (for longitude west of Greenwich, use a minus sign), or east of the central meridian in many formulas.

λ_0 — longitude east of Greenwich of the central meridian of the map, or of the origin of rectangular coordinates (for west longitude, use a minus sign), often assumed to be zero.

μ — linear scale factor (not restricted by direction if no subscripts).

μ_1 — linear scale factor along a vertical (or great circle passing through pole of projection).

μ_2 — linear scale factor along an almucantar (or small circle centered on pole of projection).

ρ — radius in plane polar coordinates.

Σ — area of portion of surface on map. Sometimes used for a sum.

Φ — planetocentric latitude.

ϕ — north geodetic or geographic latitude (if latitude is south, apply a minus sign).

ϕ_0 — central latitude or origin of y coordinates at central meridian.

ω — maximum angular deformation.

Preface

This manual discusses the theory and practical problems of map projections, their development, and their application to the design of various maps of the Earth and celestial bodies. Briefly and systematically, the authors have tried to provide the results of many years of research and of the experience of scholars from many different countries in this field, as well as a blended summary of the experience of the two authors studying the subject under widely contrasting cultures but with considerable common interest.

This experience is relevant to any nation or individual making maps, and it is appropriate that Russian and American approaches be brought together in this first joint international textbook from two countries that have been the source of the most extensive research in recent decades on the subject of map projections.

It should be noted that several of the statements given cannot be found in any other cartographic source. The book also provides in detail the information necessary in the search for and the selection, calculation, and application of projections for specific tasks and for the solution of a large group of cartographic, navigational, and other problems, using maps designed for various purposes.

This book can be used as a precise practical manual or handbook by scientists and engineers as well as cartographers in solving practical problems in cartography or in other fields of science and commerce using the techniques of mathematical cartography, maps, and space imagery. It should be useful for the training of college and university students in cartographic specialties. The mathematical approaches and the treatment of various problems should stimulate the development of creativity and related thinking of future engineers in scientific fields concerned with space and cartography.

By attempting to provide a thorough presentation of the theoretical and practical problems of map projections, the authors hope that this book will make it possible for undergraduate and graduate students, university faculty, and specialists improving their cartographic skills to study this subject with considerable thoroughness and to carry out independent research based on it.

The book was written by Lev M. Bugayevskiy* and John P. Snyder†. It is largely based on the book *Kartograficheskiye Proyektsii – Spravochnoye Posobiye* (*Cartographic Projections – A Reference Manual*), by Lev M. Bugayevskiy and Lyubov' A. Vakhrameyeva (now deceased) and published in Moscow in 1992, but parts of the book have been extensively revised by both the present authors. The

* Professor, Moscow State University of Geodesy and Cartography (MIIGAiK), Moscow, Russia.
† Retired from the US Geological Survey, Reston, Virginia. Present address: 17212 Friends House Road, Sandy Spring, Maryland 20860, USA.

second author's goal, as requested by the first author, has been to add pertinent Western material for balance and to correct some of the impressions inadvertently given in the Russian text about Western projections, without detracting from the role of the text in discussing the important Russian contributions to mathematical cartography.

The translation of all Russian materials into English was made by Drs. Igor P. Gerasimov and Sophia A. Butyugina with participation of Mr. Stephen K. Tecku. Professors Andrey M. Portnov and Kira B. Shingareva, two of the first author's colleagues at MIIGAiK, have provided valuable assistance in communications between the authors. For guidance to and liaison with the publisher, the authors thank Professors Mark Monmonier and Duane F. Marble of the USA.

Transliteration of Russian words and names into English follows the style suggested by the US Board on Geographic Names. Geometric figures are those from the Russian text, with changes in some lettering to use English instead of Russian letters and to orient the X- and Y-axes in accordance with US conventions. Most Russian map illustrations have been replotted using the second author's own computer program, except where the mathematical basis of the Russian illustration is unclear to both authors. Numerous other maps have been added to those in the Russian edition.

Owing to various problems of communication, both in transmitting information and in the fact that neither author can reasonably speak or write in the language of the other, thus requiring translation, much of the editing and many of the additions by the second author have not been thoroughly checked by the first author. On the other hand, several of the Russian sources referenced were not available to the second author, but they may be useful to a number of readers. Each of the authors has found some errors in the Russian edition as published, apart from the second author's added material. The responsibility for any remaining errors may be due to either author or to both. It is hoped that careful checking within each author's abilities has minimized this problem.

LEV M. BUGAYEVSKIY
JOHN P. SNYDER

Introduction

The successful solution to many problems related to commerce and science is based on a wide use of maps of various scales, contents, and purposes.

Maps have a number of features and advantages. Among them is a one-to-one correspondence of elements of nature and society and their representation on the map. Maps serve as a basis for measuring and obtaining various qualitative and quantitative characteristics necessary for solving scientific and commercial problems. Maps have highly important informational and cognitive properties. They can be used for objective investigation into specific problems in various fields of economics and science (navigation, land use and forest organization, organization of private and governmental facilities, etc.).

The capacity to solve these problems is based on the fact that positions on maps can be determined on a strictly mathematical basis, the study and elaboration of which is the subject of mathematical cartography. Mathematical cartography deals with the theory of map projections, map scales and their variation, the division of maps into sets of sheets, and nomenclature, as well as with problems of making measurements and carrying out the investigation of various phenomena from maps. Mathematical cartography also includes the study of map projections which make use of geodetic measurements and the development of graphical methods for solving problems of spherical trigonometry and astronomy, marine navigation and aeronavigation, and even crystallography. It is also the basis for developing the theory of methods and techniques for map design.

The main objectives of mathematical cartography are

- the development of the theory, and, above all, working out new methods of research on map projections, including the 'best' and 'ideal' ones;
- research into various map projections, their nature, properties, and capacity for practical application;
- improvement of the available map projections, their unification and standardization; development of new map projections satisfying the requirements of science and industry for the compilation of various maps, including thematic and special maps, as well as the processing of geodetic measurements and the solution of geodetic and applied engineering problems;
- developing and using the algorithm and program for a given map projection, and incorporating these and subsequent improvements of computational techniques into computer software libraries;
- development of other mathematical elements of maps (determination of scale variation and the design of maps, and division into sheets according to map purpose, contents, and other preset requirements);
- development of methods and means for performing various measurements from

maps, considering the map projection properties and including methods of reading computerized cartographic information;
- research on and solutions to problems of mathematical map design (e.g. methods of map projection transformation using various equipment including computers with peripherals; methods of making map graticules on automated and non-automated coordinatographs);
- development of technical aids to take measurements from maps;
- development of the theory and methods of automation in mathematical cartography.

The main element of a map's mathematical basis is a map projection. Its properties influence the choice of the nominal scale and design of the map, and in turn determine a close interrelationship of all the map elements.

This book is devoted to this main element, and hence to mathematical cartography, i.e. the theory of map projections.

A brief history of the development of map projections

The science of map projection was 'born' more than two thousand years ago when Greek scientists started to depict the Earth and celestial sky with the help of meridians and parallels (see the works by Anaximander, Hipparchus, Apollonius, and Eratosthenes).

The detailed work of about AD 150 by Claudius Ptolemy, *Geography*, was of great importance to cartography. Along with his descriptions of methods for map design and the determination of the Earth's dimensions, he dealt with map projection.

The Renaissance witnessed an intensive development of cartography. It was an epoch of great geographical discoveries, which were in need of accurate, reliable maps for the governing countries – for military campaigns, and for developing trade and navigation. Only a mathematical basis and the results of various surveys could lead to such maps.

An important event in the further development of map projections was the creation of geographical atlases by Dutch cartographers Abraham Ortelius and Gerardus Mercator in the late sixteenth century. Mercator is most famous, however, because of his development of the map projection bearing his name, and which is still in use for navigational charts.

At that time the trapezoidal projection and an oval projection by Peter Apian were widely adopted for maps of the world and of large regions. These projections were prototypes for pseudocylindrical projections which were developed later. Jean Cossin of Dieppe was apparently the first to suggest a new sinusoidal pseudocylindrical projection for world maps in the late sixteenth century.

Further development of map projections in Central and Western Europe

The eighteenth century marked the beginning of measurements of the ellipsoidal shape of the Earth, and of regular topographic surveys of large regions, starting with France. With these advances and the application of important new mathematics, especially logarithms and the calculus, new and more accurate maps were created. In cartography, important projections suggested or promoted by J. H. Lambert, Rigobert Bonne, J. L. Lagrange, Leonhard Euler, Guillaume De l'Isle, and others were introduced or were improvements on earlier prototypes.

The early nineteenth century witnessed the design of large-scale military maps, their mathematical basis being of great importance. In 1825 C. F. Gauss was the first to solve the general problem of the conformal transformation of one surface onto another. It became the basis for developing a whole class of conformal projections.

Map projections by G. B. Airy and A. R. Clarke in England in the 1860s utilized the principle of least squares developed by Gauss and Legendre. Simpler projections such as an equal-area conic by H. C. Albers, an equal-area elliptical pseudocylindrical by C. B. Mollweide, both in 1805 in Germany, and a cylindrical projection by James Gall in Scotland in 1855, caught the attention of commercial mapmakers. In the later nineteenth century, works appeared by N. A. Tissot, who had solved the general theory of distortion on map projections.

In the early twentieth century, Germany and England dominated the field with innovative map projections. In Germany, Max Eckert, Oswald Winkel, Hans Maurer, and Karlheinz Wagner made numerous contributions, especially of pseudocylindrical projections. A. E. Young, C. F. Close, and John Bartholomew used new approaches in England, both theoretical and practical. Numerous interpretive textbooks also appeared during the nineteenth and especially the twentieth centuries in many of the languages of Europe. It was not until the work of D. H. Maling (1960) that Russian contributions to map projections became known in any detail in the West.

Use of map projections in Russia

In the late seventeenth century, Russia produced maps with a graticule of meridians and parallels (the map of Russia by F. Godunov, G. Gerrits, I. Massa, and N. Vitsen).

Systematic surveys to meet the requirements of the fleet and to design a general map of Russia were started upon the orders of Peter I ('the Great') in the eighteenth century. The projections for these maps included cylindrical, trapezoidal (rectilinear pseudocylindrical), stereographic, and conic.

The *Atlas Vserossiyskoy Imperii* (*Atlas of the All-Russia Empire*) by I. Kirilov was published in 1734. A majority of its maps were compiled on equidistant conic projections with two standard parallels.

Further development of the theory of map projections is closely connected with the activity of Rossiyskaya Akademiya Nauk or the Russian Academy of Sciences. The work of the Geographical Department of the Academy resulted in publishing the *Atlas Rossiyskiy* (*Atlas of Russia*) in 1745. It included a general map of Russia and 19 maps of European and Asian parts of the country. All the maps were designed on the trapezoidal and equidistant conic projections.

The most significant progress in cartography in the second half of the eighteenth century was connected with the name of M. V. Lomonosov, who attached great significance to the mathematical basis of maps. Under his leadership in the

Geographical Department of the Academy of Sciences, maps of the whole world and Russia were compiled (including the region of the Arctic Ocean). They were designed on conformal cylindrical, oblique stereographic, and normal azimuthal equidistant projections. In the first volume of the *Trudy* (*Proceedings*) of the St Petersburg Academy of Sciences, dated 1777, there were three works by Leonhard Euler devoted to the problems of the theory of map projections. The general theory of conformal transformation of the surface of the sphere onto a plane, as well as the theory of equal-area projections with an orthogonal map graticule were developed for the first time. Research by F. I. Shubert followed the work of Euler in the field of mathematical cartography.

General boundary surveys provided cartographers with much detailed cartographic material in the second half of the eighteenth century. This material served as the basis for a series of maps for a hundred-sheet *Polrobnoy Karty Rossiyskoy Imperii* (*Detailed Map of the Russian Empire*). All the maps were designed to a scale of 20 versts to an inch (1:840000).

Military geodesists, cartographers, and astronomers (Shubert, A. P. Bolotov, N. Ya. Tsinger, and others) studied the problems of map design and its mathematical basis in the late eighteenth and early nineteenth centuries. The most significant products of Russian military cartography are 10 verst special maps of the western part of Russia, referred to as Shubert's map, and a 3 verst military topographic map of western Russia. They were both based on the Bonne projection. Soon thereafter a 10 verst special map of European Russia, designed on a conformal conic projection, was published, followed by a 10 verst map of western Siberia.

The theory of map projections was thoroughly considered in *Kurs Vysshey i Nizshey Geodesii* (*Course on Simple and Higher Geodesy*) written by A. P. Bolotov in 1849. He was the first in Russia to suggest a theory of conformal transformation from one surface onto another (according to Gauss). He also compared the Bonne pseudoconic projection, which was widely used for map design at that time, with a conformal conic projection. The latter was recommended for use in designing maps of Russia and its regions.

In 1848 a special commission under the Military Topographic Corps adopted the Müffling polyhedric projection for large-scale topographic maps of Russia. On this projection the Earth's surface is represented by quadrangles bounded by arcs of meridians and parallels.

Medium- and small-scale maps of Russia and its regions were based on conformal conic projections at that time. Maps of other countries, continents, and the world were based on oblique stereographic, pseudoconic, azimuthal equidistant, and conformal cylindrical projections.

A new stage in the development of Russian mathematical cartography is connected with the name of the famous Russian mathematician P. L. Chebyshev, who in 1856 formulated the theorem on the best conformal map projection for a given region. D. A. Grave proved the Chebyshev theorem in 1896 and carried out considerable research on the theory and practice of making equal-area and other map projections. Russian mathematicians A. N. Korkin and A. A. Markov also conducted research into the best map projections.

By the end of the nineteenth century, mathematical cartography, as a branch of surveying, had been included in the curricula of higher technical establishments and departments of physics and mathematics of Russian universities. In the early twentieth century D. I. Mendeleyev, F. N. Krasovskiy, A. A. Mikhaylov, N. Ya. Tsinger, and others conducted research in the field of map projection theories. An important event in the development of the theory was the book *Kartografiya (Cartography)* by V. V. Vitkovskiy, published in 1907.

A decree establishing the Vysshego Geodesicheskogo Upravleniya (VGU) (Higher Geodetic Department), signed by V. I. Lenin in 1919, was another important step in the development of cartography in Russia. The 1920s saw the solution to the problem of a mathematical basis for new topographic maps (in metric scales). These maps were designed on the Müffling projection used earlier.

In 1921 Krasovskiy developed two original equidistant conic projections designed for small-scale maps of the entire country and of the European portion. One of these projections was well suited to represent the country's territory, a projection that received wide acceptance in Russia. It was referred to as the Krasovskiy projection. The advantage of this projection was obvious: with it nearly 90 percent of the country's territory has an area distortion less than 1.5 percent; many of the maps of the USSR were designed on this projection.

It eventually became necessary to introduce a unified projection and system of orthogonal coordinates for various kinds of topographic projects and for the processing of geodetic measurements made throughout the country. Hence, geodesists and cartographers worked intensively to choose projections, and to establish scales, sectional map systems, and nomenclature for the maps of the USSR.

In 1928, at the Third Geodetic Convention, the Gauss–Krüger projection was adopted. Its introduction in terms of a system of 6° zones of longitude for the entire territory of the USSR was an important accomplishment of Soviet geodesy and cartography.

Remarkable progress was achieved in the design of various maps and atlases at the beginning of the 1930s. At that time the work on the design of school maps on oblique perspective cylindrical projections developed by M. D. Solov'ev was started. A million copies were published.

The Central Research Institute of Geodesy, Aerophotography, and Cartography (TsNIIGAiK) created a special group on mathematical cartography to carry out research in this field. Outstanding scientists N. A. Urmayev, V. V. Kavrayskiy, and Solov'ev headed this research.

It should be noted that various kinds of topographic and geodetic work are carried out through the combined efforts of military and civilian specialists within many countries, including the former USSR and the United States. The advantages of this cooperation within each country were clearly seen during the Second World War.

After the war, atlases of map projections and cartographic tables were published in the USSR. They were of great help to cartographers when choosing and calculating projections.

Besides the above scientists, research in mathematical cartography was conducted by N. M. Volkov, G. A. Ginzburg, A. P. Yushchenko, G. A. Meshcheryakov, A. S. Lisichanskiy, F. A. Starostin, T. D. Salmanova, A. V. Gedymin, A. K. Malovichko, G. I. Konusova, V. M. Boginskiy, and others. A harmonious system of methods for creating new map projections capable of producing not only individual projections but a great number of sets of various projections based on the practical needs (see Chapter 6) were developed. Among them are methods for producing projections based on solving Euler–Urmayev and Tissot–Urmayev equations, as well as methods for producing derivative projections and projections approximating sketches of map graticules.

Soviet scientists paid considerable attention to the application and further development of the Chebyshev–Grave theorems and proposals for the best conformal projections. Urmayev considered this intricate problem of mathematical cartography from a new point of view in his work devoted to methods for producing new map projections. They were further developed in the works by V. V. Kavrayskiy, P. D. Belonovskiy, A. I. Dinchenko, N. Ya. Vilenkin, L. A. Vakhrameyeva, L. M. Bugayevskiy, Lisichanskiy, Konusova, A. I. Shabanova, and A. A. Pavlov. Research dealing with the concept of best projections with various kinds of distortion was conducted by Meshcheryakov, M. A. Topchilov, Yu. M. Yuzefovich, Ye. N. Novikova, and others. For a summary of this additional work by Russian and other scientists, the reader is referred to Chapter 11.

Use of map projections in the United States

The principal contribution to map projections during the nineteenth century in the United States was the ordinary polyconic projection devised by Swiss-born F. R. Hassler in about 1820. The first director of the Survey of the Coast established in 1807 by President Thomas Jefferson, Hassler called the projection the best for mapping US coastal data. When the US Geological Survey (USGS) was formed in 1879, that bureau also adopted the polyconic projection for the thousands of topographic quadrangles to be prepared detailing the land portions of the United States. These were prepared at scales of 1:62 500 and ultimately at 1:24 000 in most regions.

The polyconic projection continued to be used by the US Coast and Geodetic Survey (USC&GS; successor to the Survey of the Coast) until the increased use of the Mercator projection in the early twentieth century. The polyconic was also used by the USGS until the 1950s, when newly mapped quadrangles began to be based on the conformal projection of the State Plane Coordinate System (SPCS) zone applying to the particular region.

Although Christopher Colles had published a perspective conic projection in the United States in 1794, this was not further used. The Lambert conformal conic projection, presented in Germany in 1772, was adopted by the USC&GS between 1918 and 1935 for maps of the entire United States, for maps of the individual states at a scale of 1:500 000, and for grid zones of the SPCS in states extending predominantly east to west.

The transverse Mercator was adopted by the USC&GS for grids in SPCS zones in states predominantly north to south, and by the Army Map Service for the Universal Transverse Mercator (UTM) grid system extending over the entire world in zones 6° wide in longitude. The Albers equal-area conic projection was applied to many other maps of the entire country or to sections by the USGS as published in the *National Atlas* in 1970.

In addition to these projections dominant in government mapping, several others were used to a smaller extent by governmental agencies, but more commonly by commercial publishers of world atlases, such as Rand McNally and Hammond. In addition, O. M. Miller and W. A. Briesemeister of the American Geographical

Society devised projections in the 1940s and 1950s which were often used for their own maps and for those published by others. These included the bipolar oblique conic conformal, the Miller oblated stereographic, the Miller cylindrical, and the Briesemeister elliptical projections.

Numerous other projections were developed by various individuals in US academic, commercial, and governmental environments. While many have been only slightly used, a few became popular. A. J. van der Grinten's circular world map projection presented in 1904 was used by the US Department of Agriculture and for 60 years by the National Geographic Society for its sheet maps and world atlases, until its replacement in 1988 by a pseudocylindrical projection developed by A. H. Robinson of the University of Wisconsin in 1963, actually for Rand McNally, but little known until the 1988 adoption. J. P. Goode of the University of Chicago promoted his interrupted projections, especially the homolosine, beginning in 1925.

Governmental research on map projections in the United States was formerly confined largely to the USC&GS during the late nineteenth century but also especially under mathematician O. S. Adams in the early decades of the twentieth century with the issuance of numerous 'Special Publications'. Included were projections involving complicated elliptic functions as well as the more straightforward Lambert and Albers conic projections applied to the Earth as an ellipsoid. F. W. McBryde also developed several pseudocylindrical projections with the USC&GS and independently. During the 1980s J. P. Snyder of the USGS wrote several papers and books on the subject, developing a few new projections. Extended academic research was most notable under W. R. Tobler of the University of Michigan during the 1960s and 1970s.

Measurement of the Earth's shape

Although the Earth was assumed to be spherical prior to the eighteenth century, Isaac Newton and others concluded that the Earth should be an oblate ellipsoid of revolution, slightly flattened at the poles. The French, under the Cassini family, were beginning a detailed survey of France and concluded, based on measurements of the meridian through Paris but within France, that the Earth was elongated at the poles. To resolve the disparity, the French Academy of Sciences sent survey teams to Peru and Lapland to measure meridians at widely separated latitudes. This established the validity of Newton's conclusions and led to numerous meridian measurements in various locations during the next two centuries.

Nineteenth-century calculations of the Earth's shape were made by geodesists such as Airy and George Everest of England in 1830, F. W. Bessel of Germany in 1841, and Clarke of England in 1866 and 1880. These values are still used in parts of the world for precision mapping; the United States began using the Clarke 1866 ellipsoid in 1880. In the twentieth century came many additional measurements, especially leading to the Hayford or international ellipsoid adopted by the International Union of Geodesy and Geophysics (IUGG) in 1924, and the Krasovskiy ellipsoid adopted by the Soviet Union in 1940.

The launching of artificial satellites by the Soviet Union and the United States led to satellite geodesy and new measurements of the Earth's shape. Most prominent have been the US Defense Mapping Agency's series of World Geodetic

Systems (WGS), the most recent being WGS 84. Independently measured is the Geodetic Reference System 1980 or GRS 80 ellipsoid, for which the values were adopted by IUGG in 1979. The GRS 80 and WGS 84 have dimensions which may be assumed to be identical for all mapping purposes. Because several of these reference ellipsoids are currently in use, the more prominent are listed with their dimensions in Appendix 6.

1

General theory of map projections

The physical surfaces of the Earth and of celestial bodies have complex shapes. In order to represent them on a plane it is necessary to move from a physical surface to a mathematical one, very close to the former. The latter can be described with corresponding equations.

The surface of the globe may be taken mathematically as a sphere, as an ellipsoid of revolution, or in particular cases as a triaxial ellipsoid. In theoretical studies, the problems of depicting more complex regular and irregular surfaces are considered.

When studying map projections, one assumes that the parameters of the approximating surfaces (see Appendix 6), initial geodetic data (and analogous values), and the methods of transformation from physical surfaces to the approximating surfaces, usually developed from the results of astronomical and gravimetrical works, are known.

1.1 Coordinate systems used in mathematical cartography

Curvilinear, three-dimensional rectangular, plane rectangular, and plane polar coordinate systems are used in the theory of map projections. Their specific expressions depend to some extent on the shape of the surface being depicted; that is why we will consider first the basic coordinate systems. They are generally associated with the transformation of the surface of the sphere and of the ellipsoid of revolution. We shall also consider coordinate systems for the triaxial ellipsoid.

1.1.1 Geographical and geocentric coordinate systems

Numerous sets or families of parametric lines can be established on the surface of an ellipsoid (or sphere). They can be assumed to be corresponding systems of curvilinear coordinates.

On the surface of an ellipsoid of revolution, geographic coordinates coincide with both geodetic and astronomical coordinates. The geodetic latitude is an angle formed by the normal to the surface of the ellipsoid at a given point and the equatorial plane. The geodetic longitude is a dihedral angle formed by the plane of a given meridian and the plane of the prime meridian.

The astronomical latitude is an angle between the direction of a plumb line (the normal to the surface of the geoid, an undulating surface corresponding to mean sea level) and the plane perpendicular to the Earth's axis of rotation. The astronomical longitude is a dihedral angle between planes of astronomical meridians.

On the surface of the sphere, spherical coordinates with the pole of the coordinate system at the geographical pole are referred to as geographical, if determined like geodetic coordinates.

Let us take families of geographical parallels of latitude ϕ and meridians of longitude λ forming a geographical coordinate system.

Draw a normal AO' to the surface (Figure 1.1) at a random point $A(\phi, \lambda)$ on an ellipsoid.

We choose two main sections out of an innumerable set of the normal sections possible on the ellipsoid: the one that coincides with the plane of meridian PAP' is called a meridian section, and the one orthogonal to the former is called the section of the first vertical.

The radii of curvature of these normal sections respectively equal (see Appendix 2)

$$M = a(1 - e^2)/(1 - e^2 \sin^2 \phi)^{3/2} \tag{1.1}$$

$$N = a/(1 - e^2 \sin^2 \phi)^{1/2} \tag{1.2}$$

where $e = [(a^2 - b^2)/a^2]^{1/2}$, a, and b are respectively the first eccentricity and the major and minor semiaxes of the ellipsoid of revolution.

Now, let us determine a spatial geocentric coordinate system $OX_G Y_G Z_G$ for which the origin coincides with the center of mass of the Earth (actually with the center of the ellipsoid of revolution), the Z_G-axis is in the direction of the Earth's North Pole, the X_G-axis is in the direction of the intersection of the Greenwich meridian and the equator, and the Y_G-axis is directed to the east (a quasigeocentric coordinate system, if its origin coincides with the center of the reference ellipsoid).

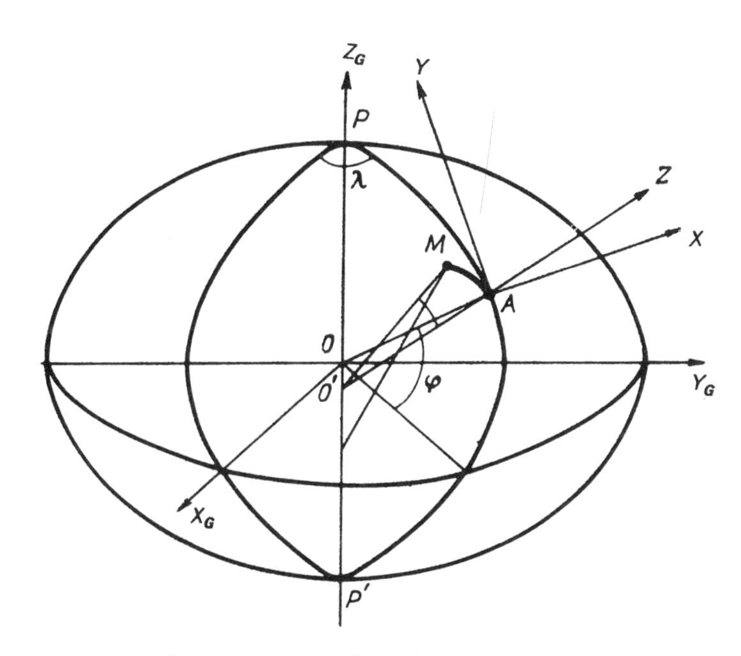

Figure 1.1 Geocentric and topocentric spatial coordinate systems.

Taking into account equation (1.2), the relationship between the geocentric and geographical coordinate systems can be written as

$$X_G = (N + h)\cos\phi\cos\lambda, \quad Y_G = (N + h)\cos\phi\sin\lambda$$
$$Z_G = [N(1 - e^2) + h]\sin\phi \tag{1.3}$$

where h is the elevation of point C' above the surface of the ellipsoid along the normal to the surface (usually it is assumed in the theory of map projections that $h = 0$).

The reverse transition from geocentric to geodetic coordinates is made using known geodetic formulas. We should note that equations (1.3) take the form of equations of an arbitrary surface at $h \neq 0$; where $h = 0$ at every point of the surface, we have the surface of the ellipsoid of revolution.

1.1.2 Topocentric horizon coordinate system

The topocentric horizon coordinate system is the system for which the origin coincides with point $Q_0(\phi_0, \lambda_0, H_0)$ lying at any arbitrary (current) point in space, the Y-axis is in the plane of the meridian of point Q and in the direction of the North Pole, the Z-axis coincides with normal O_1Q to the ellipsoid surface at point Q, and the X-axis completes the system (Figure 1.2).

The relationship between the topocentric and geocentric coordinate system can be written in the following form:

$$\begin{pmatrix} X_G \\ Y_G \\ Z_G \end{pmatrix} = A\begin{pmatrix} X \\ Y \\ Z + N_0 + H_0 \end{pmatrix} - \begin{pmatrix} 0 \\ 0 \\ e^2 N_0 \sin\phi_0 \end{pmatrix}$$
$$\begin{pmatrix} X \\ Y \\ Z \end{pmatrix} = A'\begin{pmatrix} X_G \\ Y_G \\ Z_G + e^2 N_0 \sin\phi_0 \end{pmatrix} - \begin{pmatrix} 0 \\ 0 \\ N_0 + H_0 \end{pmatrix} \tag{1.4}$$

where A is the matrix of coordinate transformation

$$A = \begin{pmatrix} -\sin\lambda_0 & -\cos\lambda_0 \sin\phi_0 & \cos\lambda_0 \cos\phi_0 \\ \cos\lambda_0 & -\sin\lambda_0 \sin\phi_0 & \sin\lambda_0 \cos\phi_0 \\ 0 & \cos\phi_0 & \sin\phi_0 \end{pmatrix} \tag{1.5}$$

Here A' is the transpose of matrix A, and N_0 is the radius of sectional curvature of the first vertical at the point of new pole $Q(\phi_0, \lambda_0)$.

Taking into account (1.3)–(1.5), the formulas for the coordinates of the topocentric horizon take the form

$$\begin{aligned} X &= (N + h)\cos\phi\sin(\lambda - \lambda_0) \\ Y &= (N + h)[\sin\phi\cos\phi_0 - \cos\phi\sin\phi_0\cos(\lambda - \lambda_0)] \\ &\quad + e^2(N_0 \sin\phi_0 - N\sin\phi)\cos\phi_0 \\ Z &= (N + h)[\sin\phi\sin\phi_0 + \cos\phi\cos\phi_0\cos(\lambda - \lambda_0)] \\ &\quad + e^2(N_0\sin\phi_0 - N\sin\phi)\sin\phi_0 - (N_0 + H_0) \end{aligned} \tag{1.6}$$

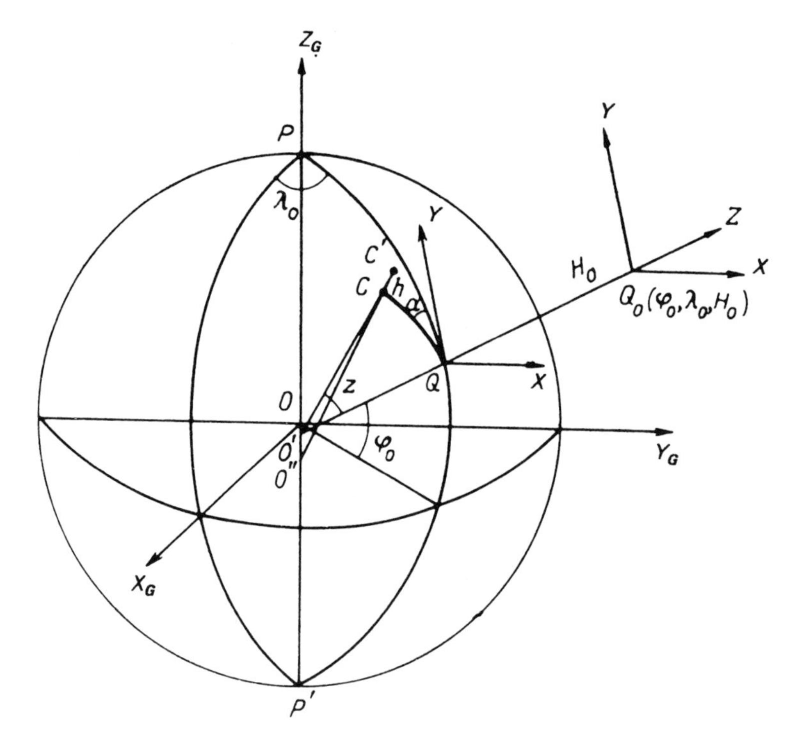

Figure 1.2 Topocentric horizon and spheroidal (or spherical) polar coordinate systems.

1.1.3 Polar spheroidal or spherical coordinate systems

In the polar spheroidal coordinate system (z, a), pole $Q(\phi_0, \lambda_0)$ is at a selected point, a is the angle between normal planes at the pole Q, and z is the angle between normal $O'Q$ and the direction to a given point on the surface of ellipsoid C as seen from point O' (see Figure 1.2).

With $CO' = N'_0$, from Figure 1.2 we can get

$$X = N'_0 \sin z \sin a, \quad Y = N'_0 \sin z \cos a, \quad Z = N'_0 \cos z - N_0 \tag{1.7}$$

Taking into account (1.2), (1.6), and (1.7), the lengths of the sides of triangle $CO'O''$ are

$$CO'' = N, \quad CO' = N'_0, \quad O'O'' = (N \sin \phi - N_0 \sin \phi_0)e^2$$

and setting $h = 0$, we have

$$N'_0 = N_0\left[1 - \frac{e^2}{4}(\sin \phi - \sin \phi_0)^2\left(1 + \frac{e^2}{4}(5 \sin^2 \phi + \sin^2 \phi_0 + 2 \sin \phi \sin \phi_0 - 4)\right) + \cdots\right] \tag{1.8}$$

$$N = N'_0\left[1 + e^2(\sin \phi - \sin \phi_0)\left(\sin \phi + \frac{e^2}{2}[\sin \phi \sin \phi_0(\sin \phi + \sin \phi_0) + (\sin \phi - \sin \phi_0)(3 \sin^2 \phi - 1)]\right) + \cdots\right]$$

Then, relating the formulas for polar spheroidal coordinates (z, a) and geographical coordinates (ϕ, λ), we have

$$\sin z \cos a = \frac{N}{N'_0} [\sin \phi \cos \phi_0 - \cos \phi \sin \phi_0 \cos(\lambda - \lambda_0)] + e^2\left(\frac{N_0}{N'_0} \sin \phi_0 - \frac{N}{N'_0} \sin \phi\right)\cos \phi_0$$

$$\sin z \sin a = \frac{N}{N'_0} \cos \phi \sin(\lambda - \lambda_0) \tag{1.9}$$

$$\cos z = \frac{N}{N'_0} [\sin \phi \sin \phi_0 + \cos \phi \cos \phi_0 \cos(\lambda - \lambda_0)] + e^2\left(\frac{N_0}{N'_0} \sin \phi_0 - \frac{N}{N'_0} \sin \phi\right)\sin \phi_0$$

To the accuracy of terms up to e^2 we can also write

$$\sin z \cos a = t_1 + e^2\tau\left((t_1 \sin \phi - \cos \phi_0) + \frac{e^2}{2} (t_1 t_2 - 2t_3 \sin \phi_0)\right) + \cdots$$

$$\sin z \sin a = t_4\left[1 + e^2\tau\left(\sin \phi + \frac{e^2}{2} t_2\right)\right] + \cdots \tag{1.10}$$

$$\cos z = t_5 + e^2\tau\left((t_5 \sin \phi - \sin \phi_0) + \frac{e^2}{2} (t_2 t_5 - 2t_3 \sin \phi_0)\right) + \cdots$$

where

$$t_1 = \sin \phi \cos \phi_0 - \cos \phi \sin \phi_0 \cos(\lambda - \lambda_0)$$

$$t_2 = \sin \phi \sin \phi_0(\sin \phi + \sin \phi_0) + (\sin \phi - \sin \phi_0)(3 \sin^2 \phi - 1)$$

$$t_3 = \sin^2 \phi - \tfrac{1}{2} \sin \phi_0(\sin \phi - \sin \phi_0) \tag{1.11}$$

$$t_4 = \cos \phi \sin(\lambda - \lambda_0)$$

$$t_5 = \sin \phi \sin \phi_0 + \cos \phi \cos \phi_0 \cos(\lambda - \lambda_0)$$

$$\tau = \sin \phi - \sin \phi_0$$

When solving problems of cartography, photogrammetry, and some problems of geodesy, in (1.10) it is sufficient to include terms of up to e^2 (Bugayevskiy and Portnov 1984).

Then from (1.10) we get

$$\sin z \cos a = t_1 + e^2\tau(t_1 \sin \phi - \cos \phi_0) + \cdots$$

$$\sin z \sin a = t_4(1 + e^2\tau \sin \phi) + \cdots \tag{1.12}$$

$$\cos z = t_5 + e^2\tau(t_5 \sin \phi - \sin \phi_0) + \cdots$$

When using spherical coordinate systems, the eccentricity e is equal to zero, and equations (1.10)–(1.12) take the form

$$\begin{aligned} \sin z \cos a &= t_1 = \sin\phi \cos\phi_0 - \cos\phi \sin\phi_0 \cos(\lambda - \lambda_0) \\ \sin z \sin a &= t_4 = \cos\phi \sin(\lambda - \lambda_0) \\ \cos z &= t_5 = \sin\phi \sin\phi_0 + \cos\phi \cos\phi_0 \cos(\lambda - \lambda_0) \end{aligned} \tag{1.13}$$

Depending on the position of the pole in the spheroidal (or spherical) polar coordinate system, we distinguish between (1) a direct, main, or simply geographical coordinate system where the pole of the accepted coordinate system coincides with the geographical pole; (2) a transverse or equatorial coordinate system where the pole of the spheroidal (or spherical) polar coordinate system used is on some point of the equator; and (3) an oblique or horizon coordinate system with its pole at a latitude which is between 0 and 90°.

Besides the above, there is a normal polar spheroidal (or spherical) coordinate system in which the coordinate lines assume the simplest form for a given representation.

In the normal coordinate system, which coincides with the main system, the pole used is at the geographical pole, and the meridians and parallels are represented in the simplest manner.

On oblique and transverse aspects of projections, the normal graticule falls where the above set of meridians and parallels would appear. For convenience, the graticule lines that correspond to meridians of the normal projection will be called verticals, and the parallels of the normal projection will be called almucantars. These terms have been used in astronomy for the spherical coordinates of the sky with respect to a local horizon. Lines of constant a are the verticals. When representing the surface of a sphere, they are arcs of great circles intersecting at the pole points of either oblique or transverse system $Q(\phi_0, \lambda_0)$. Lines of constant z are almucantars. When representing the surface of a sphere, they are small circles, perpendicular to the verticals.

As a rule, normal systems are those for which their poles are at the geographical poles.

1.1.4 Polar geodetic coordinate systems

The two polar geodetic coordinates (α, s) of point $C(\phi, \lambda)$ are (1) the length s of the geodetic line from the pole of the polar coordinate system $Q_0(\phi_0, \lambda_0)$ to this point C and (2) the azimuth α of line Q_0C at point Q_0 (Figure 1.3). In this system lines of constant α are 'straight' geodetic lines from pole Q_0 and lines of constant s a set of geodetic circles orthogonal to the first family, which are not geodetic lines but complex curves of double curvature (Morozov 1979).

1.1.5 Elliptic coordinates

A surface being mapped can be defined by three equations of the form

$$x_e = F_1(u, v), \quad y_e = F_2(u, v), \quad z_e = F_3(u, v)$$

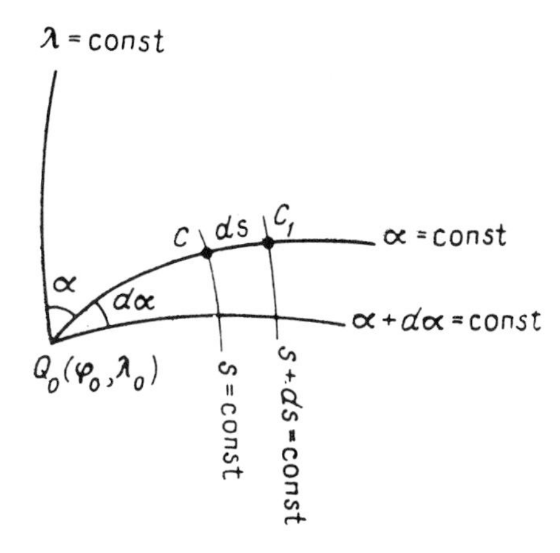

Figure 1.3 Polar geodetic coordinate system.

expressing the rectangular coordinates of the points on the surface as functions of arbitrary parameters u and v, which are curvilinear coordinates (Urmayev 1962; Vakhrameyeva *et al.* 1986).

Two systems of confocal spherical ellipses make the basis for obtaining curvilinear elliptic coordinates. Figure 1.4 shows focus F' common for both spherical ellipse MC with a second focus at point F and spherical ellipse MB with a second focus at point F'_1. The position of arbitrary point M is determined by its distances from foci $FM = a$ and $MF' = b$. If λ is the longitude of point M referred to the plane of initial meridian $PCAP_1$ perpendicular to the plane of the figure, and if ϕ is the latitude of the point, then from formulas of spherical trigonometry we have

$$\begin{aligned} \cos a &= \sin\phi \sin\phi_0 - \cos\phi \cos\phi_0 \sin\lambda \\ \cos b &= \sin\phi \sin\phi_0 + \cos\phi \cos\phi_0 \sin\lambda \end{aligned} \tag{1.14}$$

where ϕ_0 is the latitude of the pole point.

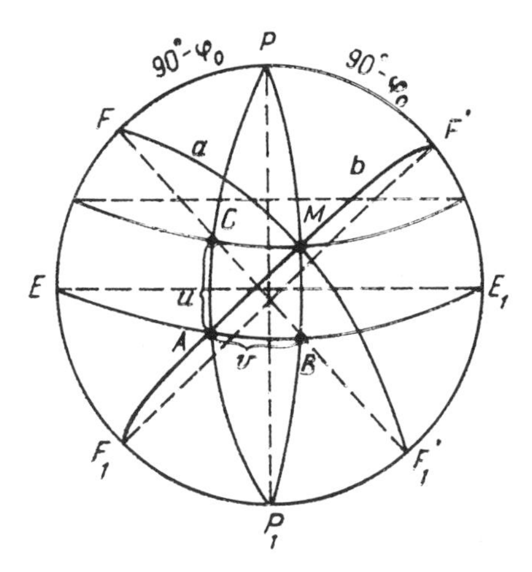

Figure 1.4 Elliptical coordinate system.

Point C is equally distant from both F and F', hence

$$FC = (a + b)/2$$

and point B is equally distant from both F' and F'_1. But $F'M = b$ and $F'_1M = \pi - a$; consequently

$$BF'_1 = \pi/2 - (a - b)/2$$

If we denote $\cup AC = u$ and $\cup AB = v$ and assume them to be elliptic coordinates, then, knowing both a and b, we can find these coordinates from formulas

$$\sin \phi_0 \sin u = \cos \frac{a + b}{2}, \quad \cos \phi_0 \sin v = \sin \frac{a - b}{2} \tag{1.15}$$

(test formulas: $\cos u \cos v = \cos \phi \cos \lambda$).

The formulas presented show that elliptic coordinates depend on the position of the foci (F, F', F_1, F'_1) of the spherical ellipses on the surface of the sphere; according to this attribute, elliptic coordinates are divided into different systems (those of Peirce, Guyou, and Adams (see section 7.4) are often related to elliptic functions).

1.1.6 Coordinate systems of the triaxial ellipsoid

Spatial rectangular, planetocentric, and geodetic coordinate systems are generally used when determining and using map projections of the triaxial ellipsoid.

The spatial rectangular coordinate system for the triaxial ellipsoid is written in the form

$$\frac{X^2}{a^2} + \frac{Y^2}{b^2} + \frac{Z^2}{c^2} = 1 \tag{1.16}$$

where a, b, c are semiaxes of the triaxial ellipsoid.

In the planetocentric system (Φ, λ), coordinate Φ is the planetocentric latitude, i.e. an angle between a radius vector from the center of the ellipsoid to a given point and the equatorial plane, and λ is the planetocentric longitude, i.e. the dihedral angle between the two planes passing through the polar axis of the ellipsoid and through the initial and current points, respectively.

The relation between spatial rectangular and planetocentric coordinates is expressed by the following formulas:

$$X = \rho(\Phi, \lambda)\cos \Phi \cos \lambda, \quad Y = \rho(\Phi, \lambda)\cos \Phi \sin \lambda$$

$$Z = \rho(\Phi, \lambda)\sin \Phi$$

where

$$\rho(\Phi, \lambda) = [(\alpha \cos \Phi \cos \lambda)^2 + (\beta \cos \Phi \sin \lambda)^2 + (\gamma \sin \Phi)^2]^{-1/2}$$

$$\alpha = 1/a, \quad \beta = 1/b, \quad \gamma = 1/c$$

The geodetic coordinate system, as applied to the triaxial ellipsoid, is treated ambiguously.

In the works of Clarke, Krasovskiy, N. A. Bespalov, and other scientists it was suggested that latitude Φ of the triaxial ellipsoid should be called the complement of the 90° angle between the normal to its surface and the axis of rotation. The definition of the meridian is given in two ways:

1. as a curve along which all normals to the ellipsoid are perpendicular to a line lying in the plane of the equator, this line being the same for a given meridian;
2. as a curve along which all tangents are directed to the north (or south).

These two definitions are not identical. The above authors, applying the first definition to the meridian, defined the line described in the second definition as a line of the north (or south) direction.

The formulas relating spatial rectangular coordinates X, Y, Z and geodetic coordinates ϕ, λ are given in the form

$$X = a \cos \phi \cos \lambda / W$$

$$Y = a(1 - e_a^2)\cos \phi \sin \lambda / W$$

$$Z = a(1 - e^2)\sin \phi / W$$

where

$$W = \sqrt{1 - e^2 \sin^2 \phi - e_a^2 \cos^2 \phi \sin^2 \lambda}$$

$$e^2 = (a^2 - c^2)/a^2, \quad e_a^2 = (a^2 - b^2)/a^2$$

e and e_a are the first polar and equatorial eccentricities, respectively.

According to another point of view, one should consider the concepts of conventional-geodetic and geodetic latitudes, geodetic longitudes and reduced latitudes as well.

The geodetic longitude is the dihedral angle between the two planes of the section passing through the polar axis of the ellipsoid and through the original and current points, respectively (meridians are section lines of these planes on the surface of the triaxial ellipsoid).

In order to introduce the notion of latitude, the following should be taken into account. Let line AK (Figure 1.5) be a normal to ellipse PDP_1 at point A. For the ellipsoid of revolution with semiaxes d and c, this normal would be a normal to its surface at point A, and at the same time angle ϕ^0 would be a geodetic latitude. However, for the triaxial ellipsoid, line AK does not appear to be normal to the surface and angle ϕ^0 does not appear to be the geodetic latitude. Hence, angle ϕ^0 between normal AK to ellipse PAP_1 at point A and line OD we shall call the false-geodetic latitude.

Angle ϕ of the intersection of a normal to the surface of the triaxial ellipsoid at point A with the equatorial plane ($Z = 0$) is called the geodetic latitude.

If we draw a circle of radius $d = OD$ in meridional plane PDP_1, then by analogy to the ellipsoid of revolution, angle u between lines OA' and OD can be called the reduced latitude of a given point on the triaxial ellipsoid.

The surface of the triaxial ellipsoid can be given in the following form by parametric equations:

$$X = d \cos u \cos \lambda, \quad Y = d \cos u \sin \lambda, \quad Z = c \sin u \tag{1.17}$$

where, according to Figure 1.5,

$$d = b(1 - k^2 \cos^2 \lambda)^{-1/2}, \quad k^2 = 1 - (b/a)^2 \tag{1.18}$$

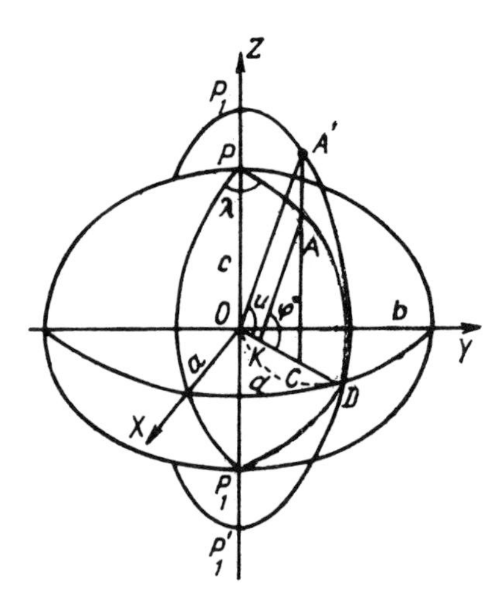

Figure 1.5 Coordinate system of the triaxial ellipsoid.

Denoting

$$p^2 = 1 - (c/d)^2 \tag{1.19}$$

we obtain the formulas for the relationship between the false-geodetic and reduced latitudes:

$$\cos^2 u = \cos^2 \phi^0/(1 - p^2 \sin^2 \phi^0)$$
$$\tan \phi^0 = \frac{d}{c} \tan u \tag{1.20}$$

On writing the equations for the normal to the surface of the triaxial ellipsoid at a given point with the equatorial plane $Z = 0$, we obtain from the formulas of analytical geometry the expressions for determining geodetic latitudes:

$$\sin \phi = d \sin^2 u/\sqrt{c^2 \cos^2 u(1 + z^2) + d^2 \sin^2 u} \tag{1.21}$$

or

$$\sin \phi = \sin \phi^0/\sqrt{1 + z^2 \cos^2 \phi^0}$$

where

$$z = -\frac{d_\lambda}{d} = \frac{k^2 \sin 2\lambda}{2(1 - k^2 \cos^2 \lambda)} \tag{1.22}$$

and

$$d_\lambda = -ab \sin 2\lambda \frac{a^2 - b^2}{2} (a^2 \sin^2 \lambda + b^2 \cos^2 \lambda)^{-3/2}$$

1.1.7 Elements of the spheroidal or spherical quadrangle: isometric coordinate systems and their relation to geographical systems

Let us draw through an arbitrary point A on the surface of the ellipsoid a line for a meridian and for a parallel and then add one more line for a meridian and for a parallel at an infinitesimal distance from the initial set.

As a result of the nature of the surface under consideration, we will get an infinitesimal convex spheroidal quadrangle *ABCD* with an accuracy of up to infinitesimal values of a higher-order infinitesimal, and which will be taken as a flat infinitesimal rectangle (Figure 1.6).

The elements of this quadrangle are:

1. an infinitesimal segment of a meridian

$$ds_m = AB = M\,d\phi \tag{1.23}$$

2. an infinitesimal segment of a parallel (an arc of finite dimensions is equal to $r(\lambda_2 - \lambda_1)$)

$$ds_p = BC = r\,d\lambda \tag{1.24}$$

3. a linear ellipsoidal element

$$ds = \sqrt{ds_m^2 + ds_p^2} = \sqrt{M^2\,d\phi^2 + r^2\,d\lambda^2} \tag{1.25}$$

4. a linear element of azimuth

$$\alpha = \arctan\left(\frac{ds_p}{ds_m}\right) = \arctan\left(\frac{r\,d\lambda}{M\,d\phi}\right) \tag{1.26}$$

5. an infinitesimal spheroidal quadrangle of area

$$dS = ds_m\,ds_p = Mr\,d\phi\,d\lambda \tag{1.27}$$

where the radius of curvature of the parallel for latitude ϕ

$$r = N \cos \phi \tag{1.28}$$

From (1.23) and (1.24) it follows that when differentials $d\phi$ and $d\lambda$ are equal, infinitesimal arc $ds_m \neq ds_p$, since $r \neq M$. This circumstance makes the use of a geodetic coordinate system difficult in a number of cases (e.g. when obtaining conformal projections).

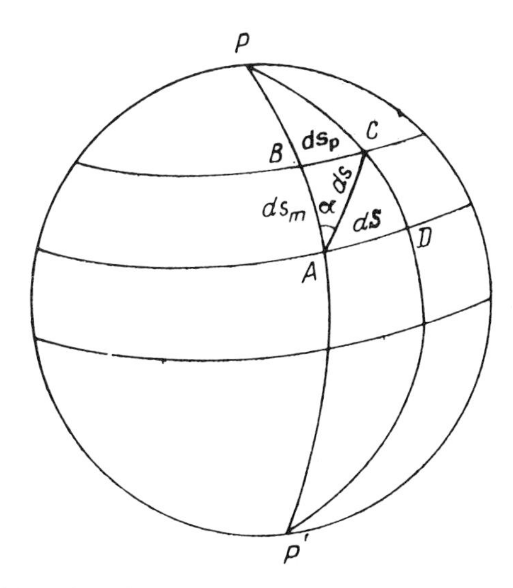

Figure 1.6 Elements of the spheroidal quadrangle.

Let us consider the coordinate system called isometric, in which the differential arguments corresponding to the infinitesimal arcs of meridians and parallels are equal to each other. To do this let us rewrite the formula of linear element (1.25) in the form

$$ds^2 = r^2 (dq^2 + d\lambda^2) \tag{1.29}$$

where

$$dq = (M/r)\, d\phi \tag{1.30}$$

Integrating (1.30) with reference to (1.1), (1.2), and (1.28), we obtain the formula for calculating an isometric latitude:

$$q = \ln U + C$$

where

$$U = \frac{\tan(45^\circ + \phi/2)}{\tan^e(45^\circ + \psi/2)} \tag{1.31}$$

$$\psi = \arcsin(e \sin \phi)$$

$$e = [1 - (b/a)^2]^{1/2} \tag{1.32}$$

the first eccentricity of the ellipsoid, and C is a constant of integration that can be made equal to zero so that $q = 0$ when $\phi = 0$. Thus, isometric latitudes of points on the surface of the ellipsoid (as a function of geodetic latitude) are determined by the formula (see Appendix 5)

$$q = \ln U \tag{1.33}$$

Isometric longitudes are equal to geodetic longitudes λ.

When transforming the surface of a sphere, the formula for isometric latitude, taking into account (1.30) and (1.33), will take the form (where subscript s refers to sphere)

$$q_s = \ln \tan(\pi/4 + \phi_s/2) \tag{1.34}$$

When polar coordinates are used, isometric latitudes will equal the following:

1. when transforming the surface of the ellipsoid

$$q = \ln \frac{\tan(z/2)}{\tan^e(\psi/2)}, \quad \psi = \arccos(e \cos z)$$

   where polar coordinates z, a are determined:

   (a) when the pole of the system is situated at the geographical pole, by the formulas

$$z = 90^\circ - \phi, \quad a = -\lambda \tag{1.35}$$

   (b) when the pole is situated at any point of the ellipsoid, by formulas (1.12) and (1.11), to an accuracy of up to e^2 terms;

2. when transforming the surface of the sphere

$$q = \ln \tan(z_s/2)$$

   where z, a are determined by formulas (1.13) (in a particular case by (1.35), when the pole of the coordinate system is situated at the geographical pole).

From the formulas obtained for isometric coordinates and equation (1.29) it is not difficult to see that, when differentials dq and $d\lambda$ are equal, the corresponding infinitesimal arc lengths of meridians and parallels ds_m and ds_p are also equal on the projection.

1.2 Definition of map projections: equations for meridians and parallels; the map graticule; conditions for transformation

Let us highlight on the ellipsoid (or sphere) a closed single connected zone Δ_1 with boundary Γ_1 and a system of curvilinear coordinates u, v; then on the plane we will have a corresponding zone Δ_2, with boundary Γ_2, for which a system of rectangular coordinates x, y is adopted. Let each point A of the ellipsoidal (or spherical) zone Δ_1 correspond to one and only one point A' of the plane zone Δ_2. With an infinitesimal displacement ds of this point A on the ellipsoid, point A' on the plane is also displaced at an infinitesimal value ds' and vice versa. Then it is possible to find a one-to-one correspondence of points on both zones and to express the relationship between these coordinate points in terms of

$$x = f_1(u, v), \quad y = f_2(u, v) \tag{1.36}$$

$$u = F_1(x, y), \quad v = F_2(x, y) \tag{1.37}$$

Here u, v can be, in particular, coordinates that are geodetic (ϕ, λ), isometric (q, λ), planetocentric (Φ, λ), or otherwise curvilinear. (Without loss of investigative generality, we will assume them to be geodetic coordinates ϕ, λ.) Functions f_1, f_2, F_1, F_2 are finite and continuous together with their first- and second-order partial derivatives (i.e. double continuous differentials); they are single valued and independent (Jacobian determinant $h = \partial(x, y)/\partial(\phi, \lambda)$) at all points of the region being mapped and should not be equal to zero (see formula (1.47)).

Equations (1.36) and (1.37) provide map projections in general form. Their properties depend on the above functions f_1, f_2, F_1, F_2. These functions can be of different forms. Hence, a set of map projections will also have various properties. This results in two definitions:

1. A map projection is a mathematically expressed method of transforming the surface of the Earth or a celestial body, assumed to be an ellipsoid, sphere, or other regular surface, onto a plane.
2. A map projection is a method of establishing one-to-one correspondence of the points on the surface being transformed and on the plane.

In this case equations (1.36) determine the so-called direct transformation of the given surface onto a plane, rectangular coordinates being expressed as functions of geodetic coordinates. Equations (1.37) determine the inverse transformation, geodetic coordinates being expressed as functions of rectangular coordinates.

Along with expressions (1.37) there are also equations for parallels and meridians. Equations for these coordinate lines can also be put in other forms. Eliminating successive longitudes and latitudes among (1.36), we have equations for parallels and meridians in an implicit form:

$$\Phi_1(x, y, \phi) = 0, \quad \Phi_2(x, y, \lambda) = 0 \tag{1.38}$$

Applying expression (1.36), we may obtain an equation for a parallel of latitude ϕ_0 and an equation for a meridian of longitude λ_0 in the parametric form

$$\begin{aligned} x &= f_1(\phi_0, \lambda), \quad y = f_2(\phi_0, \lambda) \\ x &= f_1(\phi, \lambda_0), \quad y = f_2(\phi, \lambda_0) \end{aligned} \tag{1.39}$$

The representation of meridian and parallel lines on maps on the map projection adopted is called a map graticule. The intervals between these lines depend on the purpose of the map.

The shape of the map graticule depends on the equations for the given projection:

- If $x = f_1(\lambda)$ and $y = f_2(\phi)$, then parallels and meridians are represented by two systems of parallel, mutually perpendicular, straight lines.
- In the case where $x = f_1(\phi, \lambda)$ and $y = f_2(\phi)$, parallels are represented by straight lines parallel to the X-axis, and meridians are represented by curves.
- If $x = f_1(\lambda)$ and $y = f_2(\phi, \lambda)$, parallels are represented by curves, and meridians by straight lines.
- If $x = f_1(\phi, \lambda)$ and $y = f_2(\phi, \lambda)$ parallels and meridians are represented by various curves.

The geographical pole p on maps can be represented by a point or a segment of a straight line or a curve. Its type of representation is determined by

$$\begin{aligned} &x_p = 0, \quad y_p = f_1(\phi_p) \quad \text{for a point} \\ &x_p = f_1(\phi_p, \lambda), \quad y_p = f_2(\phi_p) \quad \text{for a straight line} \\ &x_p = f_1(\phi_p, \lambda), \quad y_p = f_2(\phi_p, \lambda) \quad \text{for a curved line} \end{aligned} \tag{1.40}$$

Meridians and parallels of the map graticule can be represented on maps by straight lines symmetrical about a straight central meridian, the equator, or both these lines, or asymmetrically.

The type of graticule representation depends on the following conditions:

1. For projections symmetrical about the straight central meridian: ordinates must be even functions of the longitude

$$y(\phi, \lambda) = y(\phi, -\lambda)$$

abscissas must be odd functions of the longitude, and when $\lambda_0 = 0$ we get $x_0 = 0$

$$x(\phi, \lambda) = -x(\phi, -\lambda) \tag{1.41}$$

and the condition in which the central meridian is intersected by parallels at right angles gives

$$(\partial y/\partial \lambda)_{\lambda=0} = 0$$

2. For projections symmetrical about a straight equator: ordinates must be odd functions of the latitude

$$y(\phi, \lambda) = -y(-\phi, \lambda)$$

abscissas must be even functions of the latitude

$$x(\phi, \lambda) = x(-\phi, \lambda) \tag{1.42}$$

when $\phi_0 = 0$ we get $y = 0$, and the condition in which the equator is intersected by meridians at right angles gives

$$(\partial x/\partial \phi)_{\phi=0} = 0$$

All conditions (1.41) and (1.42) must be satisfied by projections for which map graticules are symmetrical about both the straight central meridian and the equator. In cases where at least one of these conditions is not satisfied, the map graticules for the projections being investigated are asymmetric about either one or both of these lines.

The design of the map graticule is used to determine coordinates of points, to plot points on the maps using their coordinates, to determine the relative arrangement of regions, and to solve cartographic and other mapping problems. On maps, generally of large scale, besides the map graticule, coordinate grids are often shown as a system of mutually perpendicular lines, drawn at preset intervals parallel to the axes of the rectangular coordinate system of the projection used. There may also be rectangular grids in metric, English (Imperial), and other units.

1.3 Elements for transforming an infinitesimal spheroidal (or spherical) quadrangle onto a plane

An infinitesimal convex quadrangle $ABCD$ (see Figure 1.6) of an ellipsoidal (or spherical) surface is represented by an infinitesimal oblique quadrangle $A'B'C'D'$ (Figure 1.7) on a plane; its linear element $AC = ds$ is represented by an infinitesimal segment $A'C' = ds'$ of a curve. To an accuracy of up to infinitesimal values of higher-order infinitesimals, we can assume quadrangle $A'B'C'D'$ to be an infinitesimal parallelogram and linear element ds' to be an infinitesimal straight line segment.

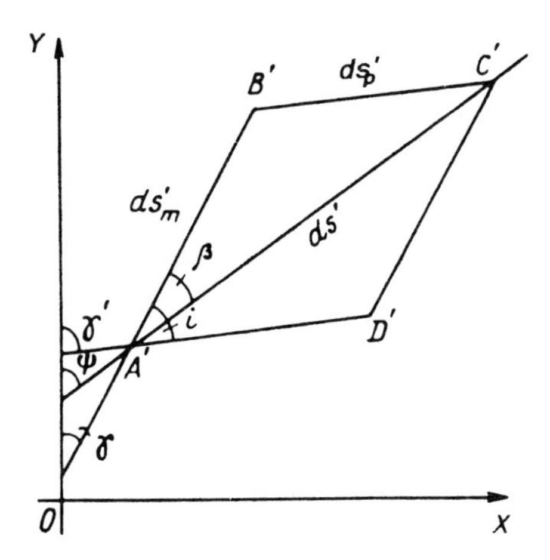

Figure 1.7 Transformation elements of an infinitesimal spheroidal quadrangle.

Then the tangents to the curves at point A' will coincide with the corresponding straight line segments forming angles γ', Ψ, and γ with the X-axis, and the elements of the transformation being investigated will receive values as shown below.

1.3.1 Linear element on a plane

$$ds' = (e\ d\phi^2 + 2f\ d\phi\ d\lambda + g\ d\lambda^2)^{1/2} \tag{1.43}$$

where e, f, g are Gauss coefficients equal to

$$e = x_\phi^2 + y_\phi^2, \quad f = x_\phi x_\lambda + y_\phi y_\lambda, \quad g = x_\lambda^2 + y_\lambda^2 \tag{1.44}$$

and $x_\phi = (\partial x/\partial \phi)$, $y_\lambda = (\partial y/\partial \lambda)$, etc.

Along the meridians (at constant λ) and parallels (at constant ϕ) we can get infinitesimal segments as follows:

$$ds'_m = \sqrt{e}\ d\phi, \quad ds'_p = \sqrt{g}\ d\lambda \tag{1.45}$$

1.3.2 Angle *i* between the meridian and parallel on the projection

$$i = \arctan(h/f) = \arccos(f/\sqrt{eg}) = \arcsin(h/\sqrt{eg}) \tag{1.46}$$

where

$$h = x_\lambda y_\phi - x_\phi y_\lambda = \sqrt{eg - f^2} \tag{1.47}$$

is considered to be only positive.

The angles may also be determined from

$$\varepsilon = \arctan(-f/h) \tag{1.48}$$

where $\varepsilon = i - 90°$, the deviation of i from a right angle on the projection;

$$\gamma_m = \arctan(x_\phi/y_\phi) \tag{1.49}$$

is the convergence of a meridian at a given point, or its angle with the direction of the Y-axis on the map, and

$$\gamma_p = \arctan(x_\lambda/y_\lambda)$$

is the angle between the direction of the parallel and the northern direction of the Y-axis at a given point.

Angle i is considered to be clockwise from (east of) north, and its quadrant is determined by the sign of f. If $f > 0$, then $i < 90°$, and the angle is in the first quadrant. If $f < 0$, then $i > 90°$, and the angle is in the second quadrant. If $f = 0$, the angle $i = 90°$ and meridians and parallels are represented by orthogonal lines. Hence, expression (1.50) is the condition for an orthogonal map graticule on the projection:

$$f = x_\phi x_\lambda + y_\phi y_\lambda = 0 \tag{1.50}$$

1.3.3 Azimuth β of linear element *ds'* on the projection

$$\cot \beta = (e/h)(r/M)\cot \alpha + (f/h) \tag{1.51}$$

or

$$\tan \beta = Mh \tan \alpha/(er + Mf \tan \alpha) \tag{1.52}$$

See also (1.61) and (1.105).

1.3.4 Area Σ representing an infinitesimal spheroidal (or spherical) quadrangle

$$d\Sigma = ds_m\, ds_p \sin i = h\, d\phi\, d\lambda \qquad (1.53)$$

1.4 Scale

In the theory of map projections, concepts and formulas for both area and linear scale are considered.

1.4.1 Linear scale

The nominal linear scale of a map must be distinguished from the local scale.

Nominal linear scale
The nominal linear scale shows the general reduction of either the whole ellipsoid (or sphere) or some part of it to represent the surface being mapped on the plane.

This scale is placed on the map but it is preserved only at some points or along some lines of the map. Changing the nominal scale does not affect the properties of the projection used, and thus it is often conveniently considered to be unity or 1.0 when carrying out projection research.

Local linear scale
The local linear scale of the representation of a given point and along a given direction is the ratio of the length of an infinitesimal segment on the projection to that of the corresponding infinitesimal segment on the surface of the ellipsoid (or sphere). The ratio of this local linear scale to the nominal linear scale of the map is called the (linear) scale factor μ:

$$\mu = ds'/ds$$

Taking into account the squares of linear elements (1.43) and (1.25), and the azimuth formula (1.26),

$$\mu^2 = \frac{e\, d\phi^2 + 2f\, d\phi\, d\lambda + g\, d\lambda^2}{M^2\, d\phi^2 + r^2\, d\lambda^2} = \frac{e}{M^2}\cos^2\alpha + \frac{f}{Mr}\sin 2\alpha + \frac{g}{r^2}\sin^2\alpha \qquad (1.54)$$

Introducing $P = e/M^2$, $Q = f/Mr$, and $R = g/r^2$, formula (1.54) for the local scale factor along any direction can be represented in the form

$$\mu^2 = P\cos^2\alpha + Q\sin 2\alpha + R\sin^2\alpha$$

In the direction of the meridian, azimuth $\alpha = 0$. Hence, taking into account (1.44), we can get from (1.54)

$$m = \mu_{\alpha=0} = \sqrt{e}/M = (1/M)(x_\phi^2 + y_\phi^2)^{1/2} \qquad (1.55)$$

i.e. the formula for the local linear scale factor m along a meridian.

Correspondingly, along a parallel the direction $\alpha = 90°$. From expression (1.54), we find

$$n = \mu_{\alpha=90°} = \sqrt{g}/r = (1/r)(x_\lambda^2 + y_\lambda^2)^{1/2} \tag{1.56}$$

i.e. the formula for the local linear scale factor n along a parallel.

The formula for the local linear scale factor along any direction (1.54), taking into account (1.55), (1.56), and (1.46), takes the form

$$\mu^2 = m^2 \cos^2 \alpha + mn \cos i \sin 2\alpha + n^2 \sin^2 \alpha \tag{1.57}$$

Taking into account (1.51), an analogous formula, but along azimuth direction β on the projection, can be recast in the form

$$\frac{1}{\mu^2} = \csc^2 i \left(\frac{1}{n^2} \sin^2 \beta + \frac{1}{m^2} \sin^2(i - \beta) \right) \tag{1.58}$$

or

$$\frac{1}{\mu^2} = P_1 \cos^2 \beta + Q_1 \sin 2\beta + R_1 \sin^2 \beta \tag{1.59}$$

where

$$P_1 = e^{-1}M^2, \quad Q_1 = -(e/h)^{-1} fM^2, \quad R_1 = (M^2 f^2 + e^2 r^2)(eh^2)^{-1}$$

From the formulas given above, it follows that the local linear scale factor in a particular direction depends on both point coordinates and azimuths of linear elements related to direction.

Extremes of local linear scale

In order to determine these values we have to differentiate either formula (1.57) or (1.54), which depend on the azimuths of linear elements on the surface of the ellipsoid, with respect to α, and set the derivative obtained (when $\alpha = \alpha_0$) to zero. Then

$$\tan 2\alpha_0 = \frac{2mn \cos i}{m^2 - n^2} = \frac{2Q}{P - R} = \frac{2Mfr}{er^2 - gM^2} \tag{1.60}$$

Applying formula (1.51) for relating azimuth, rewritten considering (1.46), (1.55), and (1.56) in the form

$$\cot \beta = \frac{m}{n} \csc i \cot \alpha + \cot i \tag{1.61}$$

we obtain a formula analogous to (1.60), but as a function of azimuth β of a linear element on the projection

$$\tan 2\beta_0 = \frac{n^2 \sin 2i}{m^2 + n^2 \cos 2i} \tag{1.62}$$

As the period for tangents is equal to π, equations (1.60) and (1.62) yield two roots each, and consequently two directions α_0, $\alpha_0 + 90°$ and β_0, $\beta_0 + 90°$ in azimuth on both the ellipsoid and the projection, along which the local linear scale factors at a given point have extreme values a and b (a is the greater of the two). These directions are orthogonal and are called base directions. In this case, if the map graticule is orthogonal, then the base directions apply to meridians and parallels.

Note that the azimuth relationship can be determined with the formulas (Meshcheryakov 1968)

$$\sin \beta = \frac{n \sin i \sin \alpha}{\mu}, \quad \cos \beta = \frac{m}{\mu} \cos \alpha + \frac{n}{\mu} \cos i \sin \alpha$$
$$\cos(i - \beta) = \frac{m}{\mu} \cos i \cos \alpha + \frac{n}{\mu} \sin \alpha, \quad \sin(i - \beta) = \frac{m}{\mu} \sin i \cos \alpha \tag{1.63}$$

From these formulas it is not difficult to find the values α_0 of azimuths of base directions of the projections through the values β_0.

In order to determine extreme local linear scale factors, it is sufficient to substitute values of azimuths α_0 and $\alpha_0 + 90°$ in formula (1.57) or (1.54), or values β_0 and $\beta_0 + 90°$ in (1.58) or (1.59).

It is more convenient, however, to use two theorems of Apollonius for conjugate diameters of an ellipse, showing the relationship of extreme local linear scale factors to the local linear scale along meridians and parallels (see Ellipse of distortion, section 1.6.1):

$$m^2 + n^2 = a^2 + b^2, \quad mn \cos \varepsilon = ab \tag{1.64}$$

Hence

$$a = \tfrac{1}{2}(\sqrt{m^2 + 2mn \cos \varepsilon + n^2} + \sqrt{m^2 - 2mn \cos \varepsilon + n^2})$$
$$b = \tfrac{1}{2}(\sqrt{m^2 + 2mn \cos \varepsilon + n^2} - \sqrt{m^2 - 2mn \cos \varepsilon + n^2}) \tag{1.65}$$

Using expressions (1.57) and (1.65), we can obtain one more formula for determining local linear scale factors along any direction:

$$\mu^2 = a^2 \cos^2 (\alpha - \alpha_0) + b^2 \sin^2 (\alpha - \alpha_0) \tag{1.66}$$

Analogous to this we can get general formulas for local linear scale factors along verticals and almucantars in cases where oblique and transverse spheroidal (or spherical) coordinate systems are used. They can be expressed, for example, by formulas (1.11), (1.12), and (1.13). Writing the square of a linear element of an ellipsoid to an accuracy of terms of e^2 for the elements in the form of

$$ds^2 = P^2(dz^2 + \sin^2 z \, da^2) \tag{1.67}$$

where

$$P = N_0\left(1 - \frac{e^2}{2} [\sin z \cos a \cos \phi_0 + \sin \phi_0 (\cos z - 1)]^2\right) \tag{1.68}$$

the formulas for the local linear scale factor along oblique verticals and almucantars have the respective forms

$$\mu_1 = \frac{1}{P} (x_z^2 + y_z^2)^{1/2}, \quad \mu_2 = \frac{1}{P \sin z} (x_a^2 + y_a^2)^{1/2} \tag{1.69}$$

For transforming the surface of a sphere,

$$\mu_1 = \frac{1}{R} (x_z^2 + y_z^2)^{1/2}, \quad \mu_2 = \frac{1}{R \sin z} (x_a^2 + y_a^2)^{1/2} \tag{1.70}$$

1.4.2 Local area scale

The local area scale factor at a given point is the ratio of the area of an infinitesimal quadrangle on the projection to that of the corresponding infinitesimal quadrangle on the surface of the ellipsoid or sphere:

$$p = d\Sigma/dS \tag{1.71}$$

Taking into account (1.53) and (1.27) we have

$$p = h/(Mr) \tag{1.72}$$

Hence, taking into account (1.46), (1.47), (1.55), (1.56), and (1.64) we get

$$p = (x_\lambda y_\phi - x_\phi y_\lambda)/Mr \tag{1.73}$$

$$p = mn \sin i = mn \cos \varepsilon = ab \tag{1.74}$$

If we substitute the new variable S from the equation $dS = Mr\, d\phi$ for latitude ϕ, then equation (1.73) takes the form

$$p = x_\lambda y_S - x_S y_\lambda \tag{1.75}$$

For oblique or transverse coordinate systems we must take into account (1.12) and (1.11):

$$p = \mu_1 \mu_2 \cos \varepsilon \tag{1.76}$$

where

$$\varepsilon = -\arctan\left(\frac{x_z x_a + y_z y_a}{x_a y_z - y_a x_z}\right)$$

1.5 *Conditions for conformal, equal-area, and equidistant transformation of an ellipsoidal (or spherical) surface onto a plane*

Map projections can be conformal, equal area (or equivalent), and arbitrary (in particular cases, 'equidistant' according to the nature of their distortion). While creating these projections it is necessary that their equations satisfy the corresponding conditions of transformation.

1.5.1 Conditions for conformal mapping

Conformal projections are those on which there is no distortion of local angles and azimuths of linear elements, i.e. on which the identity

$$\beta \equiv \alpha \tag{1.77}$$

holds. From equation (1.51) it follows that this identity holds only in the case where

$$f = 0 \quad \text{and} \quad (e/h)(r/M) = 1 \tag{1.78}$$

Consequently, taking into account (1.46) we get

$$\sqrt{e}/M = \sqrt{g}/r \quad \text{and} \quad f = 0 \tag{1.79}$$

Expressions (1.78) and (1.79) can be regarded as two pairs of conditions for conformality, from which it follows that on these projections local linear scale does not depend on direction (the shape or identity of infinitesimal figures holds) and that on them the map graticule is orthogonal. Applying equations (1.44), (1.55), (1.56), and (1.30), we get the following pair of conditions for conformality:

$$m = n, \quad \varepsilon = 0 \tag{1.80}$$

For specific surfaces,

$$(1/M^2)(x_\phi^2 + y_\phi^2) = (1/r^2)(x_\lambda^2 + y_\lambda^2), \quad x_\phi x_\lambda + y_\phi y_\lambda = 0 \tag{1.81}$$

$$x_\lambda = (r/M)y_\phi, \quad y_\lambda = -(r/M)x_\phi \tag{1.82}$$

to represent the ellipsoid on a plane;

$$x_\lambda = y_\phi \cos\phi \quad \text{and} \quad y_\lambda = -x_\phi \cos\phi \tag{1.83}$$

to represent the surface of a sphere; and

$$x_\lambda = y_q \quad \text{and} \quad y_\lambda = -x_q \tag{1.84}$$

to represent the sphere and ellipsoid in the isometric coordinate system.

Each of the last three formula groups are also called Cauchy–Riemann conditions (they preserve the sign combination in which $h > 0$).

1.5.2 Conditions for equal-area transformation

When deriving map projections we consider the transformation of a single connected closed region Δ_1 of an ellipsoid (or sphere) with boundary Γ_1 onto region Δ_2 of the plane bounded by Γ_2.

The areas of these regions by parts using (1.27) and (1.53) are determined by formulas

$$S = \iint_S Mr \, d\phi \, d\lambda \tag{1.85}$$

$$\Sigma = \iint_\Sigma h \, d\phi \, d\lambda \tag{1.86}$$

where double integrals are used to compute areas S and Σ, respectively.

Equal-area or equivalent projections are those in which areas S and Σ of the above regions on an ellipsoidal (or spherical) surface and on a plane, respectively, are identical, i.e.

$$\Sigma \equiv S \tag{1.87}$$

Then in view of (1.85) and (1.86) the condition of equivalency takes the form

$$h = Mr \tag{1.88}$$

Applying equations (1.46), (1.47), (1.55), and (1.56), this condition can be expressed by the following formulas:

$$x_\lambda y_\phi - x_\phi y_\lambda = Mr \tag{1.89}$$

to represent an ellipsoid on a plane;

$$x_\lambda y_\phi - x_\phi y_\lambda = R^2 \cos \phi \tag{1.90}$$

to represent the surface of a sphere; and

$$mn \sin i = mn \cos \varepsilon = 1 \tag{1.91}$$

$$p = ab = 1 \tag{1.92}$$

1.5.3 Conditions for equidistant transformation

On so-called equidistant projections, lengths are preserved along one of the base directions. These projections are very often considered together with an orthogonal map graticule.

Projections equidistant along meridians or verticals
Here the identities to maintain are

$$m \equiv 1 \quad \text{and} \quad \mu_1 \equiv 1 \tag{1.93}$$

Hence, subject to (1.55), (1.69), and (1.70), the conditions desired take the form

$$x_\phi^2 + y_\phi^2 = M^2 \quad \text{and} \quad x_z^2 + y_z^2 = P^2 \tag{1.94}$$

where M and P are determined, respectively, by (1.1) and (1.68).

Projections equidistant along parallels or almucantars
For these projections the corresponding identities

$$n \equiv 1 \qquad \mu_2 \equiv 1 \tag{1.95}$$

are to hold. Hence, subject to (1.56), (1.69), and (1.70), the equidistant conditions for this situation take the form

$$x_\lambda^2 + y_\lambda^2 = r^2 \quad \text{and} \quad x_a^2 + y_a^2 = P^2 \sin^2 z \tag{1.96}$$

where r and P are determined, respectively, by (1.28) and (1.68).

1.6 Distortion on map projections

An analysis of projection distortion makes it possible to appreciate the advantages of various projections and to apply the results found to practice and research.

It is usually noted that there are two kinds of distortion on maps:

1. distortion of length on all projections except that on conformal and equal-area projections distortion results, respectively, from the variation in local scale only

from point to point on a conformal projection and from maintaining the correct area scale at each point on an equal-area projection;

2. distortion of length of finite straight segments and the angles between them, resulting from the curvature of the plotted geodetic lines when taking measurements from maps.

In some cases the advantages of map projections depend not so much on the value and character of the distortion but on their other properties, e.g. the shape of the map graticule, the method of representing positional lines, the appearance of sphericity, and so on. This leads to the need to consider distortion more generally as the value of elements characterizing actual projection properties in comparison with ideal properties, e.g. the importance of representing geodetic lines or loxodromes on the projection as straight lines and so on. Taking into account values like these makes it possible to develop general criteria for choosing and creating projections satisfying all the necessary requirements in the best possible way.

It should be noted that Euler and Tissot made great contributions to the development of distortion theory. These problems were intensively dealt with in the Soviet Union in the works of V. V. Kavrayskiy, N. A. Urmayev, Meshcheryakov, Konusova, and others. In other countries, O. S. Adams in the United States and A. E. Young in England performed substantial research in this field.

1.6.1 Ellipse of distortion; maximum angular deformation

Assume an infinitesimal quadrilateral $ABCD$ on an ellipsoid (or sphere). To a high degree of accuracy, we can assume it to be a flat infinitesimal rectangle (Figure 1.8). We assume the transformation of this quadrilateral $A'B'C'D'$ onto a plane to be an infinitesimal parallelogram with the same degree of accuracy (Figure 1.9). Let every point on the surface of the ellipsoid, e.g. point A, have coordinate systems $\eta A \zeta$ and

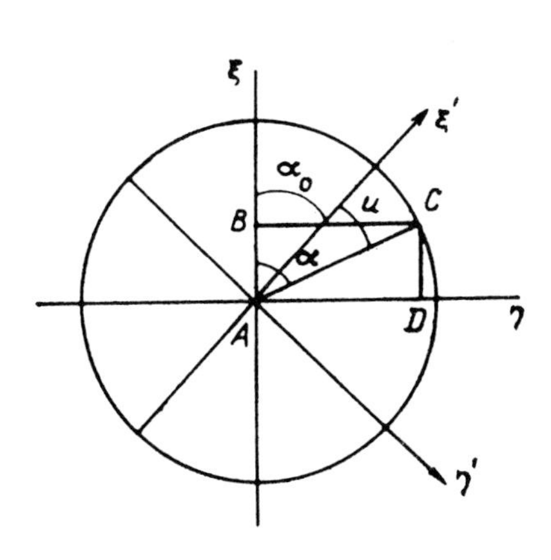

Figure 1.8 The infinitesimal circle and quadrangle on the ellipsoid.

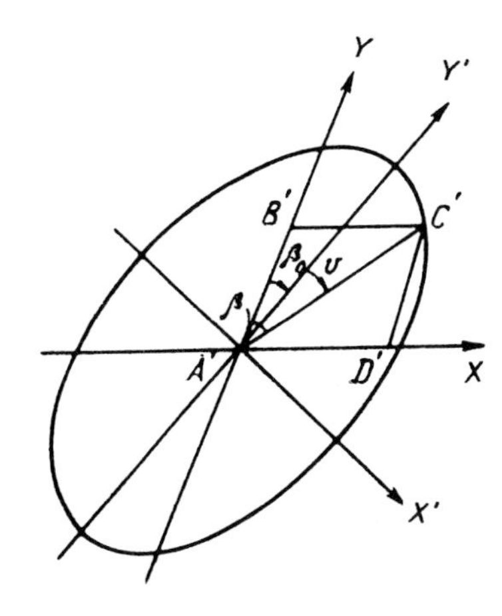

Figure 1.9 Geometry of a distortion ellipse.

$\eta'A\zeta'$, and let every corresponding point on the plane have coordinate systems $xA'y$ and $x'A'y'$, where the axes are in the directions of meridians, parallels, and base lines. Now let us draw a circle of radius $R = AC$ with its center at point A on the surface of the ellipsoid (or sphere) (see Figure 1.8); then

$$\eta^2 + \xi^2 = R^2 \tag{1.97}$$

Taking into account values of the local linear scale factors along meridians

$$m = A'B'/(AB) = y/\xi$$

and parallels

$$n = A'D'/(AD) = x/\eta$$

equation (1.97) for the plane takes the form

$$x^2/(n^2R^2) + y^2/(m^2R^2) = 1 \tag{1.98}$$

It follows that in the general case an infinitesimal circle (1.97) on the surface of the ellipsoid (or sphere) is represented by an infinitesimal ellipse (1.98) on the plane. It also follows from (1.98) that in a particular case, i.e. conformal and semiconformal projections, where local linear scale factors along meridians and parallels are equal ($m = n$), an infinitesimal circle on the surface of the ellipsoid (or sphere) is represented by a similar infinitesimal circle on the plane.

We should note that it is more convenient to use finite rather than infinitesimal values to interpret distortion geometrically. Proceeding from this, the ellipse of distortion or the Tissot indicatrix or indicator is made a finite ellipse (1.98), the transformation of a circle of radius R (1.97).

In order to draw the indicatrix, it is sufficient to calculate the values of local linear scale factors m, n or a, b and those of angles i (or ε) and β at the given point, and then to plot straight segments proportional to the values of the local scale factors on the meridians, parallels, and base directions on the projection.

Now, from Figures 1.8 and 1.9 we can write the values for the distortion of angles with base directions on both the ellipsoid (or sphere) and projection

$$u = \arctan(\eta'/\xi'), \quad v = \arctan(y'/x')$$

We can also write the deflections of angles on the projection

$$\Delta u = u - v \tag{1.99}$$

and the formula relating these angles

$$v = \arctan\left(\frac{b}{a} \tan u\right) \tag{1.100}$$

It is now possible to recast the azimuths of linear elements in the form

$$\alpha = u + \alpha_0 \quad \text{and} \quad \beta = v + \beta_0 \tag{1.101}$$

Substituting $\omega/2 = u_0 - v_0$, for the maximum angular distortion or deformation, according to Vitkovskiy, we obtain formulas for their calculation:

$$\sin\frac{\omega}{2} = \frac{a-b}{a+b}, \quad \cos\frac{\omega}{2} = \frac{2\sqrt{ab}}{a+b}, \quad \tan\frac{\omega}{2} = \frac{a-b}{2\sqrt{ab}} \tag{1.102}$$

In doing so, the values of the angles at which distortion ω is a maximum can be obtained from the equations

$$u_0 = \frac{\pi}{4} + \frac{\omega}{4} = \arctan\left(\frac{a}{b}\right)^{1/2}$$

$$v_0 = \frac{\pi}{4} - \frac{\omega}{4} = \arctan\left(\frac{b}{a}\right)^{1/2} \tag{1.103}$$

$$\alpha_{\omega=\max} = \alpha_0 + u_0, \quad \beta_{\omega=\max} = \beta_0 + v_0$$

1.6.2 Distortion of azimuth

The difference

$$\Delta\alpha = \beta - \alpha \tag{1.104}$$

between the azimuth on the projection and on the ellipsoid (or sphere) is called the distortion of azimuth (Urmayev 1962). The value of this difference depends on the direction.

Let us rewrite equation (1.61) in the following form:

$$\tan \beta = \frac{n \sin i \tan \alpha}{m + n \cos i \tan \alpha} \tag{1.105}$$

An analysis of this equation shows that when $\alpha = \beta = 0$ azimuths are not distorted, and when $\beta = \alpha$ directions of non-distorted azimuths are equal:

$$\tan \alpha = \frac{n \sin i - m}{n \cos i} = \frac{m - n \cos \varepsilon}{n \sin \varepsilon} \tag{1.106}$$

After differentiating (1.104) and (1.105) we can get

$$d\alpha/d\beta = \mu^2/p = 1 \tag{1.107}$$

and in view of (1.57) we can obtain a formula for determining the directions of the maximum distortion of azimuths:

$$(n^2 - p)\tan^2 \alpha + 2mn \cos i \tan \alpha + (m^2 - p) = 0 \tag{1.108}$$

It can be seen from (1.107) that it is along these directions that there is no linear distortion on equal-area projections.

Now, taking into account equations (1.99), (1.101), and (1.104), it is easy to write a formula relating the distortion of azimuths and angles:

$$\beta - \alpha = (\beta_0 - \alpha_0) - (u - v)$$

or

$$\Delta\alpha = (\beta_0 - \alpha_0) - \Delta u$$

Hence, it follows that even if there is no angular distortion Δu, there is distortion of azimuths $\Delta\alpha = (\beta_0 - \alpha_0)$ and, vice versa, if there is no distortion of azimuths, there is angular distortion of the same value $\Delta u = \beta_0 - \alpha_0$.

1.6.3 Linear distortion on the projection

It is necessary to distinguish relative linear distortion at a given point along a certain direction, relative linear distortion at a given point along all directions, and root-mean-square or arithmetic-mean distortion within the boundaries of the entire region which is being mapped.

The values of relative linear distortion at a given point along a given direction are considered to be equal to

$$v_1 = \mu - 1, \quad v_2 = \ln \mu, \quad v_3 = 1 - (1/\mu), \quad v_4 = \tfrac{1}{2}(\mu^2 - 1) \tag{1.109}$$

All these values differ from each other only by small values of second- or higher-order infinitesimals with respect to these values.

The general value of relative linear distortion at a given point along all directions may be expressed by formulas suggested by various scientists and because of that are given their names.

- the Airy criteria

$$\varepsilon_{A_1}^2 = \tfrac{1}{2}[(a-1)^2 + (b-1)^2]$$
$$\varepsilon_{A_2}^2 = \frac{1}{2}\left[\left(\frac{a}{b} - 1\right)^2 + (ab-1)^2\right] \tag{1.110}$$

- the Airy–Kavrayskiy criterion

$$\varepsilon_{A-K}^2 = \tfrac{1}{2}(\ln^2 a + \ln^2 b) \tag{1.111}$$

- the Jordan criterion

$$\varepsilon_J^2 = \frac{1}{2\pi}\int_0^{2\pi} (\mu - 1)^2 \, d\alpha \tag{1.112}$$

- the Jordan–Kavrayskiy criterion

$$\varepsilon_{J-K}^2 = \frac{1}{2\pi}\int_0^{2\pi} \ln^2 \mu \, d\alpha \tag{1.113}$$

- the Klingach criterion

$$\varepsilon_K^2 = \left[P_\omega\left(\frac{a}{b} - 1\right)^2 + P_p(ab-1)^2\right] \Big/ (P_\omega + P_p) \tag{1.114}$$

  where with the help of weight factors P_ω and P_p we can establish the desired balance between angular and area distortion;
- the Konusova criterion

$$\alpha = \arctan\left[\left(\frac{a}{b} - 1\right) \Big/ (ab - 1)\right] \tag{1.115}$$

  with the help of which we can evaluate or preset projection distortion: $\alpha = 0$ for conformal projections; $\alpha = \pi/2$ for equal-area projections; and $0 < \alpha < \pi/2$ for arbitrary projections, depending on the type of distortion.

To assess the advantages of map projections, besides those stated above, other criteria have been suggested and used by Weber, Friedrich Eisenlohr, Yu. S. Frolov, and others.

Note that in the case of projections with an orthogonal map graticule, the extreme local scale factors a and b take on the values of local linear scale factors m and n for all the criteria given.

Linear distortion within the entire region being mapped is assessed with the help of the criteria of minimax or variations. A criterion of the minimax type is that of Chebyshev, according to which the ratio of the maximum value of the local linear scale factor μ_{max} to the minimum value μ_{min} is determined (within the boundaries of the region being represented) for the projection under consideration. When the criteria of variations or least squares are used for the projection being studied (within the boundaries of the entire region being represented), the value of one of the functions is

$$E^2 = (1/F) \int_F \varepsilon^2 \, dF \tag{1.116}$$

where ε^2 is determined by one of the criteria (1.110)–(1.115) or others analogous to them. To an accuracy sufficient for practice and research, the value of this function can be found in the following way. The region being mapped is divided into k small plots, for which midpoint values of ε^2 are determined by one of the formulas (1.110)–(1.115). Then we determine

$$E^2 = \frac{1}{k} \sum_{i=1}^{k} \varepsilon_i^2 \tag{1.117}$$

or

$$E = \left(\frac{1}{k} \sum_{i=1}^{k} \varepsilon_i^2 \right)^{1/2} \tag{1.118}$$

1.6.4 Area distortion on the projection

Relative area distortion is determined from the equation

$$v_p = p - 1 = ab - 1 = mn \cos \varepsilon - 1 \tag{1.119}$$

For a projection with an orthogonal map graticule we can write

$$v_p = mn - 1 \tag{1.120}$$

For conformal projections we will get

$$v_p = m^2 - 1 \tag{1.121}$$

1.6.5 Ratio of angular and area distortion on the projection

Let us rewrite the formulas for maximum angular (1.102) and area (1.119) types of distortion in the form

$$\sin \frac{\omega}{2} = \frac{a - b}{a + b} = \frac{v_1 - v_2}{2(v_1 + v_2)}$$

$$v_p = ab - 1 = (1 + v_1)(1 + v_2) - 1$$

where $v_1 = a - 1$ and $v_2 = b - 1$ are the values of linear distortion along the base directions.

From these formulas we can get the approximate relationships

$$\omega \approx v_1 - v_2 \tag{1.122}$$

$$v_p \approx v_1 + v_2 \tag{1.123}$$

For conformal projections $\omega = 0$, $a = b$, and $v_1 = v_2$. From (1.123), $v_p = 2v$, i.e. the area distortion or area scale error is about twice the linear scale error at a given point on these projections.

On equal-area projections $v_p = 0$ and from (1.123) and (1.122) we have $v_2 = -v_1$ and $\omega = 2v$, i.e. the maximum angular deformation (in radians) is about twice the maximum linear scale error at a given point.

On equidistant projections we can find, respectively, that if $b = 1$, then $v_2 = 0$, $\omega = v_1 = v$, and $v_p = v_1 = v$; if $a = 1$, then $v_1 = 0$, $\omega = -v_2 = v$, and $v_p = v_2 = v$; i.e. both the angular and area types of distortion are approximately equal on these projections.

Correlating the distortion given makes it possible to arrive at three conclusions:

1. the decrease of angular distortion on a projection inevitably results in an increase of area distortion on this projection and vice versa;
2. in the case where neither angular nor area distortion is desirable, it is advisable to use projections similar to equidistant ones;
3. map projections should be chosen so that not only will they satisfy the condition of minimal distortion, but also the distortion characteristics should lead to maps with optimal conditions for the purpose, permitting the solution of problems from the map.

1.6.6 Central line and central point

A line and a point are called central if the distortion of length, angles, and areas there is at a minimum.

It is known that every function close to its minimum changes no faster than its arguments. Consequently, in the vicinity of the central line or the central point, the distortion on a projection also changes slowly.

Hence, the choice of the criterion for a projection to map a specific region is to be carried out so that the central point of the projection is approximately at a midpoint of the region being mapped, and the central line is in the middle and in the direction of the greatest expanse of this region.

1.6.7 Distortion and corrections due to curvature of geodetic lines as represented on the projection

In Figure 1.10, let s_{12} be a curved segment of the representation of a geodetic line, with d_{12} its chord, β_{12} the azimuth on the projection of geodetic line 1–2 at point 1, γ the convergence of the meridian at point 1, $\alpha_{\partial_{12}}$ the direction angle of chord 1–2, and δ_{12} a correction to the azimuth for the curvature of the geodetic lines as

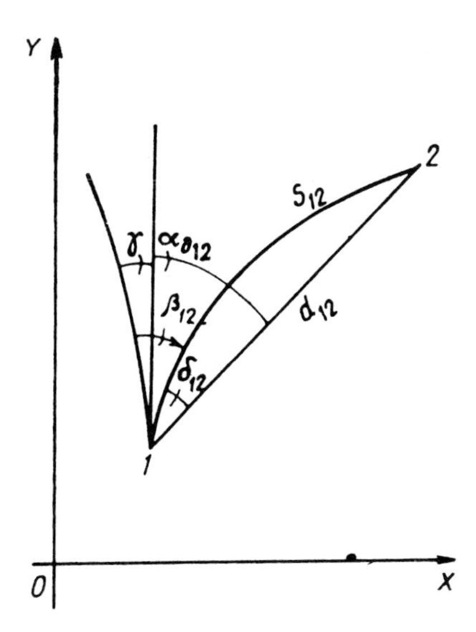

Figure 1.10 Azimuth and directional angle on the projection.

represented. Then we can write

$$\beta_{12} = \alpha_{\partial_{12}} + \gamma - \delta_{12}$$

The values of the direction angle of chord 1–2 and of the convergence of the meridian at the point on the projection can be easily determined by the formulas

$$\alpha_{\partial_{12}} = \arctan\left(\frac{x_2 - x_1}{y_2 - y_1}\right), \quad \gamma = \arctan\left(\frac{x_\phi}{y_\phi}\right)_1$$

where x_1, y_1, x_2, y_2 are rectangular coordinates of points 1 and 2 of the given segment on the projection, and x_ϕ and y_ϕ are partial derivatives at point 1. The main difficulty in finding azimuth β_{12} on the projection is in determining correction δ_{12}. For distance $s_{12} < R$, the radius of a sphere for a celestial body under investigation, we can use the following formulas:

$$\begin{aligned} \delta_{12} &= \frac{1}{2} k_1 s_{12} + \frac{1}{6}\left(\frac{dk}{ds}\right)_1 s_{12}^2 + \frac{1}{24}\left(\frac{d^2k}{ds^2}\right)_1 s_{12}^3 + \cdots \\ s - d &= \frac{1}{24} k_1^2 s_{12}^3 + \frac{1}{24} k_1 \left(\frac{dk}{ds}\right)_1 s_{12}^4 + \cdots \end{aligned} \tag{1.124}$$

where k_1, $(dk/ds)_1$, and $(d^2k/ds^2)_1$ are the curvature of the representation of the geodetic line and its derivatives at the first point.

In the general case, the curvature on a plane of the transformation of a geodetic line from the surface of an ellipsoid or sphere may be determined by the formula (Meshcheryakov 1968)

$$k = \frac{1}{r^2 p}\left(\frac{\partial(r, n)}{\partial q} \cos(i - \beta) - r \frac{\partial m}{\partial \lambda} \cos \beta - \frac{\partial i}{\partial q} \sin(i - \beta)\right) - \frac{p}{r^2 \mu^3}\left(\frac{\partial r}{\partial q} \sin \alpha\right)$$

However, in practice, measurements of angles and distance are generally taken from maps based on conformal projections. In these cases the curvatures of the conformal representation of geodetic line k_1 on the plane (at point 1) can be determined from the formula

$$k = \frac{1}{rm}\left(\frac{\partial \ln m}{\partial q}\sin\alpha - \frac{\partial \ln m}{\partial \lambda}\cos\alpha\right) \tag{1.125}$$

where q, λ are isometric coordinates. Formula (1.125) is general. Using it, we can obtain formulas for specific conformal projections. Some of these formulas are given in courses on spheroidal geodesy (Morozov 1979).

1.7 Transformation of one type of surface onto other types: that of the ellipsoid of revolution onto the surface of a sphere

1.7.1 General transformation of one type of surface onto other types

Suppose we have two regular surfaces s and σ, for which two curvilinear coordinate systems (u, v) and (ν, ϑ) are given, respectively. Suppose that a singly connected closed region Δ_1 is designated on the first surface and its corresponding region Δ_2 on the second surface.

We impose the condition that every point of the first surface corresponds to one and only one point of the second one. For an infinitesimal motion ds of a given point of the first region, the corresponding point in the second region moves an infinitesimal distance $d\sigma$ and vice versa.

Then, in a general form, the equations for transforming region Δ_1 of surface s onto region Δ_2 of surface σ can be written in the following way:

$$\nu = f_1(u, v), \quad \vartheta = f_2(u, v) \tag{1.126}$$

where f_1, f_2 are single-valued functions, continuous with their first- and second-order partial derivatives, and Jacobian $\partial(\nu, \vartheta)/\partial(u, v)$ is nowhere equal to zero in the region being represented.

It is known that in order to transform one surface onto another, it is sufficient to define these mutually represented surfaces in terms of their first quadratic forms, which can be written in the form

$$ds^2 = E\,du^2 + 2F\,du\,dv + G\,dv^2 \tag{1.127}$$

$$d\sigma^2 = E'\,d\nu^2 + 2F'\,d\nu\,d\vartheta + G'\,d\vartheta^2 \tag{1.128}$$

where E, F, G, E', F', G' are Gaussian coefficients on both surfaces

$$E = X_u^2 + Y_u^2 + Z_u^2, \quad G = X_v^2 + Y_v^2 + Z_v^2$$
$$F = X_u X_v + Y_u Y_v + Z_u Z_v \tag{1.129}$$

$$E' = x_\nu^2 + y_\nu^2 + z_\nu^2, \quad G' = x_\vartheta^2 + y_\vartheta^2 + z_\vartheta^2$$
$$F' = x_\nu x_\vartheta + y_\nu y_\vartheta + z_\nu z_\vartheta \tag{1.130}$$

However, applying equations (1.127) and (1.128) in the geometric form to transform complex surfaces is very difficult. This problem is solved much more simply if we use an isometric coordinate system (Meshcheryakov 1968).

Assume that squares of linear elements of the surface in isometric form are as follows:

- on the first surface

$$ds^2 = P^2(d\xi^2 + d\eta^2) \tag{1.131}$$

- on the second surface

$$d\sigma^2 = T^2(dx^2 + dy^2) \tag{1.132}$$

Then the local linear scale factors in any direction, when transforming the first surface onto the second one, are expressed by the formula

$$\mu^2 = \frac{T^2}{P^2}(e \cos^2 \alpha + f \sin 2\alpha + g \sin^2 \alpha) \tag{1.133}$$

Now it is easy to obtain, analogous to the methods described in sections 1.3 and 1.4, equations for transforming surfaces (Meshcheryakov 1968). In order to use these equations it is first necessary to transform from geometric to isometric forms.

When transforming the surfaces of an ellipsoid of revolution or sphere, it is not difficult to solve this problem. For a triaxial ellipsoid and more complex surfaces, however, the original curvilinear coordinate systems are not orthogonal. Therefore, it is difficult to reduce differential forms of type (1.127) and (1.128) to the isometric type, e.g. (1.131) and (1.132). So far this problem has only been partially clarified in the mathematical and specialized literature.

Carl Jacobi, in considering a general method for creating the conformal transformation of the triaxial ellipsoid onto a surface, touched partially the problem of obtaining isometric coordinates.

In 1825 Gauss developed the general theory of conformal transformation of one kind of surface onto another, by which he showed the process of obtaining various surfaces by using isometric coordinates. Meshcheryakov (1968) considered a method for creating isometric coordinates by both the Gauss method and according to the theory of harmonic functions. In the former case, it is necessary to solve the differential equations

$$E\,du + (F \pm i\sqrt{EG - F^2})\,dv = 0 \quad (i = \sqrt{-1}) \tag{1.134}$$

In the latter case, in order to obtain isometric coordinates it is necessary to develop the harmonic function ϕ on the surface and then another conjugate harmonic function ψ.

The conditions of conjugation require a whole set of equations with first-order partial derivatives, where the partial derivatives are connected by the Beltrami condition

$$\psi_u = -\frac{E_{\phi_v} - E_{\phi_u}}{\sqrt{EG - F^2}}, \quad \psi_v = \frac{G_{\phi_u} - F_{\phi_v}}{\sqrt{EG - F^2}}$$

The Cauchy–Riemann condition is a particular case of the Beltrami condition.

The solution of the given problem by the second method is reduced, finally, to integrating equations (1.134).

This method gives only a general solution of the problem; specific formulas transforming complex linear elements of the surface from the geometric form to isometric equivalents have not been developed.

We will consider one of the methods for determining the isometric coordinates for the surface of a triaxial ellipsoid in section 7.11.

1.7.2 Principal concepts in transforming the ellipsoid of revolution onto the surface of a sphere

Applying (1.126), the equations for transforming the ellipsoid onto the surface of the sphere can be rewritten in the form

$$\phi' = f_1(\phi, \lambda), \quad \lambda' = f_2(\phi, \lambda) \tag{1.135}$$

where ϕ, λ, ϕ', λ' are geographical coordinates of the ellipsoid and sphere, respectively; f_1, f_2 represent functions subject to the above limitations.

Various methods for these transformations have been developed, e.g. using geodetic transformations (the Bessel method among them) and using correspondence along normals.

The simplest is a method in which we neglect the polar oblateness and suppose that latitudes and longitudes of both the sphere and the ellipsoid are equal, i.e. $\phi' = \phi$ and $\lambda' = \lambda$. In this case, in order to reduce distortion, the radius of a sphere substituted for the ellipsoid is determined as either the mean radius of curvature along the average parallel ϕ_0 of the region being mapped (see Appendix 3)

$$R = \sqrt{M_0 N_0}$$

or as a mean radius of curvature along parallels ϕ_s and ϕ_n bounding this region

$$R = \sqrt{R_s R_n}$$

In some cases the volume of the sphere is taken to be equal to that of the Earth ellipsoid; then

$$R = \sqrt[3]{a^2 b}$$

Here M_0 and N_0 are radii of curvature of the meridian section and of the first vertical, respectively, determined from formulas (1.1) and (1.2); a, b are semiaxes of the ellipsoid of revolution, and e is the first eccentricity.

In other applications, the sphere is given a radius which provides the same surface area as the ellipsoid:

$$R = a\left[0.5\left(1 - \frac{1 - e^2}{2e} \ln \frac{1 - e}{1 + e}\right)\right]^{1/2}$$

This method of transformation can be used for designing small-scale maps, where it is possible to neglect the effect of distortion due to the ellipsoid. In mathematical cartography, conformal, equal-area, and equidistant transformations are more often used.

In addition, transformations preserving the length of the central meridian are used, as well as methods of perspective transformation of the ellipsoid onto the surface of the sphere. In the latter methods, while preserving the accuracy of calculation up to terms of e^2 (which is sufficient to solve the great majority of problems in mathematical cartography and photogrammetry) verticals a and almucantars z are also represented orthogonally on the surface of the sphere.

In all the methods used in practice it is assumed that the equatorial planes of both the ellipsoid of revolution and the sphere as well as their centers coincide, ellipsoidal parallels are represented by parallels on the sphere, the central meridians coincide and are of 0° longitude, the longitudes of the other meridians are proportional (i.e. the meridians and parallels of the ellipsoid of revolution are represented orthogonally on the surface of the sphere), and, consequently, the base directions on the transformations coincide with those of the meridians and parallels.

Let us write the squares of linear elements ds^2 for the ellipsoid and $d\sigma^2$ for the sphere in the form

$$ds^2 = M^2\, d\phi^2 + r^2\, d\lambda^2$$

$$d\sigma^2 = R^2\, d\phi'^2 + R^2 \cos^2 \phi'\, d\lambda'^2$$

where R is the radius of the sphere and $R \cos \phi'$ is the radius of curvature of a parallel of latitude on the sphere.

Then the formulas for linear scale factors take the following form: in any direction

$$\mu^2 = \frac{R^2\, d\phi'^2 + R^2 \cos^2 \phi'\, d\lambda'^2}{M^2\, d\phi^2 + N^2 \cos^2 \phi\, d\lambda^2} \tag{1.136}$$

and along the directions of meridians and parallels, respectively,

$$m = \frac{R\, d\phi'}{M\, d\phi} \tag{1.137}$$

$$n = \frac{R \cos \phi'\, d\lambda'}{N \cos \phi\, d\lambda} = \alpha' \frac{R \cos \phi'}{N \cos \phi} \tag{1.138}$$

where $\alpha' = d\lambda'/d\lambda$, the proportionality coefficient for the meridians of longitude.

1.7.3 Conformal transformation of the surface of the ellipsoid onto a sphere

From conformality terms $m = n$, $\varepsilon = 0$, and (1.137)–(1.138), we obtain a differential equation, the integration of which gives

$$q' = \alpha q + \ln C, \quad \lambda' = \alpha\lambda \tag{1.139}$$

where q', q are isometric latitudes on the surface of the sphere and of the ellipsoid of revolution, determined by formulas (1.34) and (1.31)–(1.33), respectively. Constants α and C are parameters for which the prescribed conditions result in various types of transformations.

Mollweide method

Suggested in 1807, this method is characterized by the following initial conditions. The linear scale is preserved along the equator, and the parallels and meridians of the ellipsoid and the sphere are equal along the equator and the central meridian, respectively: if $\phi = 0$ then $\phi' = 0$; if $\lambda = 0$ then $\lambda' = 0$. The constant parameters have values $\alpha = C = 1$.

After expanding equations (1.139) into a Taylor series, taking into account (1.31)–(1.34), we can obtain a formula relating the given geodetic latitude ϕ and the 'conformal' latitude ϕ' of the transformation

$$\phi' = \phi - A \sin 2\phi + B \sin 4\phi - C \sin 6\phi + \cdots \tag{1.140}$$

where

$$A = \left(\frac{e^2}{2} + \frac{5}{24} e^4 + \frac{3}{32} e^6\right) = 0.003\,356\,554 + 692.339''$$

$$B = \left(\frac{5}{48} e^4 + \frac{7}{80} e^6 + \cdots\right) = 0.000\,004\,694 = 0.968'' \tag{1.141}$$

$$C = \left(\frac{13}{480} e^6 + \cdots\right) = 0.000\,000\,008 = 0.002''$$

The greatest difference between latitudes, $\phi' - \phi$, is $11'32.34''$ at approximately parallel $\phi = 45°$ (numerical values are given for the GRS 80 ellipsoid).

The formulas for linear and areal scale factors may be written as follows:

$$m = n = \frac{R}{a}\left(1 + \frac{e^2}{2} \sin^2 \phi\right), \quad p = m^2 = \frac{R^2}{a^2}(1 - e^2 \sin^2 \phi) \tag{1.142}$$

where $R = a = 6\,378\,137$ m.

Maximum linear distortion $v_m = 0.3$ percent occurs at the poles; the greatest difference between spheroidal and spherical latitudes, $11'$, is approximately along the parallel of latitude $\phi = 45°$.

Gauss' methods

The first method was suggested in 1822, the second in 1844. In the first method the initial conditions are: (1) the scale is equal to that along the central parallel of the region being represented, if $\lambda = 0$ and $\lambda' = 0$; (2) along the central parallel of the region spheroidal and spherical latitudes are equal ($\phi_0 = \phi'_0$); and (3) the radius R of the sphere is equal to N_0, i.e. to the radius of curvature of the normal section along the parallel of latitude ϕ_0.

The constant parameters of equations (1.139) take the form

$$\alpha = 1, \quad C = \left(\frac{1 + e \sin \phi_0}{1 - e \sin \phi_0}\right)^{e/2}$$

In the second method the initial conditions are: if $\lambda = 0$ and $\lambda' = 0$, then $m_0 = 1$, $(dm/d\phi)_0 = 0$, and $(d^2m/d\phi^2)_0 = 0$ apply to the central point of the region.

V. P. Morozov suggested the following specific formulas for the above transformation in 1969 and 1979.

For the first method

$$\phi' = \phi_0 + b + P_{03} b^3 - P_{04} b^4 - P_{05} b^5, \quad \lambda' = \lambda \tag{1.143}$$

where

$$b = \frac{s - s_0}{N_0}, \quad P_{03} = \frac{\eta_0^2}{6}, \quad P_{04} = \frac{\eta_0^2 \tan \phi_0}{24}(3 + 4\eta_0^2)$$

$$P_{05} = \frac{\eta_0^2}{120}(4 - 3\tan^2 \phi_0 + 3\eta_0^2 - 24\eta_0^2 \tan^2 \phi_0 + 4\eta_0^4 - 24\eta_0^4 \tan^2 \phi_0) \tag{1.144}$$

$$\eta_0^2 = e'^2 \cos^2 \phi_0$$

s, s_0 are the arc lengths of the meridians from the equator to the given parallel and to the central parallel of the region, respectively, determined by the formula (see Appendix 4)

$$s = \frac{a}{1 + n'}\left[\left(1 + \frac{n'^2}{4} + \frac{n'^2}{64} + \cdots\right)\phi - \left(\frac{3}{2}n' - \frac{3}{16}n'^3 - \cdots\right)\sin 2\phi \right.$$
$$\left. + \left(\frac{15}{16}n'^2 - \frac{15}{64}n'^4 + \cdots\right)\sin 4\phi - \left(\frac{35}{48}n'^3 + \cdots\right)\sin 6\phi + \cdots\right] \tag{1.145}$$

where $n' = (a - b)/(a + b)$, $e'^2 = e^2/(1 - e^2)$, a, b are semiaxes of the ellipsoid of revolution, and e' is the second eccentricity of the ellipsoid of revolution.

For the second method

$$\phi' = \phi_0^* + b - P_{04} b^4 - P_{05} b^5 + \cdots, \quad \lambda' = P_0 \lambda \tag{1.146}$$

where

$$b = \frac{s - s_0}{R}, \quad R = \sqrt{M_0 N_0}, \quad P_0 = \sqrt{1 + \eta_0^2 \cos^2 \phi_0}$$

$$P_{04} = \frac{\eta_0^2 \tan \phi_0^*}{6}, \quad P_{05} = \frac{\eta_0^2}{30}(1 - 6\eta_0^2 \tan^2 \phi_0^*) \tag{1.147}$$

$$\tan \phi_0^* = \frac{\tan \phi_0}{V_0}, \quad V_0 = \sqrt{1 + \eta_0^2}$$

1.7.4 Equal-area and equidistant transformations of the ellipsoid onto the surface of a sphere

Equal-area transformation

This requires that $p = mn = 1$. Applying (1.137) and (1.138) we get a differential equation, the integration of which gives for 'authalic' latitude ϕ''

$$\sin \phi'' = \alpha \frac{a^2}{R^2}(1 - e^2)(\sin \phi + \tfrac{2}{3}e^2 \sin^3 \phi + \tfrac{3}{5}e^4 \sin^5 \phi + \cdots) + C \tag{1.148}$$

where α, C are constant parameters and R the radius of the equivalent sphere, the type of equal-area transformation depending on how they are determined. By locating ϕ at latitude ϕ'' on the sphere, true area for the ellipsoid is maintained on the sphere.

For example, using the following initial conditions, at the equator and at the pole with latitude $\phi''_0 = \phi_0 = 0$, $\phi''_{90} = \phi_{90} = 90°$, all longitudes $\lambda'' = \lambda$. Then $\alpha = 1$, $C = 0$, and, further, to the accuracy of terms to e^4

$$R = a(1 - \frac{e^2}{6} - \frac{17}{360} e^4 + \cdots) \tag{1.149}$$

$$\phi'' = \phi - A_1 \sin 2\phi + B_1 \sin 4\phi - \cdots \tag{1.150}$$

where

$$A_1 = \frac{e^2}{3} + \frac{31}{180} e^4 + \cdots, \quad B_1 = \frac{17}{360} e^4 + \cdots \tag{1.151}$$

Using the elements of the GRS 80 ellipsoid we have $A_1 = 461.864''$, $B_1 = 0.436''$, $R = 6\,371\,007$ m. The greatest difference of latitudes $\phi'' - \phi$ is 7′41.9″ approximately along the parallel $\phi = 45°$. Linear scale factors and maximum angular deformation to an accuracy of terms of e^2 are equal to

$$n = 1 - \frac{e^2}{6} \cos^2 \phi + \cdots, \quad m = 1 + \frac{e^2}{6} \cos^2 \phi + \cdots, \quad \sin \frac{\omega}{2} = \frac{e^2}{6} \cos^2 \phi + \cdots \tag{1.152}$$

Areas of maximum linear scale and angular deformation are along the equator ($\phi = 0$) and are equal to $n_e = 0.999$, $m_e = 1.001$, $\omega = 3.84'$.

Transformations equidistant along meridians

Here the condition is $m = 1$. Taking into account (1.137) we obtain a differential equation, the integration of which gives (in radians)

$$\phi''' = s/R + C \tag{1.153}$$

where ϕ''' is a 'rectified' latitude with spacing on the sphere at true distances of ϕ on the ellipsoid; s is the arc length of the meridian from the equator to the given parallel and is determined by equation (1.145), C is a constant (taken to be zero), and R is the radius of the sphere.

On imposing the condition that the arc lengths of the meridian from the equator to the poles on the sphere and on the ellipsoid are equal, we obtain

$$R = \frac{a}{1 + n'} \left(1 + \frac{n'^2}{4} + \frac{n'^4}{64} + \cdots\right) \tag{1.154}$$

Using the GRS 80 ellipsoid, $R = 6\,367\,449$ m. Along parallels, the linear and area scale factors as well as the maximum angular deformation (ω' in minutes), to an accuracy of terms with e^2, can be found by the formulas

$$n = p = 1 - \frac{e^2}{4} \cos 2\phi + \cdots, \quad \omega' = \frac{e^2}{4} \rho' \cos 2\phi + \cdots \tag{1.155}$$

where ρ' is the conversion factor from radians to minutes ($180 \times 60/\pi$).

Transformations equidistant along parallels

Applying the condition $n = 1$ and (1.138) we can get a differential equation and, after its integration, the expression

$$\cos \phi^{\text{IV}} = \frac{a}{\alpha R}\left(1 + \frac{e^2}{2} \sin^2 \phi + \frac{3}{8} e^4 \sin^4 \phi + \cdots\right)\cos \phi \tag{1.156}$$

We can develop a number of these transformations depending on the values of α, R, and the given original parallel, where $\alpha = \lambda'/\lambda$. In particular, if we take the initial conditions to be such that the equatorial and polar latitudes are $\phi_0^{\text{IV}} = \phi_0 = 0$, $\phi_{90}^{\text{IV}} = \phi_{90} = 90°$, and the longitudes $\lambda^{\text{IV}} = \lambda$, then $\alpha = 1$, $R = a$.

Applying (1.156) we get

$$\tan \phi^{\text{IV}} = (1 - e^2)^{1/2} \tan \phi \tag{1.157}$$

i.e. latitude ϕ^{IV} for this transformation is equal to reduced latitude u.

The formulas for the linear and area scale factors and the maximum angular deformation, all along meridians, will take the form

$$m = p = 1 + \frac{e^2}{2} \cos^2 \phi + \frac{e^4}{8} (3 - 2 \sin^2\phi - \sin^4 \phi) + \cdots \tag{1.158}$$

$$\omega' = \frac{e^2}{2} \rho' \cos^2 \phi + \cdots \tag{1.159}$$

1.7.5 Methods of transformation in which the meridians and parallels on the ellipsoid do not coincide with those on the sphere

Conformal transformation preserving scale along the central meridian

The initial conditions are as follows: the transformation is symmetrical about the central meridian; the longitude of the central meridian is $\lambda_0 = \lambda_0' = 0$; the region being transformed is narrow and extends along the central meridian; the latitudes of the equator and pole are, respectively, equal to $\phi_0' = \phi_0 = 0$, $\phi_{90}' = \phi_{90} = 90°$; the transformation is conformal and arc lengths along the central meridian are preserved.

After expanding equations (1.135) into a Taylor series in powers of $l = (\lambda - \lambda_0)$, and taking into account the above conditions, we obtain the latitudes and longitudes of the given transformation:

$$\begin{aligned} \phi' &= \phi_m' + a_2 l^2 + a_4 l^4 + \cdots \\ \lambda' &= a_1 l + a_3 l^3 + a_5 l^5 + \cdots \end{aligned} \tag{1.160}$$

where

$$\phi'_m = \phi - \left(\frac{3}{2}n' - \frac{9}{16}n'^3 + \cdots\right)\sin 2\phi + \left(\frac{15}{16}n'^2 - \cdots\right)\sin 4\phi$$
$$- \left(\frac{35}{48}n'^3 - \cdots\right)\sin 6\phi + \cdots$$

$$a_1 = \left(1 + \frac{n'}{2} + \frac{n'^2}{4} - \frac{7}{96}n'^3 + \cdots\right) + \left(\frac{n'}{2} - \frac{n'^2}{8} - \frac{5}{12}n'^3 + \cdots\right)\cos 2\phi$$
$$- \left(\frac{3}{8}n'^2 + \frac{5}{96}n'^3\right)\cos 4\phi + \left(\frac{7}{24}n'^3 \cdots\right)\cos 6\phi + \cdots$$

$$a_2 = \left(\frac{n'}{2} + \frac{21}{32}n'^2 + \frac{25}{64}n'^3 + \cdots\right)\sin 2\phi + \left(\frac{n'}{8} + \frac{3}{32}n'^2 - \frac{15}{32}n'^3 + \cdots\right)\sin 4\phi$$
$$- \left(\frac{5}{32}n'^2 + \frac{41}{192}n'^3 + \cdots\right)\sin 6\phi + \cdots$$

$$a_3 = \left(\frac{n'}{24} + \frac{31}{96}n'^2 + \cdots\right) + \left(\frac{n'}{6} + \frac{37}{48}n'^2 + \cdots\right)\cos 2\phi$$
$$+ \left(\frac{n'}{8} + \frac{15}{32}n'^2 + \cdots\right)\cos 4\phi + \left(\frac{n'^2}{48} + \cdots\right)\cos 6\phi + \cdots$$

$$a_4 = \left(\frac{7}{96}n' + \frac{9}{16}n'^2 + \cdots\right)\sin 2\phi + \left(\frac{n'}{12} + \frac{83}{128}n'^2 + \cdots\right)\sin 4\phi$$
$$+ \left(\frac{n'}{32} + \frac{13}{48}n'^2 + \cdots\right)\sin 6\phi + \cdots$$

$$a_5 = \left(\frac{n'}{240} + \cdots\right) + \left(\frac{17}{480}n' + \cdots\right)\cos 2\phi + \left(\frac{n'}{16} + \cdots\right)\cos 4\phi$$
$$+ \left(\frac{n'}{32} + \cdots\right)\cos 6\phi + \cdots$$

$$n' = (a - b)/(a + b)$$

Negative perspective transformation of the ellipsoid onto the surface of a sphere
Let the surface of the sphere be tangent to the ellipsoid of revolution at a given point $Q_0(\phi_0, \lambda_0)$, which is the pole of the polar spheroidal coordinate system (Figure 1.11).

Let us introduce the following symbols:

$$O'Q_0 = N_0, \quad O'C = N'_0, \quad S_n O' = D$$

Hence,

$$\sin z_{sp} = \frac{N'_0 \sin z}{D + N'_0 \cos z}\left(\cos z_{sp} + \frac{D}{R}\right) \tag{1.161}$$

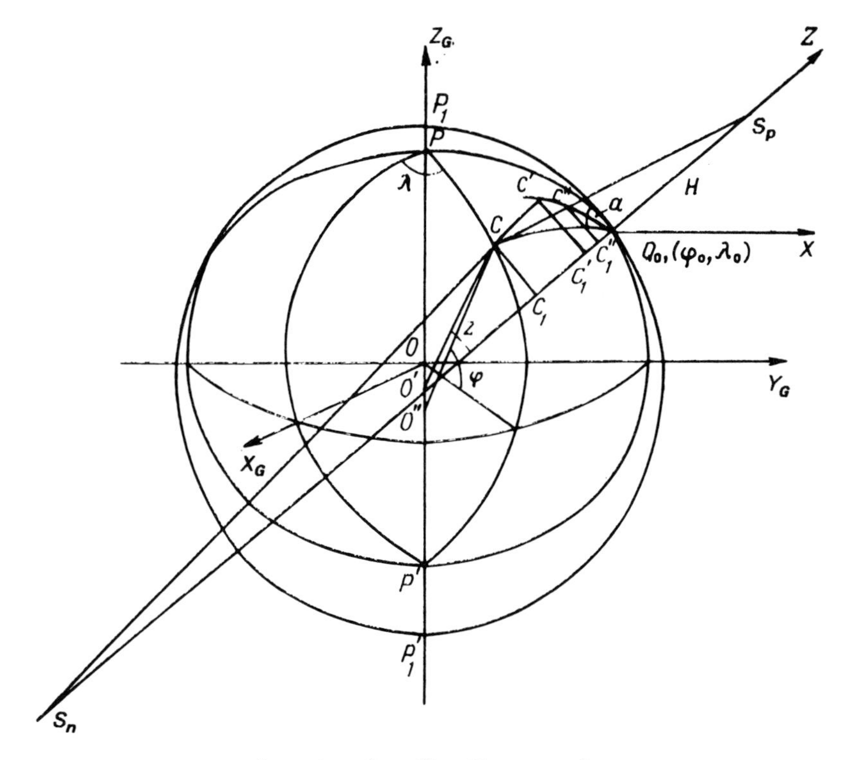

Figure 1.11 Perspective transformation of an ellipsoid onto a sphere.

where z, z_{sp} are the arc distances $Q_0 C$ on ellipsoid and sphere, respectively. After expanding this equation into a Taylor series in powers of $\Delta z = (z_{sp} - z)$, we obtain the formula relating spherical and spheroidal latitudes to an accuracy of terms to e^2:

$$z_{sp} = z - \frac{e^2}{2} [\sin z \cos a \cos \phi_0 + \sin \phi_0 (\cos z - 1)]^2$$

$$\times \frac{D \sin z}{N_0 + D \cos z} \tag{1.162}$$

where z, a are determined from equations (1.11)–(1.12), and N_0' is determined from (1.8).

To the same degree of accuracy we obtain the local linear scale factors μ_1 along verticals and μ_2 along almucantars from the formulas

$$\mu_1 = \left(1 + \frac{e^2}{2} [\sin z \cos a \cos \phi_0 + \sin \phi_0 (\cos z - 1)]^2\right) \frac{dz_{sp}}{dz} \tag{1.163}$$

$$\mu_2 = \left(1 + \frac{e^2}{2} [\sin z \cos a \cos \phi_0 + \sin \phi_0 (\cos z - 1)]^2\right) \frac{\sin z_{sp}}{\sin z} \tag{1.164}$$

It follows from formulas (1.161)–(1.164) that we can get a set of various perspective transformations, depending on the position of the point of view (value D). For example, when $D = 0$ (projecting from the center of the sphere) we get $z_{sp} = z$,

$$\mu_1 = \mu_2 = 1 + \frac{e^2}{2} [\sin z \cos a \cos \phi_0 + \sin \phi_0 (\cos z - 1)]^2, \quad p = \mu^2 \quad (1.165)$$

i.e. to an accuracy of its terms to e^2, this perspective transformation of the surface of the ellipsoid onto that of a sphere tangent to the ellipsoid at a given point is a conformal projection.

This makes it possible to generalize V. V. Kavrayskiy's deduction about the properties of a central perspective projection and to note that any perspective projection of the ellipsoid onto the surface of a sphere provides a transformation similar to a conformal projection (to an accuracy of its terms to e^2), if the point of view is the center of the sphere and does not depend on the distance of the center from the center of the ellipsoid along the axis of revolution or on the position of the pole of the polar spheroidal coordinate system.

Positive perspective transformation of the ellipsoid onto the surface of a sphere
Let us denote $S_p O' = D$, $S_p Q_0 = H$ (Figure 1.11); hence,

$$\sin z_{sp} = \frac{N'_0 \sin z}{D - N'_0 \cos z} \left(\frac{D}{R} - \cos z_{sp} \right) \quad (1.166)$$

After expanding this equation in powers of $\Delta z = (z_{sp} - z)$, we obtain, to an accuracy of its terms to e^2,

$$z_{sp} = z + \frac{e^2}{2} [\sin z \cos a \cos \phi_0 + \sin \phi_0 (\cos z - 1)]^2 \frac{D \sin z}{N_0 - D \cos z} \quad (1.167)$$

$$\mu_1 = 1 + \frac{e^2}{2} \tau \left(\tau + \frac{2\tau_1 D \sin z}{N_0 - D \cos z} + \frac{D\tau(N_0 \cos z - D)}{(N_0 + D \cos z)^2} \right) + \cdots$$

$$\mu_2 = 1 + \frac{e^2}{2} \tau^2 \left(1 + \frac{D}{N_0 - D \cos z} \right) \quad (1.168)$$

$$p = \mu_1 \mu_2, \quad \omega = 2 \arcsin \left(\frac{\mu_1 - \mu_2}{\mu_1 + \mu_2} \right) \quad (1.169)$$

$$\tau = \sin z \cos a \cos \phi_0 + \sin \phi_0 (\cos z - 1)$$

$$\tau_1 = \cos z \cos a \cos \phi_0 - \sin z \sin \phi_0$$

Projections with double or triple transformation
The formulas for map projections transforming the ellipsoid directly onto a plane are very often awkward. Furthermore, their application to regions far from the pole, with complex outlines extending in random directions, does not provide us with the possibility of obtaining minimal distortion properly distributed within the boundaries of the regions being mapped. In these cases so-called double or triple transformation may be used.

In order to achieve this we

1. determine the desired type of distortion to be obtained with the transformation, and, following that,
2. transform the corresponding surface of the ellipsoid onto the sphere in appropriate fashion; from geodetic coordinates of the points on the surface of the ellipsoid, we obtain spherical coordinates ϕ',λ' for these points, with the pole of this coordinate system at the geographical pole;
3. choose or determine from equations (1.170)–(1.171) the pole of the new polar spherical coordinate system, and from the values of points ϕ', λ' and formula (1.13) calculate polar coordinates z, a;
4. denote $\phi'' = 90° - z$, $\lambda'' = -a$, and substitute them in place of spherical geographical coordinates ϕ', λ', calculating rectangular coordinates x, y and the characteristics of the appropriate map projection in accordance with the degree and nature of the distortion using known formulas. To obtain a projection with triple transformation, the rectangular coordinates x, y obtained are additionally transformed in accordance with the given conditions. Then we can obtain the final values of the rectangular coordinates and the projection characteristics.

1.8 Classification of map projections

Map projections can be classified according to various features:

- by the properties of the transformation (distortion characteristics);
- by the shape of the normal graticule of meridians and parallels;
- by the orientation or aspect of the map graticule depending on the location of the pole of the coordinate system adopted;
- by the form of differential equations defining the map projection;
- by the method of obtaining the projection, etc.

Most commonly, the projection classification is based on three principal features, which are the first three listed above, i.e. the distortion characteristics, the shape of the normal graticule, and the graticule orientation.

Soviet scientists developed new classifications of projections based on various principles. For example, Meshcheryakov (1968) suggested that projections be classified by the form of differential equations and gave the basis for the so-called genetic classification. This classification is complete enough and interesting but not easily understood, as it is not related to the shape of the graticule of meridians and parallels. Methods of projection and, consequently, their classification by this feature are given adequately in Chapter 6.

1.8.1 Classification of map projections by distortion characteristics

Based on distortion characteristics, projections are divided into conformal, equal-area (or equivalent), and arbitrary categories.

On conformal projections the similarity of infinitesimal parts of the representation is preserved and, hence, the local linear scale does not depend on direction

($m = n = a = b = \mu$). Angular distortion is absent ($\omega = 0$), and the area scale is equal to the square of the linear scale ($p = a^2$).

The condition of conformality is characterized by the equations considered in section 1.5.1. When transforming regions of finite dimensions using these projections, the change in local linear scale characterizes the distortion of the final outlines. Areas can be greatly distorted using these projections.

On equal-area projections the relationship of areas on the territory to be mapped and on the plane remains constant. This constant area relationship applies not only to infinitesimal plots but also to plots of finite areas.

On these projections the local area scale factor p is a constant (most often $p = 1$). For all ellipsoids of revolution the condition of equivalency is of the form (see equation (1.47) for h)

$$h = Mr$$

For the surface of a sphere it is

$$h = R^2 \cos \phi$$

Extreme linear scale factors are inversely proportional to each other:

$$a = 1/b \quad \text{and} \quad b = 1/a$$

The maximum angular deformation on these projections is preferably calculated from formulas with tangents, which are in the following form for equal-area projections:

$$\tan(\omega/2) = (a - b)/2 \quad \text{or} \quad \tan(\pi/4 + \omega/4) = a$$

where a and b are extreme linear scale factors at a given point.

Projections arbitrary in distortion characteristics are neither conformal nor equal area. Both areas and angles are distorted on these projections, i.e. p varies, and ω is not zero.

Among arbitrary projections we should distinguish equidistant projections where the extreme linear scale along one of the main directions remains constant, i.e. $a = 1$ or $b = 1$. On these projections, respectively, $p = b$ or $p = a$.

To compute the maximum angular deformation it is advisable to use the general formula (1.102)

$$\sin(\omega/2) = (a - b)/(a + b)$$

If the graticule of the equidistant projection is orthogonal, then the base directions coincide with meridians and parallels, and these projections are, consequently, called equidistant along meridians or equidistant along parallels. In practice, however, the term 'equidistant projection' applies to equidistance along meridians or verticals (see section 1.8.3), unless stated otherwise.

With the development of the theory and practice of mathematical cartography, concepts of projection properties and the ability to assess them changed and were made more precise. Extensive utilization of arbitrary projections gave rise to requirements for the criteria used in assessing these projections.

Until recently, distortion characteristics have often been assessed only intuitively, without any quantitative evaluation. Konusova's research proved that distortion characteristics can be defined as a relationship of various types of distortion.

She suggested vector $\bar{\rho}$ as a common index of the value and character of distortion at any point on a projection; for projections which are distorted in area $\bar{\rho} = (p - 1)$, and in shape $\bar{\rho} = (\omega - 1)$, where $\omega = a/b$.

1.8.2 Classification of map projections by the shape of the normal graticule of meridians and parallels

The normal graticule of meridians and parallels is obtained when a normal system of polar spherical coordinates is used (the pole of this coordinate system coincides with the geographical pole); projections with such a graticule are called normal projections.

The classification of projections by the shape of the normal graticule of meridians and parallels is the most convenient, simplest, and most easily understood. Suggested by the outstanding Russian cartographer V. V. Kavrayskiy in the 1930s, this classification played a positive role in both using known and creating new variants of map projections, and in their application to the design of various maps. In its present form Kavrayskiy's classification does not embrace many new map projections, whether with constant or variable curvature of parallels.

A further development is a new classification of projections made by the Department of Map Design and Compilation of the Moscow Engineering Institute of Geodesy, Aerophotography, and Cartography (MIIGAiK) (Starostin *et al.* 1981). According to the authors' idea, this classification embraces all possible sets of map projections and consists of two subsets. The first includes projections with parallels of constant curvature, and the second subset includes projections with parallels of various curvatures.

The first subset, depending on the shape of parallels, can be divided into three families: in the first, the parallels are straight; in the second, they are concentric circles; and in the third, they are eccentric circles. Each family is subdivided into classes according to the shape of meridians.

The first family (with straight parallels) consists of four classes:

1. Cylindrical projections, the general formulas for which are

$$x = \beta\lambda, \quad y = f(\phi)$$

   where β is a projection parameter.
2. Generalized cylindrical projections:

$$x = f_1(\lambda), \quad y = f_2(\phi)$$

   This class can be divided into two subclasses, one with the graticule symmetrical about the central meridian, and the other with the graticule asymmetrical.
3. Pseudocylindrical projections:

$$x = f_1(\phi, \lambda), \quad y = f_2(\phi)$$

   This class of projections can also be divided into two subclasses with respect to symmetry of the graticule about the central meridian.
4. Cylindrical–conic projections in which parallels are represented by a set of straight lines and meridians by concentric circles.

The second family (with concentric parallels) consists of five classes:

1. Conic projections:

$$\rho = f(c, \phi), \quad \delta = \alpha\lambda$$
$$x = \rho \sin \delta, \quad y = q - \rho \cos \delta$$

   where α and c are projection parameters, and q is constant. There is an interruption at the pole of the projection.
2. Generalized conic projections where all the formulas mentioned apply, except that the formula for the polar angle is $\delta = f_2(\lambda)$. There is also an interruption at the pole point.

   Projections of these two classes can be divided into two subclasses in view of the symmetry of the graticule about the central meridian.
3. Pseudoconic projections:

$$\rho = f_1(\phi), \quad \delta = f_2(\phi, \lambda)$$
$$x = \rho \sin \delta, \quad y = q - \rho \cos \delta$$

   where q is constant. This class also includes two subclasses: in the first subclass the graticule is symmetrical about the straight central meridian, and in the second one it is asymmetrical and the central meridian can be represented either by a straight or by a curved line.
4. Azimuthal projections:

$$\rho = f_1(\phi), \quad \delta = \lambda$$
$$x = \rho \sin \delta, \quad y = \rho \cos \delta$$

5. Pseudoazimuthal projections on which parallels are concentric circles and meridians of longitude are generally curved lines, although particular meridians may be straight. The general equations for these projections are

$$\rho = f_1(z), \quad \delta = f_2(z, a) = a + f_3(z) \sin ka$$
$$x = \rho \sin \delta, \quad y = \rho \cos \delta$$

   where k is an integer. This class is subdivided into two subclasses: in one the map graticule is symmetrical about meridians of longitude 0° and 180°, and in the other one the graticule is asymmetrical.

Two classes of projections are referred to the third family (with eccentric parallels).

1. Polyconic projections in a broad sense:

$$\rho = f_1(\phi), \quad \delta = f_2(\phi, \lambda)$$
$$x = \rho \sin \delta, \quad y = q - \rho \cos \delta$$
$$q = f_3(\phi)$$

   This class includes four subclasses. The basis for the division lies in the symmetry of the map graticule about a straight meridian, about the equator, about both the meridian and equator, or with asymmetry.

2. Polyconic projections in a narrow sense:

$$\rho = N \cot \phi, \quad \delta = f(\phi, \lambda)$$

$$x = \rho \sin \delta, \quad y = q - \rho \cos \delta$$

$$q = ks + N \cot \phi$$

   where s is the length of the meridian arc and k is a coefficient. This class includes two subclasses: the basis for these subdivisions is the symmetry of the graticule about the central meridian.

The second subset of projections consists of three families. They are divided in accordance with the representation of the pole and the form of the equations.

The first family (with no interruptions on the projection in the vicinity of the pole) includes two classes.

1. Polyazimuthal projections, where parallels are represented by ellipses, and meridians by a set of straight or curved lines radiating from the center of the ellipses.

   The general equations for these projections are

$$\rho = f_1(\phi, \lambda), \quad \delta = f_2(\phi, \lambda) = \lambda + \Phi(\phi)\sin k\lambda$$

$$x = \rho \sin \delta, \quad y = \rho \cos \delta$$

2. Generalized polyazimuthal projections on which parallels are curved lines of arbitrary curvature, and meridians are a set of straight or curved lines radiating from the pole point. The general equations are of the same form as those for the class of polyazimuthal projections.

The second family (with an interruption in the vicinity of the pole) includes four classes of projections which can be called generalized polyconic. These have, respectively, parallels that are elliptical, parabolic, hyperbolic, and of any curvature.

The general equations for these projections are

$$\rho = f_1(\phi, \lambda), \quad \delta = f_2(\phi, \lambda)$$

$$x = \rho \sin \delta, \quad y = q - \rho \cos \delta$$

$$q = f_3(\phi)$$

All classes of generalized polyconic projections, except projections with hyperbolic parallels, are divided into four subclasses according to the symmetry of the graticule. Projections with hyperbolic parallels are subdivided into two subclasses (according to the same property).

The third family includes two classes of polycylindrical projections on which there is also an interruption at the pole, but the equations for these projections are expressed only in terms of rectangular coordinates (which are inherent to cylindrical projections):

$$x = f_1(\phi, \lambda), \quad y = f_2(\phi, \lambda)$$

Meridians and parallels on these projections are represented by curved lines of arbitrary curvature. Classes of projections are distinguished by the method of obtaining rectangular coordinates. To the first class belong projections for which rectangular coordinates are given in an analytical form; to the second class belong projections with rectangular coordinates given in the form of a table. Each class of cylindrical projections includes four subclasses divided according to the symmetry of the graticule.

The above classification of projections includes not only known projections but also every projection that can be created in the future, except projections for anamorphous maps (see section 7.10).

There have also been other classifications of projections using the shape of the meridians and parallels. Most detailed was that by Maurer (1935) in Germany. He adapted the pattern that Swedish botanist Carolus Linnaeus applied to plants and animals in the eighteenth century, and he listed numerous subclasses of map projections.

1.8.3 Classification of projections by the orientation of the map graticule, depending on the location of the pole of the coordinate system adopted

Depending on the orientation of the map graticule, map projections can be divided into direct, transverse, and oblique aspects of the projections. The basis for this classification lies in the latitude ϕ_0 of pole Q of the coordinate system used.

When $\phi_0 = 90°$ we have the direct aspect, when $\phi_0 = 0°$ we have a transverse aspect, and when $0° < \phi_0 < 90°$, an oblique aspect. Therefore, on direct aspects, pole Q of the coordinate system coincides with geographical pole P, and the graticule of meridians and parallels is simplest, or normal.

On oblique and transverse aspects, the shape of meridians and parallels is generally more complex. On these projections a normal graticule is obtained with an additional graticule consisting of arcs of verticals and almucantars. Verticals on projections of the sphere are arcs of great circles which cross at the points representing the poles of the oblique (or transverse) system Q (see also section 1.1.3).

The location of verticals on the region to be mapped is determined by azimuth α, which is equal to the bihedral angle between the planes of the current and initial great circles. The initial vertical is the one coinciding with the meridian of the oblique or transverse coordinate system, i.e. with longitude λ_0.

Almucantars are small circles perpendicular to the verticals. Their locations on the region to be mapped are determined by coordinate z, the zenith distance, which is equal to the arc of the vertical from the pole of coordinate system Q to the current almucantar.

The graticule of verticals and almucantars can be regarded as analogous to that of meridians and parallels, respectively, for which geographical pole P is replaced by the pole of the oblique or transverse coordinate system Q.

The transformation from geographical coordinates to the polar spherical coordinates of an oblique or transverse system is accomplished with formulas considered in section 1.1.3.

1.8.4 Choosing poles for oblique and transverse polar spherical coordinates $Q(\phi_0, \lambda_0)$

When determining the pole for oblique and transverse aspects, there can be three cases. In the first case, which can be applied to a majority of azimuthal projections, pole Q is conceptually made to coincide with the central point of the region being mapped. The coordinates of the pole are determined directly from a map or globe or computed as an average value of the latitudes and longitudes of points situated at the edges of the territory being represented.

In the second case, which is utilized in oblique and transverse cylindrical projections, the coordinates of pole Q are determined in accordance with the position of the arc of the great circle located at a distance of 90° from the pole (the equator of an oblique or transverse system). On transverse projections this great circle coincides with a meridian.

On these projections $\phi_0 = 0$ and $\lambda_0 = \lambda_{\text{cen}} \mp 90°$, where λ_{cen} is the longitude of the central meridian. If the longitude is counted from the central meridian, then $\lambda_0 = \mp 90°$.

On oblique projections, when determining the coordinates of pole Q, it is necessary to solve two spherical triangles. First angle u_1 is found with the formula

$$\tan u_1 = \tan(\lambda_2 - \lambda_1)\cos\chi \csc(\chi - \phi_1)$$

where χ is an additional angle determined by the formula

$$\tan\chi = \tan\phi_2 \sec(\lambda_2 - \lambda_1)$$

We then obtain ϕ_0 and λ_0 with the following formulas:

$$\sin\phi_0 = \cos\phi_1 \sin u_1, \quad \tan(\lambda_0 - \lambda_1) = \csc\phi_1 \cot u_1 \tag{1.170}$$

In the third case the coordinates of the pole of an oblique or transverse system are determined by considering the position of the small circle passing through the center of the region to be mapped. This method should be applied, for example, when making oblique conic projections.

Knowing the direction of this small circle, we need to find the point of intersection of great circles orthogonal to the small circle and drawn through three of its points with coordinates ϕ_1, λ_1, ϕ_2, λ_2, and ϕ_3, λ_3. Applying the formulas relating polar and geographical coordinates considered in section 1.1.3, we get

$$\begin{aligned}\tan(\lambda_0 - \lambda_1) = {} & \{(\sin\phi_1 - \sin\phi_3)[\cos\phi_1 - \cos(\lambda_2 - \lambda_1)] \\ & - (\sin\phi_1 - \sin\phi_2)[\cos\phi_1 - \cos\phi_3 \cos(\lambda_3 - \lambda_1)]\}/[(\sin\phi_1 \\ & - \sin\phi_3)\cos\phi_2 \sin(\lambda_2 - \lambda_1) - (\sin\phi_1 - \sin\phi_2)\cos\phi_3 \sin(\lambda_3 - \lambda_1)]\end{aligned}$$

$$\tan\phi_0 = \frac{\cos\phi_2 \cos(\lambda_0 - \lambda_2) - \cos\phi_1 \cos(\lambda_0 - \lambda_1)}{\sin\phi_1 - \sin\phi_2} \tag{1.171}$$

For increased reliability, latitude ϕ_0 can be determined again from another pair of points, e.g. the first and third, or the second and third.

2

Map projections with straight parallels

2.1 Cylindrical projections

2.1.1 General formulas for cylindrical projections

The normal graticule for cylindrical projections has the simplest form: meridians are represented by equally spaced straight parallel lines, and parallels by straight parallel lines orthogonal to meridians. These projections can be conformal, equal area, or arbitrary (in particular cases they can be equidistant along meridians or arbitrary with a given distribution of distortion).

On cylindrical projections the surface being mapped is taken to be that of an ellipsoid or a sphere. The base directions on these projections coincide with meridians and parallels; therefore, scales along meridians and parallels are the extreme values for each point.

General formulas for normal ellipsoidal cylindrical projections have the form

$$x = \beta\lambda, \quad y = f(\phi)$$

$$m = \frac{dy}{M\,d\phi}, \quad n = \frac{\beta}{r}, \quad p = mn = \frac{\beta\,dy}{Mr\,d\phi}$$

$$\sin\frac{\omega}{2} = \frac{a-b}{a+b} \quad \text{or} \quad \tan\left(\frac{\pi}{4} + \frac{\omega}{4}\right) = \sqrt{\frac{a}{b}}$$

where a and b are extreme linear scale factors.

When constant parameter β is obtained, a condition is imposed so that the nominal scale is preserved along standard parallels of latitude $\pm\phi_k$. From this condition

$$n_k = \beta / r_k = 1$$

we get

$$\beta = r_k$$

i.e. β equals the radius of the standard parallels or parallels of true scale. If $\phi_k = 0$, then $\beta = a$, where a is the semimajor axis of the ellipsoid.

Distortion on normal cylindrical projections is a function of latitude only; therefore, isocols (lines of equal distortion) coincide with parallels and are straight lines.

2.1.2 Conformal cylindrical projections

On normal conformal cylindrical projections ordinate y is determined so that the linear scale is independent of direction; thus $m = n$. Substituting values for the scale in the given formula, we get

$$y = \beta \int [(M\, d\phi)/(N \cos \phi)] + C$$

It is known that

$$\int [(M\, d\phi)/(N \cos \phi)] = \ln U$$

Then

$$y = \beta \ln U + C$$

The constant of integration C can be made zero so that $y = 0$ when $\phi = 0$ at the equator.

Then the projection formulas take the form

$$x = r_k \lambda, \quad y = r_k \ln U$$

$$m = n = r_k/r$$

where U is determined from equations (1.31)–(1.32) (see also Appendix 5), and λ is in radians (1 radian $= 180°/\pi = 57.295\,779\,5°$).

Conformal cylindrical projections are known as Mercator projections. They possess a special property for showing loxodromes and are often used in making charts (Figure 2.1). A loxodrome (also called a rhumb line) is a line intersecting meridians at one and the same angle. On conformal cylindrical projections, loxodromes are represented by straight lines.

In marine navigation, distances are measured in nautical miles (1 n.m. = 1852 m). Ordinate y on a Mercator projection (when $\phi_0 = 0$) is called the meridional distance. The formulas are

$$x = a\lambda, \quad y = D' = a \ln U$$

where $a = 360 \times 60/(2\pi) = 3437.747$ nautical miles.

The projection of a sphere with one standard parallel (the equator) is the simplest conformal cylindrical projection, a strictly Mercator projection. The formulas for this projection are

$$x = R\lambda, \quad y = R \ln \tan(\pi/4 + \phi/2)$$

$$m = n = \sec \phi, \quad p = \sec^2 \phi, \quad \omega = 0$$

where ϕ and λ are in radians. Angles will regularly be calculated in radians in subsequent formulas, except when the degree (°) sign is indicated, or when the function allows calculation in degrees or radians.

An analysis of the formulas shows that scale changes most slowly near the equator; hence, a normal cylindrical projection is advantageous when designing maps of an equatorial zone, symmetrical about the equator and extending along a latitude (including large-scale charts). Angle α at which a loxodrome intersects meridians is computed by the formula

$$\tan \alpha = (x_B - x_A)/(y_B - y_A) = (\lambda_B - \lambda_A)/(\ln U_B - \ln U_A)$$

where A and B denote points through which the loxodrome is drawn.

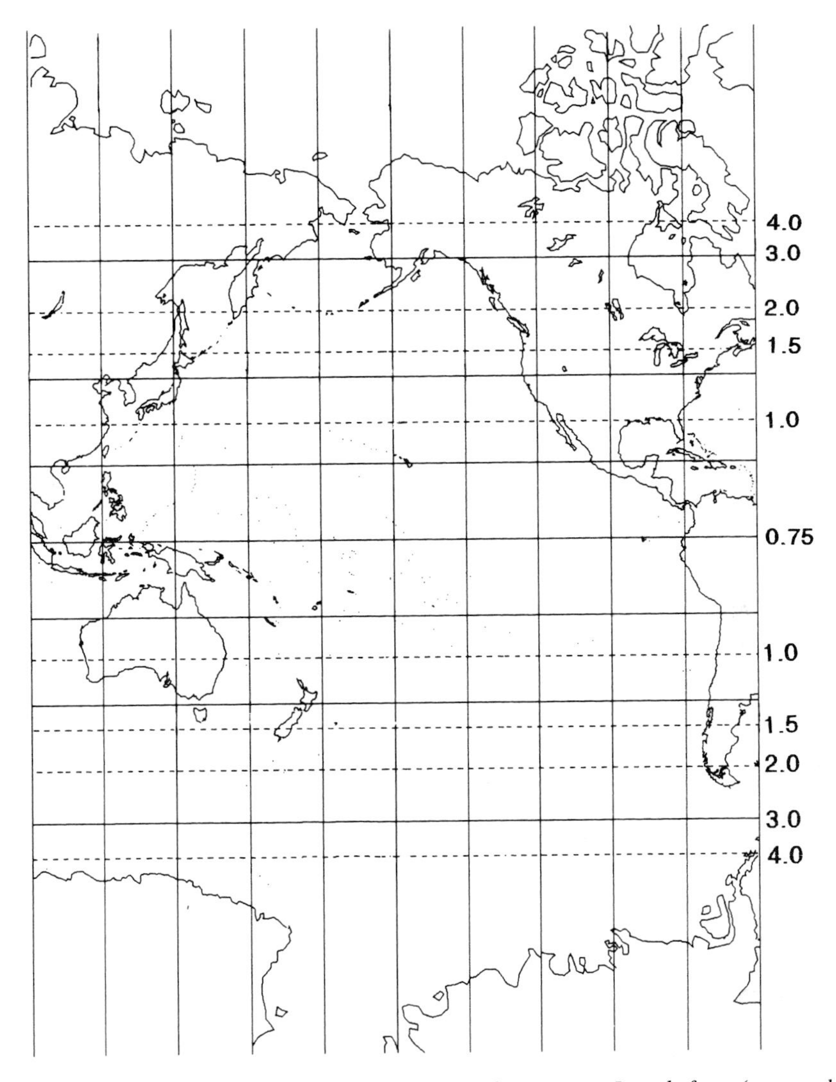

Figure 2.1 Normal Mercator conformal cylindrical projection. Isocols for p (area scale factors). 20° graticule.

A loxodrome is not the shortest distance between two points; its length can be determined from the formulas s_{le} for an ellipsoid and s_{lg} for a sphere:

$$s_{le} = \sec\alpha\,(s_B - s_A), \quad s_{lg} = R\sec\alpha\,(\phi_B - \phi_A)$$

where $(\phi_B - \phi_A)$ is in radians.

2.1.3 Equal-area cylindrical projections

On equal-area cylindrical projections, usually called cylindrical equal-area, the ordinate is determined so that the area relationship is preserved:

$$p = mn = 1$$

Incorporating the local area scale factor,

$$y = \frac{1}{\beta}\int Mr\,d\phi + C = \frac{1}{r_k}S + C$$

where C is a constant of integration which can be set to zero to make $y = 0$ when $S = \phi = 0$ on the equator; S is the area of a spheroidal quadrangle from the equator to a given parallel for a longitude difference of 1 radian and is determined by the formula (see Appendix 5)

$$S = b^2(\sin\phi + \tfrac{2}{3}e^2\sin^3\phi + \tfrac{3}{5}e^4\sin^5\phi + \tfrac{4}{7}e^6\sin^7\phi + \cdots) \qquad (2.1)$$

where $e^2 = 0.006\,694\,380$, $b = 6\,356\,752.3$ m (for the GRS 80 ellipsoid).

These projections are usually used for designing general-purpose small-scale maps, so that the surface being mapped is assumed to be a sphere. Then

$$y = (R/\cos\phi_k)\sin\phi + C$$

The general formulas for the cylindrical equal-area projections of a sphere are

$$x = R\lambda\cos\phi_k, \quad y = R\sec\phi_k\sin\phi$$

$$m = \cos\phi/\cos\phi_k, \quad n = \cos\phi_k/\cos\phi$$

$$p = 1, \quad \tan(\pi/4 + \omega/4) = a = n$$

where a is the maximum linear scale factor.

The cylindrical equal-area projection (Figure 2.2) preserving scale along the equator, or $\beta = R$ in the general formulas for cylindrical projections, was presented by J. H. Lambert in 1772, and is sometimes given his name; it may also be called an isocylindrical projection. The coordinates can be obtained from the following formulas:

$$x = R\lambda, \quad y = R\sin\phi$$

$$m = \cos\phi, \quad n = \sec\phi$$

$$p = 1, \quad \tan(\pi/4 + \omega/4) = \sec\phi$$

Other cylindrical equal-area projections have been presented by James Gall of Scotland in 1855, Walter Behrmann (1910) of Germany, and Trystan Edwards of England in 1953. The only change is in ϕ_k, which is 45°, 30°, and about 51°, respectively. In addition Arno Peters of Germany reinvented Gall's version in about 1970.

2.1.4 Cylindrical projections equidistant along meridians

On normal cylindrical projections equidistant along meridians, the ordinate is determined so that the true scale along all meridians is preserved, i.e.

$$m = dy/(M\,d\phi) = 1$$

Integrating this equation and letting the constant of integration $C = 0$, we obtain the following formulas for equidistant cylindrical projections, also called equirectangular projections.

When transforming the ellipsoid:

$$x = r_k\lambda, \quad y = s$$

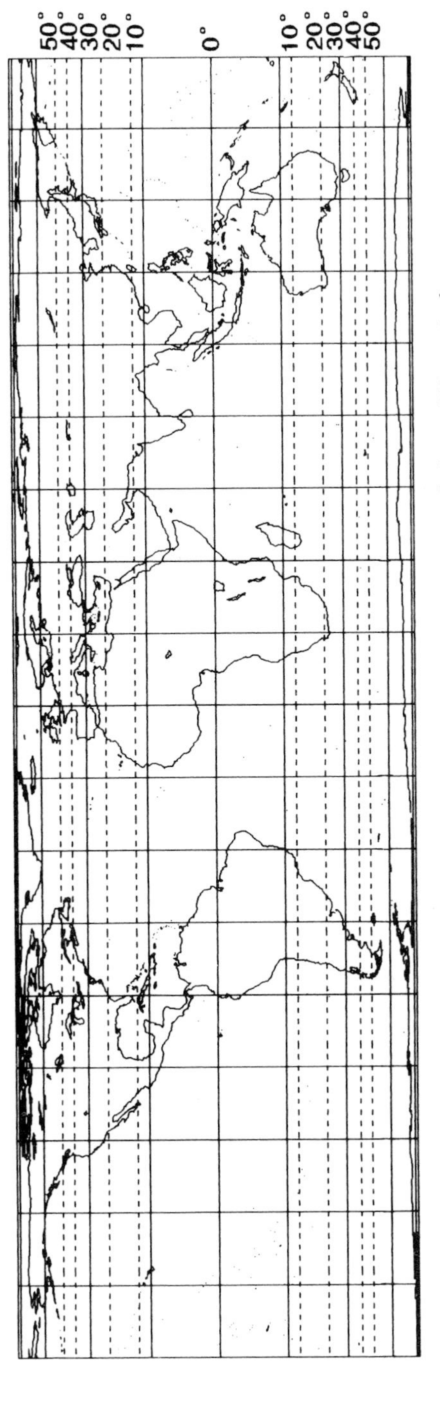

Figure 2.2 *Lambert cylindrical equal-area projection, with the equator as standard parallel. Isocols for ω. 20° graticule.*

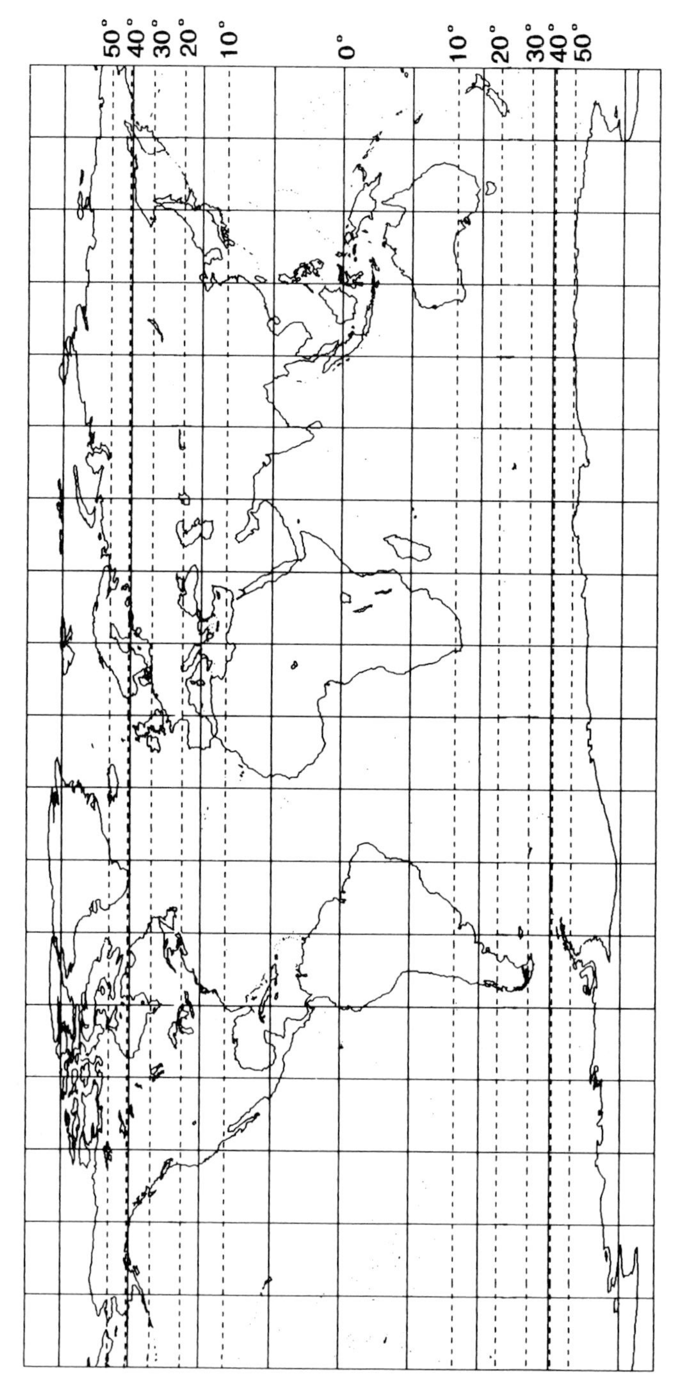

Figure 2.3 Plate carrée projection. Isocols for ω. 20° graticule.

$$m = 1, \quad n = p = r_k/r, \quad \sin(\omega/2) = (a - b)/(a + b)$$

where s is determined from equation (1.145).

When transforming the surface of the sphere:

$$x = r_k \lambda, \quad y = R\phi,$$

$$m = 1, \quad n = p = \cos\phi_k \sec\phi$$

When $\phi_k = 0$ we have a square equidistant projection or the plate carrée (square chart, Figure 2.3):

$$x = R\lambda, \quad y = R\phi$$

$$n = p = \sec\phi, \quad \sin(\omega/2) = \tan^2(\phi/2)$$

2.1.5 Arbitrary cylindrical projections

Cylindrical projections with a given distribution of distortion were suggested by N. A. Urmayev (1950), who was the first to formulate an inverse problem of mathematical cartography, i.e. writing equations of projections with a given distortion or scale. On cylindrical projections the scale along meridians is a function of the latitude. If we accept a unit radius for the sphere, then $m = dy/d\phi$; giving m values at certain points, ordinate y can be determined by integrating

$$y = \int m \, d\phi$$

For convenience and simplification of the solution to the problem, the scale can be represented in the form of a polynomial with even powers of latitude

$$m = a_0 + a_2\phi^2 + a_4\phi^4 + \cdots$$

where a_0, a_2, a_4 are coefficients which can be determined either by the solution of three simultaneous equations with three unknowns (if three values of scale factor m are given), or by interpolation.

Let us consider, as an example, deriving an arbitrary cylindrical projection as suggested by N. A. Urmayev. Along parallels of latitude $\phi_0 = 0$, $\phi_1 = 60°$, and $\phi_2 = 80°$, the corresponding scale factors are made $m_0 = 1.0$, $m_1 = 1.5$, $m_2 = 2.0$. Let us build a table of differences, where the latitude ϕ is converted to an argument a expressed (when the graticule interval $\Delta\phi = \Delta\lambda = 10°$) as the square of the tens of degrees, the scale factors m being a function of a (see Table 2.1).

Function value $f(a)$ for other arguments can be obtained from the Newton interpolation formula with divided differences:

$$f(a) = f(a_0) + (a - a_0)f_{01} + (a - a_0)(a - a_1)f_{012} + \cdots$$

where f_{01} and f_{012} are the first and second divided differences

$$f_{01} = \frac{f(a_1) - f(a_0)}{a_1 - a_0}, \quad f_{12} = \frac{f(a_2) - f(a_1)}{a_2 - a_1}$$

$$f_{012} = \frac{f_{12} - f_{01}}{a_2 - a_0} = \left(\frac{f(a_2) - f(a_1)}{a_2 - a_1} - \frac{f(a_1) - f(a_0)}{a_1 - a_0}\right) \Big/ (a_2 - a_0)$$

Table 2.1 Calculations for an Urmayev cylindrical projection

Point number	ϕ	Argument a	$f(a) = m$	Differences: First f_{01}, f_{12}	Second f_{012}
0	0°	0	1.0		
				+1/72	
1	60°	36	1.5		+1/16 128
				+1/56	
2	80°	64	2.0		

On setting $z = \phi/\text{arc } 10°$, we get

$$m = 1 + \frac{z^2}{72} + \frac{z^2(z^2 - 36)}{16\,128} = 1 + \frac{188z^2}{16\,128} + \frac{z^4}{16\,128}$$

and after integrating

$$y = z + \frac{188z^3}{48\,384} + \frac{z^5}{80\,640}$$

The final values of the rectangular coordinates for the projection have the form

$$X = \mu_0 R 100\lambda \text{ arc } 1° \quad \text{and} \quad Y = \mu_0 R 100 y$$

where X and Y are expressed in centimeters.

An example of the map graticule, with isocols for ω, is shown in Figure 2.4.

2.1.6 Oblique and transverse cylindrical projections

These projections are obtained using the following steps:

- transformation from the ellipsoid to the surface of a sphere (for large-scale maps), with spherical radius R determined by the method of transformation corresponding to the type of projection distortion selected;
- determination of pole coordinates $Q(\phi_0, \lambda_0)$;
- transformation from geodetic coordinates to polar spherical coordinates for the oblique or transverse systems;
- computation of projection coordinates, scale factors, and maximum angular deformation on the projection corresponding to the type of distortion.

These steps are carried out with the formulas and methods shown in sections 1.1.3, 1.5, 1.7.2, and 1.8.4.

Transverse cylindrical projections are recommended for maps of regions extending along meridians, and oblique projections are for regions extending along great circles of arbitrary orientation (see also section 5.1.2).

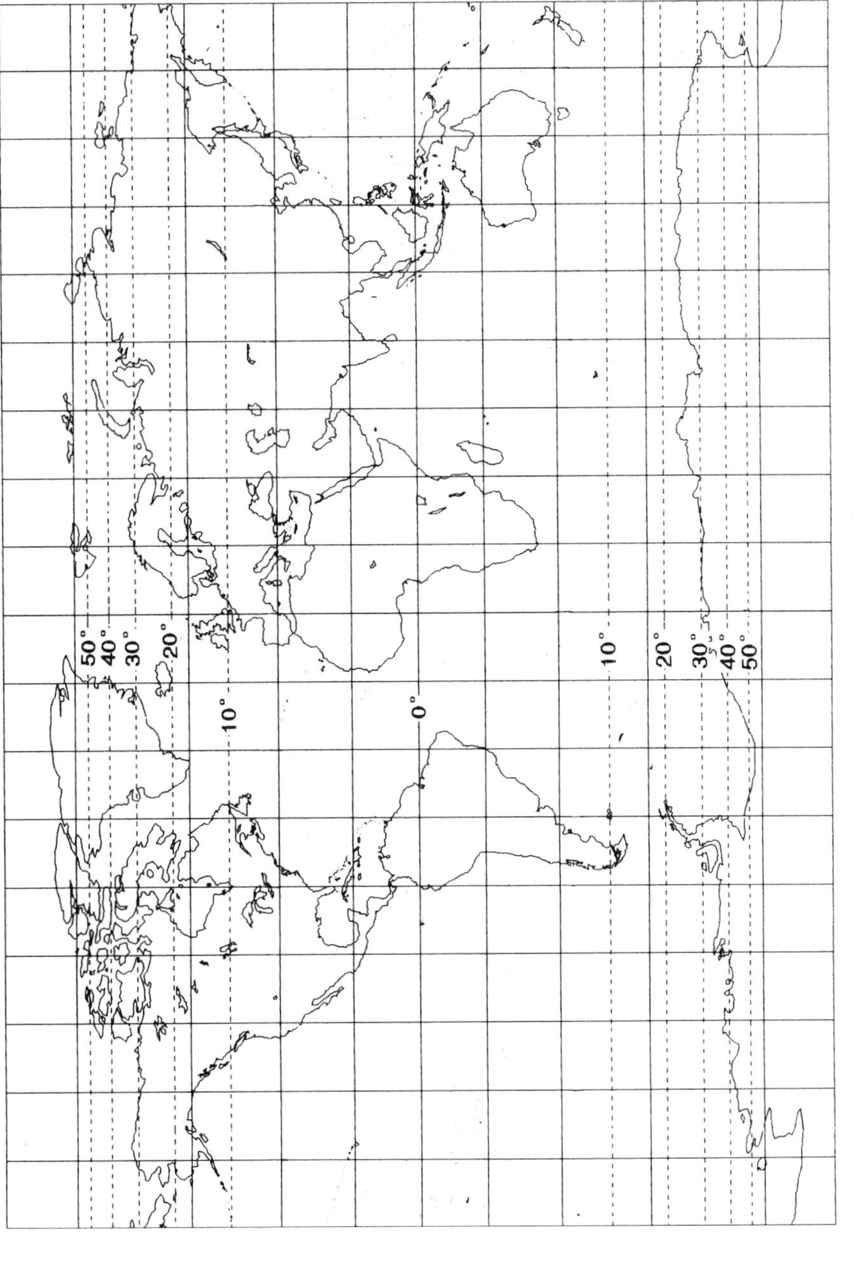

Figure 2.4 Normal Urmayev cylindrical projection (second variant), with isocols for ω. 20° graticule.

2.1.7 Perspective cylindrical projections

Cylindrical projections can be derived not only with analytical methods, but also geometrically; they possess the general properties of cylindrical projections.

Imagine a cylinder the axis of which coincides with a polar axis of the surface such as the sphere being mapped. This cylinder can intersect the sphere or be tangent to it along the equator. Let us project arcs of meridians onto the cylinder in the form of separate rays in the plane of each meridian at the given intervals of the map graticule.

For such a projection we mentally move the point of view in the equatorial plane from one meridian to another, the distance from the center of the sphere being the same for each point. Joining the projected points of meridians and parallels of the same values with lines, and flattening or developing laterally the surface of the cylinder, we obtained a map graticule for a normal perspective cylindrical projection on the plane. If one of the meridians is assumed to be the Y-axis, and either the equator or the southernmost parallel of latitude is assumed as the X-axis, the formula for rectangular coordinates of this projection will have the form

$$x = \beta\lambda, \quad y = f(\phi)$$

Meridians and parallels will be represented by two systems of perpendicular lines: distances between meridians are determined in the same way as for normal cylindrical projections, and distances between parallels by the method of perspective projection.

When making normal perspective cylindrical projections, the cartographic surface can be assumed to be an ellipsoid of revolution, but for convenience in applying these projections to general-purpose small-scale maps, we will consider only projections of the sphere.

The value of ordinate y is determined geometrically. In Figure 2.5 point A is projected from the point of view g onto point A' on the original cylinder, intersecting the globe at point A_k, which is at the parallel of latitude ϕ_k. The point of view is at distance D from the center of the sphere. If the origin of rectangular coordi-

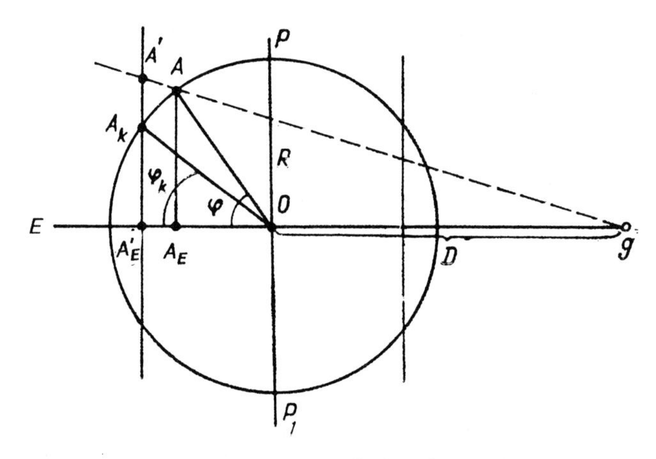

Figure 2.5 Diagram for the normal perspective–cylindrical projections.

nates coincides with point A'_E on the equator, then segment $A'A'_E$ will be the ordinate of the transformed point A.

By taking into account the similarity of gAA_E and $gA'A'_E$ as well as the properties common to cylindrical projections,

$$x = \beta\lambda, \quad y = CR\frac{\sin\phi}{k + \cos\phi}, \quad k = D/R$$

$$m = y_\phi/R = C(1 + k\cos\phi)/(k + \cos\phi)^2, \quad n = \beta/r$$

$$p = mn, \quad \sin(\omega/2) = (a - b)/(a + b)$$

where a and b are the extreme linear scale factors equal to m and n, respectively or not, as the graticule is orthogonal.

The cylinder is secant when $\phi_k \neq 0$ (thus $C = k + \cos\phi_k$, $\beta = R\cos\phi_k$) and is tangent when $\phi_k = 0$ (thus $C = k + 1$, $\beta = R$).

Perspective cylindrical projections are arbitrary in distortion characteristics. They differ from one another according to the latitude ϕ_k of the parallel along which the cylinder intersects the sphere, and the distance D of the point of view g from the center O of the sphere.

Let us consider a few particular cases.

Central cylindrical projection

Here $k = 0$, projected from the center of the sphere, with $\phi_k = 0$ (the cylinder is tangent):

$$x = R\lambda, \quad y = R\tan\phi, \quad m = \sec^2\phi, \quad n = \sec\phi$$

$$p = \sec^3\phi, \quad \sin(\omega/2) = \tan^2(\phi/2)$$

The transverse aspect is called the Wetch projection. Distortion of any form is too great for practical usage.

Braun projection

Here $k = 1$, projected stereographically, with $\phi_k = 0$ (tangent cylinder) (Braun 1867; Snyder 1993, pp. 110–11, 123–4):

$$x = R\lambda, \quad y = 2R\tan(\phi/2), \quad m = \sec^2(\phi/2), \quad n = \sec\phi$$

$$p = \sec^2(\phi/2)\sec\phi, \quad \sin(\omega/2) = (1 - \cos\phi)/(1 + 3\cos\phi)$$

Gall stereographic projection

Here $k = 1$, projected stereographically, normally with $\phi_k = 45°$, or modified with another value (the cylinder is secant, Figure 2.6):

$$x = R\lambda\cos\phi_k, \quad y = R(1 + \cos\phi_k)\tan(\phi/2)$$

$$m = \tfrac{1}{2}(1 + \cos\phi_k)\sec^2(\phi/2), \quad n = \cos\phi_k\sec\phi, \quad p = mn$$

$$\sin(\omega/2) = |\cos\phi - \cos\phi_k|/[\cos\phi\,(1 + 2\cos\phi_k) + \cos\phi_k]$$

The Gall projection modified with $\phi_k = 30°$ is used for world maps in Volume 1 of *Bol'shoy Sovetskiy Atlas Mira* (*BSAM*) (*Great Soviet World Atlas*, 1937).

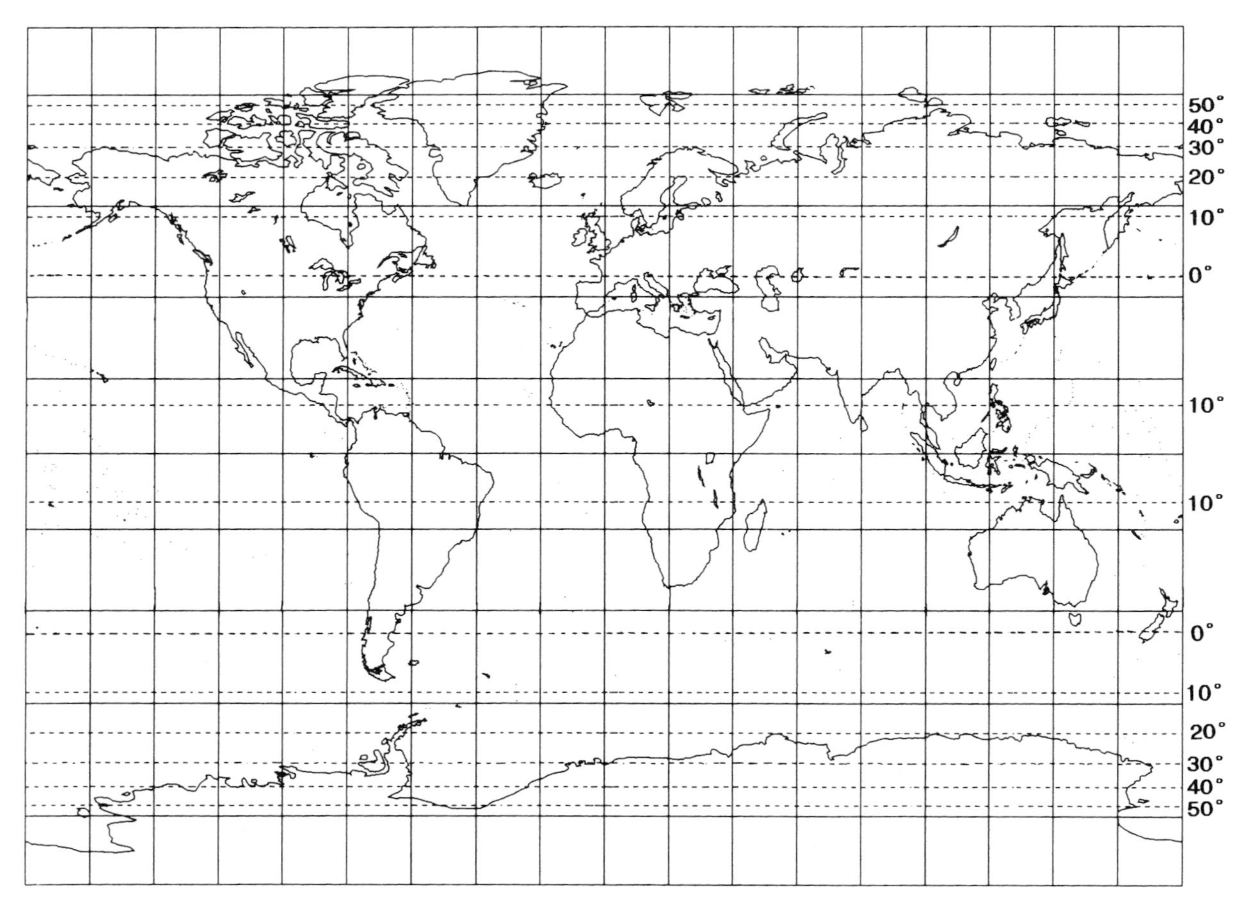

Figure 2.6 Gall stereographic projection, with isocols for ω. 20° graticule.

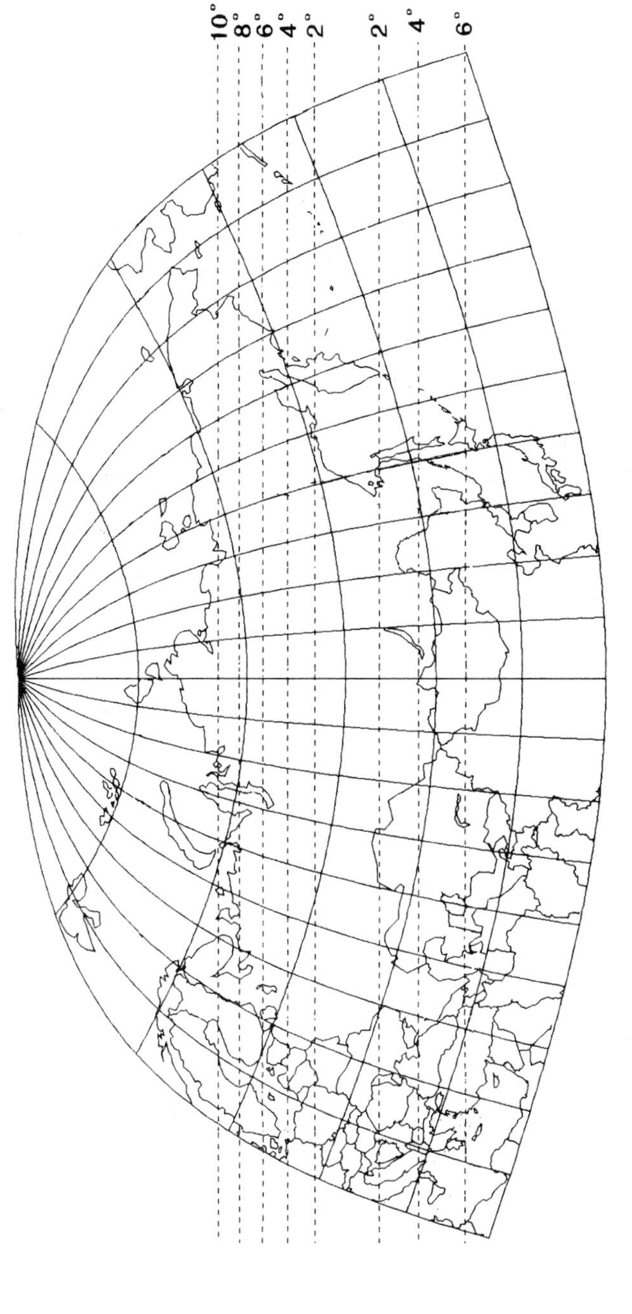

Figure 2.7 Oblique cylindrical projection (Solov'ev's variant 1), with isocols for ω. 10° graticule.

Distortion on normal perspective cylindrical projections depends only on the latitude; therefore, isocols coincide with parallels and are straight lines. The projections have two parameters k and ϕ_k which affect the arrangement of the graticule (by changing the distances between meridians and parallels) and the distribution of distortion.

On oblique perspective cylindrical projections the graticule of meridians and parallels does not coincide with a normal one. Meridians and parallels are curves, and verticals and almucantars are represented by two systems of mutually perpendicular straight lines. These projections can be prepared by the method mentioned above, but the moving point of view, from which the projection occurs onto the cylinder, is not in the equatorial plane, but in the plane of the almucantar with zenith distance $z = 90°$. General formulas for oblique perspective cylindrical projections can be obtained by a simpler method from the equations for normal perspective cylindrical projections, substituting ϕ for $(90° - z)$ and λ for a, and setting $\mu_1 = m$, $\mu_2 = n$.

Oblique perspective cylindrical projections were created for the design of educational maps. There are four values in the projections considered, ϕ_0, λ_0, k, and z_k, which affect the appearance of the graticule and the distribution of distortion.

One variant of oblique perspective cylindrical projections, developed by Solov'ev (Figure 2.7), was widely used in the former Soviet Union for the design of school maps to represent polar regions such as the Arctic Ocean ($\phi_0 = 75°$, $\lambda_0 = -80°$, $k = 1$ and $z_k = 45°$). A variant by TsNIIGAiK ($\phi_0 = 25°$, $\lambda_0 = -80°$, $k = 3$, $\phi_k = 10°$; $C = 1 + \cos \phi_k$) was used to design the map of the USSR in the 1954 *Atlas Mira* (*World Atlas*).

Combinations of perspective cylindrical projections with negative and positive transformations

In 1985–6 Mara Daskalova and Marin Andreyev suggested a projection combining perspective cylindrical projections with negative and positive transformations. Let C_1, C_2 be the points of projection (Figure 2.8), M' and M'' be the points of negative (n) and positive (p) transformation of point M from a sphere onto a wrapped cylinder; $C_1O = d$, $C_2O = D$. Then the rectangular coordinates of the projection are

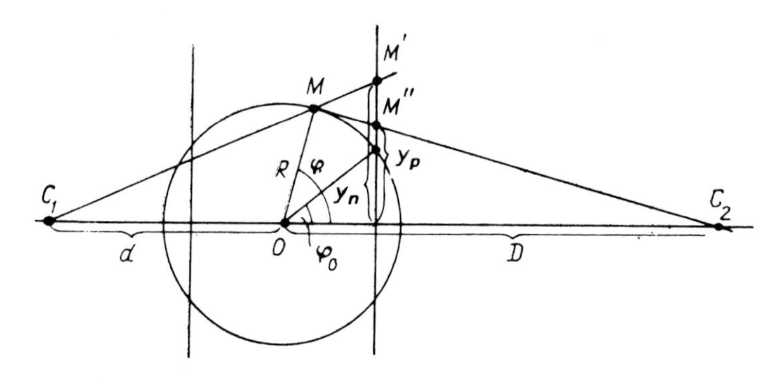

Figure 2.8 Combined perspective cylindrical projection.

Table 2.2 Distortion values for some cylindrical projections

Parameter	Latitude			
	0°	30°	60°	90°
1	2	3	4	5
		Conformal		
m	1.000	1.155	2.000	∞
n	1.000	1.155	2.000	∞
p	1.000	1.333	4.000	∞
ω	0°00′	0°00′	0°00′	Indeterminate
		Equal area		
m	1.000	0.866	0.500	0.000
n	1.000	1.155	2.000	∞
p	1.000	1.000	1.000	1.000
ω	0°00′	16°26′	73°44′	180°00′
		Equidistant		
m	1.000	1.000	1.000	1.000
n	1.000	1.155	2.000	∞
p	1.000	1.155	2.000	∞
ω	0°00′	8°14′	38°57′	180°00′
		Braun perspective		
m	1.000	1.072	1.333	2.000
n	1.000	1.155	2.000	∞
p	1.000	1.238	2.667	∞
ω	0°00′	4°16′	23°04′	180°00′

equal to

$$x = R \cos \phi_0(\lambda - \lambda_0), \quad y = \frac{k_1 y_n + k_2 y_p}{k_1 + k_2}$$

where

$$y_n = R \sin \phi \frac{d + R \cos \phi_0}{d + R \cos \phi}, \quad y_p = R \sin \phi \frac{D - R \cos \phi_0}{D - R \cos \phi}$$

and k_1, k_2 are constant coefficients, selected to suit various conditions and affecting the projection properties.

The authors suggested that these coefficients be defined using the Airy criterion for conditions providing minimum distortion within the limits of the region being mapped.

2.1.8 Characteristics of cylindrical projections

The characteristics of map projections, including cylindrical, can be determined from an analysis of the projection properties. Various criteria establishing the advantages of a projection can be utilized, e.g. values of linear, area, and angular distortion, the shape of the map projection graticule, the peculiarities of the positional lines such as loxodromes, the sphericity, and other features, the relative importance of each in a particular case depending on the purpose of the map being designed.

On cylindrical projections, as contrasted with all others, the normal map graticule is of the simplest form; on all of them, the ordinates for a given parallel are equal.

Only the Mercator projection shows all loxodromes as straight lines. This accounts for its wide application in creating navigational charts, in spite of the fact that on this projection the curvature of plotted geodetic lines is often greater than on other conformal projections.

Table 2.2 lists the distortion values for some cylindrical projections (Ginzburg and Salmanova 1964, p. 139): m is the scale factor along meridians, n is the scale factor along parallels, p is the area scale factor, and ω is the maximum angular deformation.

2.2 Pseudocylindrical projections

2.2.1 Principal concepts and general formulas

On pseudocylindrical projections parallels are represented by straight lines, and meridians by curves (sometimes by straight lines) symmetrical about the straight central meridian (Figure 2.9).

The general equations for these projections are

$$x = f_1(\phi, \lambda), \quad y = f_2(\phi)$$

It is quite possible to represent the entire globe with a pseudocylindrical projection and to repeat its extreme eastern and western portions at the opposite sides of the map. The geographical poles can be shown either by points or lines parallel to the equator. Meridians are of various shapes and are often represented by either ellipses or sinusoids, but it is possible to have pseudocylindrical projections with the meridians in the shape of parabolas, hyperbolas, or other lines.

The graticule on pseudocylindrical projections is not orthogonal except along the equator and along the central meridian:

$$\tan \varepsilon = -f/h = -x_\phi x_\lambda / y_\phi x_\lambda = -x_\phi / y_\phi$$

Hence,

$$x_\phi = -y_\phi \tan \varepsilon$$

Applying formulas for the general theory of map projection (see section 1.3) we obtain

$$m = y_\phi(\sec \varepsilon)/M, \quad n = x_\lambda(\sec \phi)/N$$

$$p = mn \cos \varepsilon = x_\lambda y_\phi / Mr, \quad \tan(\omega/2) = \tfrac{1}{2}\sqrt{(m^2 + n^2 - 2p)/p}$$

On these projections, since the meridians and parallels are not orthogonal except along the central meridian and equator, the scale factors along the meridians and parallels are generally not the extreme values.

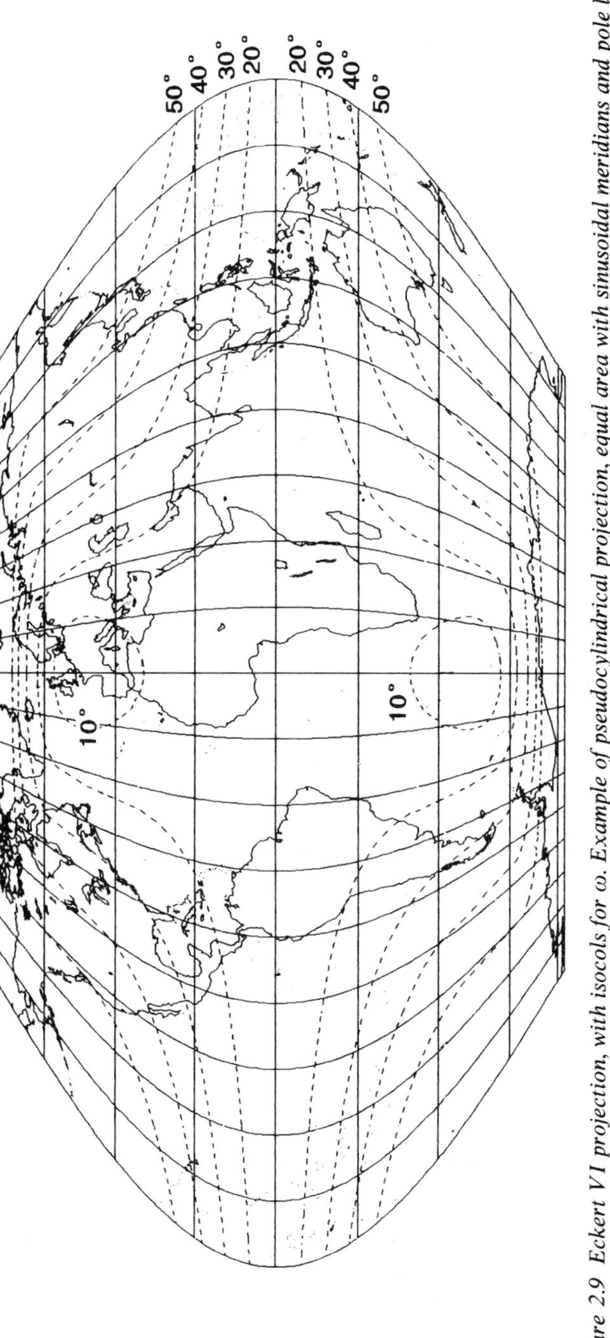

Figure 2.9 *Eckert VI projection, with isocols for ω. Example of pseudocylindrical projection, equal area with sinusoidal meridians and pole lines. 20° graticule.*

Pseudocylindrical projections, as a rule, are used for designing small-scale world maps; therefore, the surface to be mapped is usually considered to be a sphere of radius R. A rare exception is the mapping grid for Ecuador, which is based on the sinusoidal projection for the ellipsoid.

Pseudocylindrical projections can be subdivided into equal-area and arbitrary ones, according to the distortion characteristics, the former being more widely used.

If the area scale is

$$p = x_\lambda y_\phi / Mr = k^2$$

where k is a constant, then

$$x_\lambda = k^2 Mr / y_\phi$$

After integrating

$$x = \frac{k^2 Mr}{y_\phi} \lambda + f(\phi)$$

where $f(\phi)$ is an arbitrary function of latitude ϕ. When $\lambda = 0$ and $x = 0$, the Y-axis coincides with the central meridian; hence $f(\phi) = 0$.

Most commonly, $k = 1$; then

$$x = \frac{Mr}{y_\phi} \lambda \quad \text{or, for a sphere,} \quad x = \frac{R^2 \cos \phi}{y_\phi} \lambda \tag{2.2}$$

2.2.2 Equal-area pseudocylindrical projections

Soviet scientists V. V. Kavrayskiy and N. A. Urmayev suggested a generalized method of creating equal-area pseudocylindrical projections with meridians in the form of either sinusoids or ellipses. Urmayev developed a theory of equidistant hyperbolic projections.

The rectangular coordinates for pseudocylindrical projections in a parametric form, based on the representation of the meridians as well as the geographic poles, are:

- for projections with sinusoidal meridians,

$$x = (A \cos \alpha + B)\lambda, \quad y = C\alpha \tag{2.3}$$

- for projections with elliptical meridians,

$$x = (A \cos \alpha + B)\lambda, \quad y = C \sin \alpha \tag{2.4}$$

where A, B, and C are parameters characterizing the graticule size and shape, and α is a parametric angle which is a function of ϕ.

In order to derive these equal-area pseudocylindrical projections we use equation (2.2) and impose two conditions:

1. when $\phi = 0$ then $\alpha = 0$, and when $\phi = \pi/2$ then $\alpha = \pi/2$;
2. $x_p = y_p = x_e/2$, i.e. the length of the pole p is equal to the length of the central meridian and half the length of the equator e. If the pole is represented by a point, then $B = x_p = 0$. The length of the pole may also be made other propor-

tions of the length of the equator; for example, McBryde and Thomas (1949) in the United States used 1/3, and TsNIIGAiK (below) uses 0.6.

To obtain arbitrary pseudocylindrical projections, conditions other than equivalence using equation (1.90) may be imposed.

Equal-area sinusoidal pseudocylindrical projections with the poles in the form of points

Based on the above conditions, we find that

$$x = A\lambda \cos \alpha, \quad y = \alpha C, \quad y_p = x_e/2 \tag{2.5}$$

To represent the entire sphere, map semiaxes are as follows:

$$x_e = A\pi, \quad y_p = C(\pi/2)$$

then $C(\pi/2) = A(\pi/2)$; thus $C = A$, and from equations (2.5)

$$x = C\lambda \cos \alpha \tag{2.6}$$

Equating values for x in (2.6) and (2.2) we integrate the equation obtained and find

$$C^2 \sin \alpha = R^2 \sin \phi + C_1$$

where the integration constant $C_1 = 0$, so that $\phi = 0$ when $\alpha = 0$.

If $\phi = \pi/2$, then $\alpha = \pi/2$, and

$$C^2 = R^2, \quad \sin \alpha = \sin \phi, \quad \alpha = \phi$$

Using these results we can provide the equations for an equal-area sinusoidal pseudocylindrical projection often simply known as the sinusoidal, but also as the Sanson–Flamsteed projection* (Figure 2.10):

$$x = R\lambda \cos \phi, \quad y = R\phi, \quad \tan \varepsilon = \lambda \sin \phi$$

$$p = 1, \quad n = 1, \quad m_0 = 1, \quad m = \sec \varepsilon, \quad \tan(\omega/2) = (\tan \varepsilon)/2$$

Local linear scale factors along meridians and angular deformation are functions of latitude and longitude; isocols are in the shape of hyperbola-like curves symmetrical about both the equator and the central meridian.

The Sanson–Flamsteed projection was used in atlases for small-scale general world maps and for maps of South America and Africa. For many years it was considered to be one of the best projections for maps of these continents. Now it is often replaced by the oblique and equatorial aspects of the Lambert azimuthal equal-area projection.

Equal-area sinusoidal pseudocylindrical projection with poles as lines

On this projection the geographical poles are represented by straight lines parallel to the parallels of latitude, their lengths being normally half that of the equator (see Figure 2.9), although the length can be varied.

---

* Named after Nicolas Sanson d'Abbeville, a French geographer (1600–67), and John Flamsteed, an English astronomer (1646–1719), although Jean Cossin used the projection for a world map in 1570, and Mercator and others also used it before Sanson and Flamsteed.

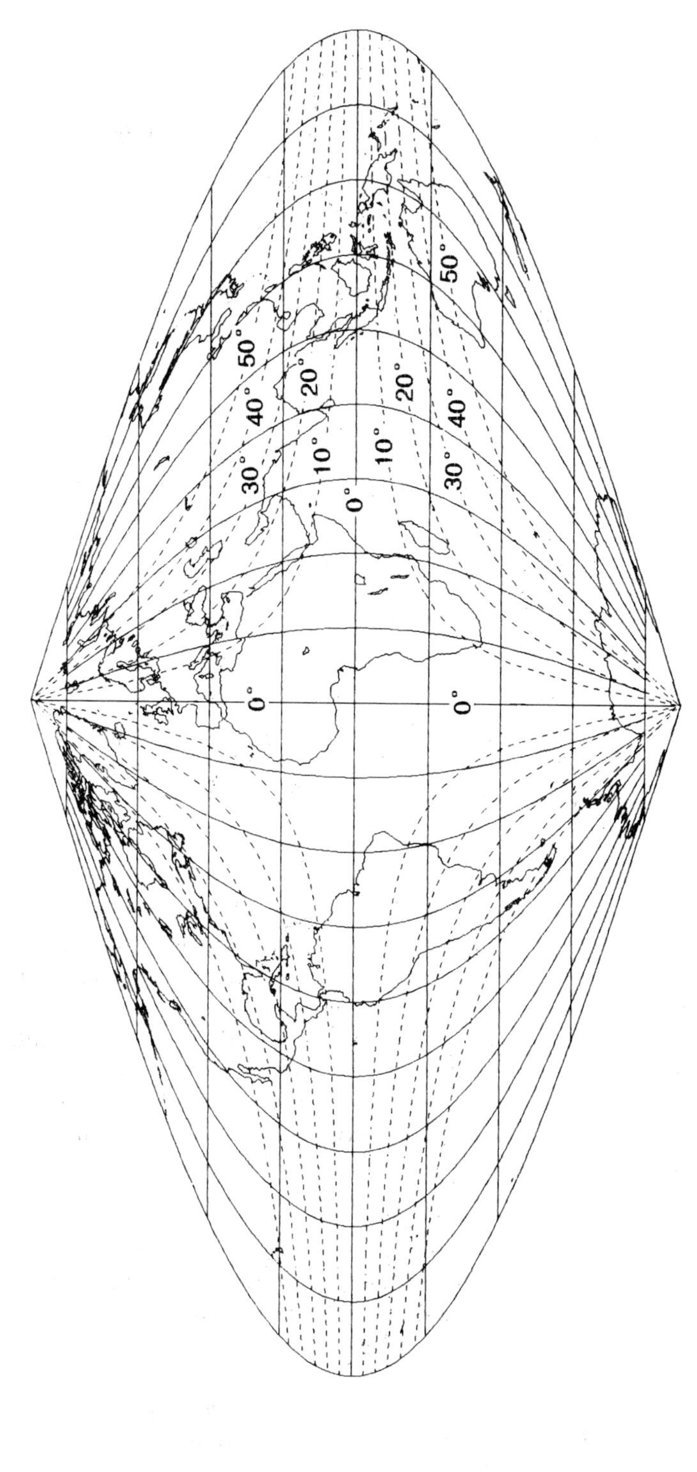

Figure 2.10 Sanson–Flamsteed sinusoidal equal-area pseudocylindrical projection. Isocols for ω. 20° graticule

Based on these conditions

$$x = (A \cos \alpha + B)\lambda, \quad y = \alpha C$$

If α and ϕ are equal at 0° and 90°,

$$C\frac{\pi}{2} = B\pi = \frac{A+B}{2}\pi, \quad A = B = \frac{C}{2}$$

then

$$x = (C/2)(\cos \alpha + 1)\lambda \tag{2.7}$$

Equating the values of abscissas (2.7) and (2.2), we get a differential equation; from integration,

$$(C^2/2)(\alpha + \sin \alpha) = R^2 \sin \phi \tag{2.8}$$

The constant of integration is made equal to zero, so that $\phi = 0$ when $\alpha = 0$.

When $\phi = \pi/2$ and $\alpha = \pi/2$,

$$\frac{C^2}{2}\left(\frac{\pi}{2} + 1\right) = R^2$$

Hence,

$$C = 2R/\sqrt{\pi + 2} \quad \text{and} \quad A = B = R/\sqrt{\pi + 2}$$

Substituting into (2.8),

$$\alpha + \sin \alpha = \frac{\pi + 2}{2} \sin \phi$$

where α is expressed in radians.

The latter equation is transcendental and can be solved by a method of successive approximations such as the Newton–Raphson method.

The formulas for an equal-area sinusoidal pseudocylindrical projection with the above ratios of length are as follows (this projection is called the Eckert VI projection and was presented in 1906, but it is a particular case of a pseudocylindrical projection):

$$x = \frac{2R\lambda}{\sqrt{\pi + 2}} \cos^2\left(\frac{\alpha}{2}\right), \quad y = \frac{2R}{\sqrt{\pi + 2}} \alpha$$

$$\alpha + \sin \alpha = \frac{\pi + 2}{2} \sin \phi, \quad \tan \varepsilon = \frac{\lambda}{2} \sin \alpha$$

$$m = \frac{\sqrt{\pi + 2}}{2} \sec^2\left(\frac{\alpha}{2}\right) \cos \phi \sec \varepsilon, \quad n = \frac{2}{\sqrt{\pi + 2}} \cos^2\left(\frac{\alpha}{2}\right) \sec \phi$$

$$p = 1, \quad \tan\left(\frac{\omega}{2}\right) = \tfrac{1}{2}\sqrt{m^2 + n^2 - 2}$$

On this projection the linear distortion along parallels is a function only of latitude; therefore, the isocols showing this type of distortion coincide with parallels. Linear distortion along meridians and angular deformation depend on both latitude and

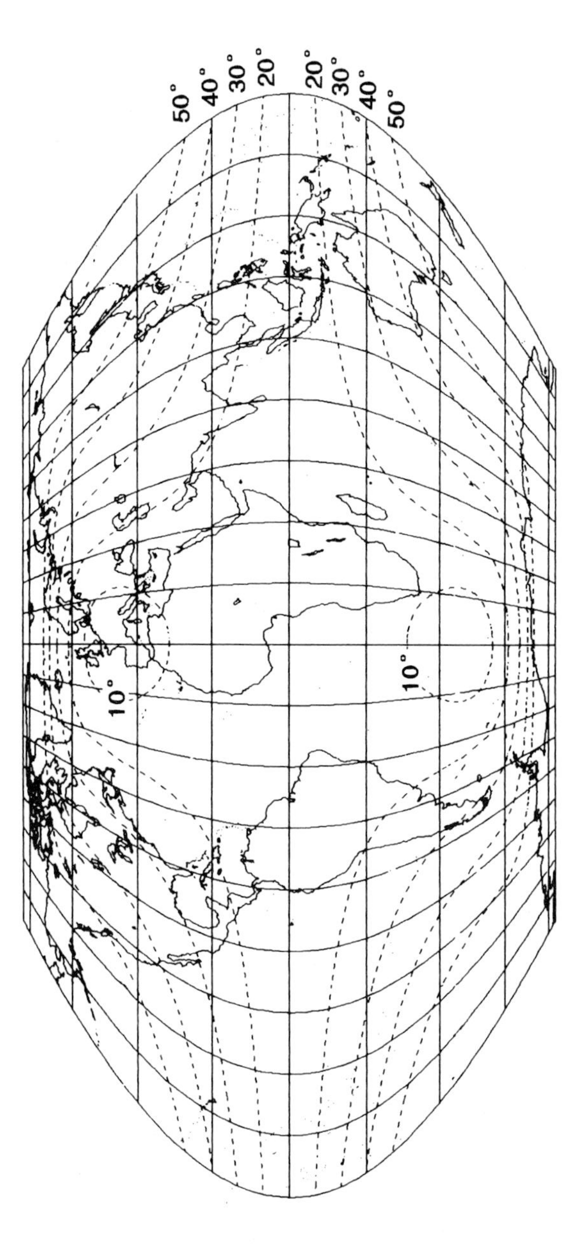

Figure 2.11 Wagner I equal-area sinusoidal pseudocylindrical projection, with isocols for ω. Kavrayskiy later independently presented this projection. 20° graticule.

longitude; isocols characterizing these types of distortion are complex curves symmetrical about the central meridian and equator.

The Eckert VI projection was widely used in Soviet geographical atlases before the Second World War.

Wagner's and Kavrayskiy's equal-area sinusoidal pseudocylindrical projection

This projection (Figure 2.11) is similar to the equal-area sinusoidal projection with a pole line described above, but only a portion of the sinusoid is used rather than the full sinusoid used above, thus eliminating iteration in the computation of coordinates. It was used by Karl Rosén for world maps in the Swedish *Nordisk Världsatlas* in 1926, and was developed in the form described below by Karlheinz Wagner in Germany in 1932 and independently by V. V. Kavrayskiy in 1936. To create this projection, two additional conditions are imposed: parameter $B = 0$, and latitude function $\alpha = \pi/3$ when $\phi = \pi/2$.

In view of the conditions imposed and the equality of ordinates at a given latitude, we can obtain a differential equation

$$\tfrac{2}{3}C^2 \cos\alpha \, d\alpha = R^2 \cos\phi \, d\phi$$

After integrating,

$$\tfrac{2}{3}C^2 \sin\alpha = R^2 \sin\phi \qquad (2.9)$$

where the constant of integration is equal to zero, so that when $\phi = 0$, $\alpha = 0$. If $\phi = \pi/2$, then $\alpha = \pi/3$; hence, substituting into (2.9),

$$C^2 = R^2\sqrt{3} \quad \text{and} \quad \sin\alpha = (\sqrt{3}/2)\sin\phi$$

The formulas for this equal-area sinusoidal projection are as follows:

$$x = \tfrac{2}{3}\sqrt[4]{3}\,\lambda\cos\alpha, \quad y = R\sqrt[4]{3}\,\alpha$$

$$\sin\alpha = (\sqrt{3}/2)\sin\phi, \quad \tan\varepsilon = \tfrac{2}{3}\lambda\sin\phi$$

$$m = (\sqrt[4]{27}/2)\sec\alpha\cos\phi\sec\varepsilon, \quad n = (2/\sqrt[4]{27})\cos\alpha\sec\phi$$

$$p = 1, \quad \tan(\omega/2) = \tfrac{1}{2}\sqrt{m^2 + n^2 - 2}$$

The curvature of meridians on this non-iterative projection changes more slowly than on the Eckert VI projection.

Equal-area elliptical pseudocylindrical projection with poles as points

This projection is called the Mollweide projection and was presented in 1805. The graticule of this projection is shown in Figure 2.12. All the meridians are ellipses, except the central meridian, which is a straight line, and the meridian with longitude $\lambda = \pm 90^\circ$, which is a circle. The circle and the straight line are limiting forms of the ellipse.

The projection graticule makes it possible to have a rounded portrayal of the entire surface of the Earth; it has been used for maps of oceans and for general maps in atlases even to the present.

Based on the conditions imposed, we can write

$$x = A\lambda\cos\alpha, \quad y = C\sin\alpha$$

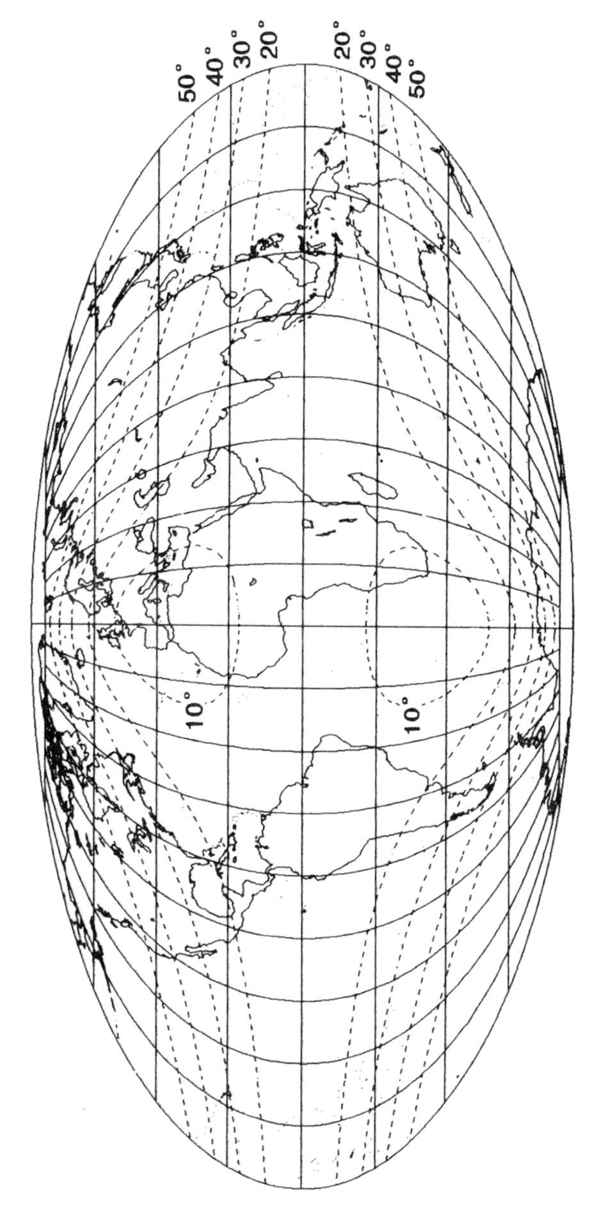

Figure 2.12 Mollweide equal-area elliptical pseudocylindrical projection with poles shown as points. Isocols for ω. 20° graticule.

$$A = 2C/\pi, \quad C = A\frac{\pi}{2} \tag{2.10}$$

$$x = (2C/\pi)\lambda \cos \alpha$$

Equating abscissas from equations (2.10) and (2.2), we obtain a differential equation

$$(C^2/\pi)(1 + \cos 2\alpha)\, d\alpha = R^2 \cos \phi\, d\phi$$

and, after integrating,

$$(C^2/2\pi)(2\alpha + \sin 2\alpha) = R^2 \sin \phi$$

The constant of integration is equal to zero, so that $\phi = 0$ when $\alpha = 0$. When $\phi = \pi/2$ and $\alpha = \pi/2$, it is found that

$$C^2 = 2R^2, \quad \text{and thus} \quad 2\alpha + \sin 2\alpha = \pi \sin \phi$$

This equation is transcendental; it is usually solved by the method of successive approximations.

Formulas for the Mollweide projection are therefore as follows:

$$x = (2\sqrt{2}/\pi)R\lambda \cos \alpha, \quad y = \sqrt{2}R \sin \alpha$$

$$2\alpha + \sin 2\alpha = \pi \sin \phi, \quad \tan \varepsilon = (2\lambda/\pi)\tan \alpha$$

$$m = (\pi/2\sqrt{2})\sec \alpha \cos \phi \sec \varepsilon, \quad n = (2\sqrt{2}/\pi)\cos \alpha \sec \phi$$

$$p = 1, \quad \tan(\omega/2) = \tfrac{1}{2}\sqrt{m^2 + n^2 - 2}$$

For this projection, isocols of maximum angular deformation are nearly the same shape as those for the Wagner/Kavrayskiy equal-area sinusoidal projection, but on the Mollweide projection the angular distortion decreases somewhat at the equator ($\omega = 12°$), and it increases ($\omega = 80°$) at the outer meridians.

On the Mollweide projection the shearing of continents is pronounced at high latitudes near the outer meridians.

Urmayev equal-area pseudocylindrical projections

N. A. Urmayev has described two sets of pseudocylindrical projections. First, with sinusoidal meridians: for world maps the pole is represented by a point. The projection equations are given in the form

$$a = 2/\sqrt[4]{27}, \quad b = \sin \phi \sin \alpha$$

$$n = \frac{a \cos \alpha}{\cos \phi}, \quad m = \frac{1}{a} \sec \varepsilon \sec \alpha \cos \phi$$

$$\tan \varepsilon = a^2 b^2 \lambda \sin \phi$$

where a is the linear scale along the equator.

Taking various values of a and b, we may obtain various pseudocylindrical equal-area projections.

Second, his projections with hyperbolic meridians: the projection equations take the form

$$x = C(1 - 2k \sec \alpha)\lambda, \quad y = A \tan \alpha$$

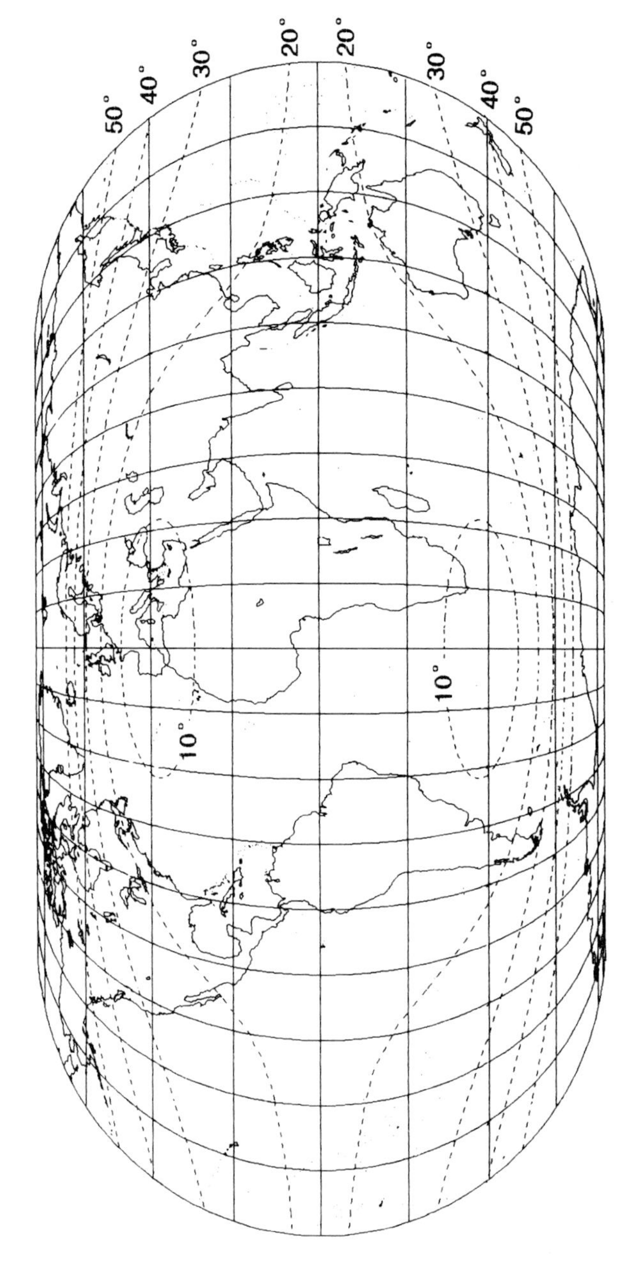

Figure 2.13 Eckert IV equal-area elliptical pseudocylindrical projection with pole lines. Isocols for ω. 20° graticule.

where α is determined by iteration from the following:

$$\frac{R^2}{AC} \sin \phi = \tan \alpha - k\left[\tan \alpha \sec \alpha + \ln \tan\left(\frac{\pi}{4} + \frac{\alpha}{2}\right)\right]$$

and A, C, and k are constant parameters.

Hojovec equal-area pseudocylindrical projections

Formulas providing rectangular coordinates for a world projection using a unit radius ($R = 1$) are given by Vladislav Hojovec (1984) in the form

$$x = \lambda \frac{\cos \phi}{a + b\phi^2}, \quad y = a\phi + \tfrac{1}{3}b\phi^3$$

or

$$x = \lambda \cos \phi, \quad y = \phi + au^2 + bu^4 + cu^6 + \cdots$$

where $u = \lambda \cos \phi$, and a, b, c are constant coefficients determined from additional conditions.

Eckert IV equal-area elliptical pseudocylindrical projection

On this projection (Figure 2.13), also presented by Eckert (1906), the length of the straight central meridian and that of the pole line are equal to half of the length of the equator. As on the Mollweide projection, the meridians are complete semi-ellipses.

Projection formulas have the form

$$x = \frac{2\lambda R}{\sqrt{\pi(\pi + 4)}}(1 + \cos \alpha), \quad y = 2R\sqrt{\frac{\pi}{\pi + 4}} \sin \alpha$$

The parametric angle α is determined by iteration from the transcendental equation

$$2\alpha + 4 \sin \alpha + \sin 2\alpha = (\pi + 4)\sin \phi$$

Wagner IV equal-area elliptical pseudocylindrical projection

First presented in 1932, this projection (Figure 2.14) by Wagner (1949, pp. 190–4) differs from Eckert IV in that each meridian is less than a semiellipse. The poles are only $\sqrt{3}/2$ or 0.866 of the distance from the equator to the vertical limits of the ellipse if it were completed. Like the Eckert IV, the poles are lines half the length of the equator and equal to the length of the central meridian. The shearing of shapes near the poles is thereby reduced. The formulas, requiring iteration, may be written as

$$x = (2\sqrt{3A/B}/\pi)R\lambda \cos \alpha, \quad y = 2\sqrt{A/B}\,R \sin \alpha$$

$$2\alpha + \sin 2\alpha = [(4\pi + 3\sqrt{3})/6]\sin \phi$$

where $A = 2\pi\sqrt{3}$ and $B = 4\pi + 3\sqrt{3}$. This was reintroduced by Putniņš as his P_2' projection in 1934.

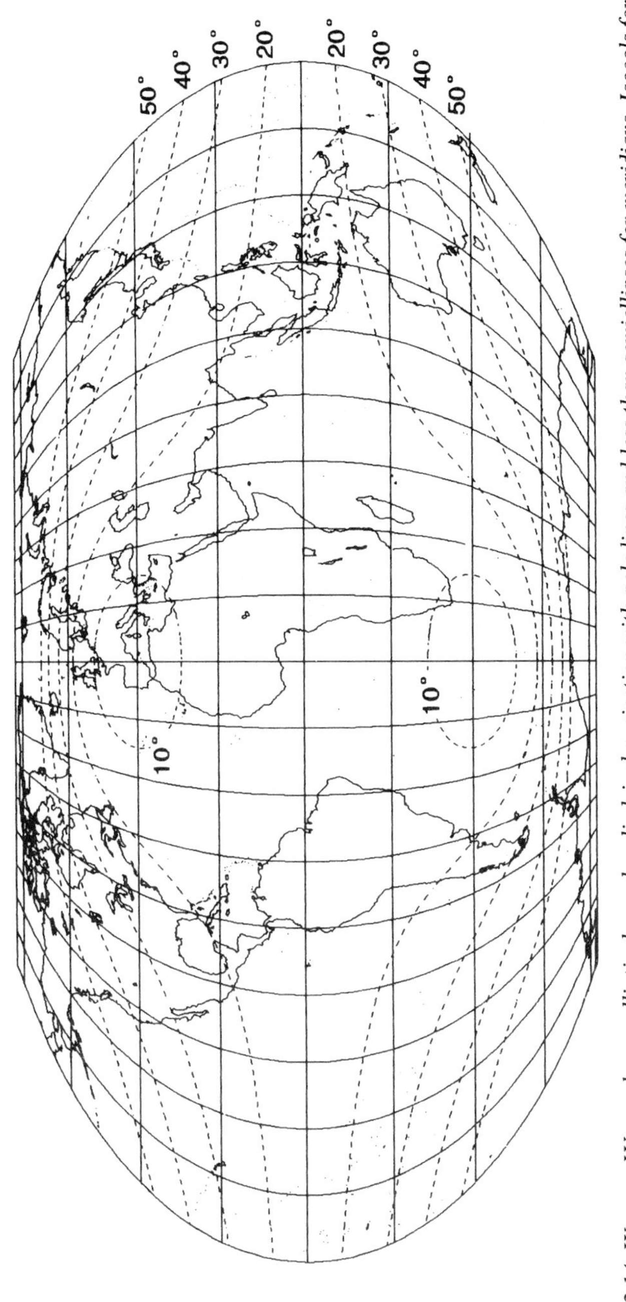

Figure 2.14 Wagner IV equal-area elliptical pseudocylindrical projection with pole lines and less than semiellipses for meridians. Isocols for ω. 20° graticule.

2.2.3 Pseudocylindrical projections with arbitrary distortion

It is possible to obtain many arbitrary projections, depending on the conditions. In particular, if it is required that the scale along the central meridian be preserved on the projection, we may obtain general formulas for this set of projections:

$$x = n_e R f(\phi, \lambda), \quad y = R\phi, \quad \tan \varepsilon = -(1/R)x_\phi$$

$$n = x_\lambda/(R \cos \phi), \quad m_0 = 1, \quad m = \sec \varepsilon, \quad p = n$$

$$\tan (\omega/2) = \sqrt{(m^2 + n^2 - 2p)/p}$$

where n_e is the linear scale factor along the equator.

Let us consider some projections similar to this and some others.

Kavrayskiy's elliptical pseudocylindrical projection

The projection is elliptical; therefore,

$$y = C \sin \alpha$$

Let us impose the condition that the scale along the central meridian is to be preserved without distortion on this projection. Then

$$y = R\phi, \quad \sin \alpha = \frac{R\phi}{C}, \quad \cos \alpha = \frac{1}{C}\sqrt{C^2 - R^2\phi^2}$$

After introducing $\cos \alpha$ into the formula

$$x = A\lambda \cos \alpha$$

we obtain

$$x = (A\lambda/C)\sqrt{C^2 - R^2\phi^2} \tag{2.11}$$

If we let $x_p = x_e/2$, from equation (2.11) we can get

$$C = R(\pi/\sqrt{3})$$

One of the meridians is a circle; for this meridian $C = A\lambda_1$, where λ_1 is its longitude. Let us substitute this expression into formula (2.11):

$$x = \frac{\lambda R}{\lambda_1}\sqrt{\frac{\pi^2 - 3\phi^2}{3}}$$

It is possible to find the meridian longitude λ_1 which is a circle, provided that along the parallels with a given latitude $\pm\phi_k$, the scale factor $n_k = 1$, i.e. the scale along these parallels has no distortion on the map:

$$\lambda_1 = \frac{\pm\sqrt{\pi^2 - 3\phi_k^2}}{\sqrt{3} \cos \phi_k}$$

If we select a value for longitude λ_1, we can determine the latitude $\pm\phi_k$ which has no distortion of length. V. V. Kavrayskiy selected $\lambda_1 = \pm 120°$ for this projection. Using this condition, the two parallels of true length $\phi_k = \pm 35°31'34''$.

Note also that, depending on the value of the longitude of the circular meridian, projection variants can be derived with different values for the local scale factor along the equator. For example, if $\lambda_1 = \pm\pi/\sqrt{3} = \pm 103°55'23''$, then $n_e = 1$, a

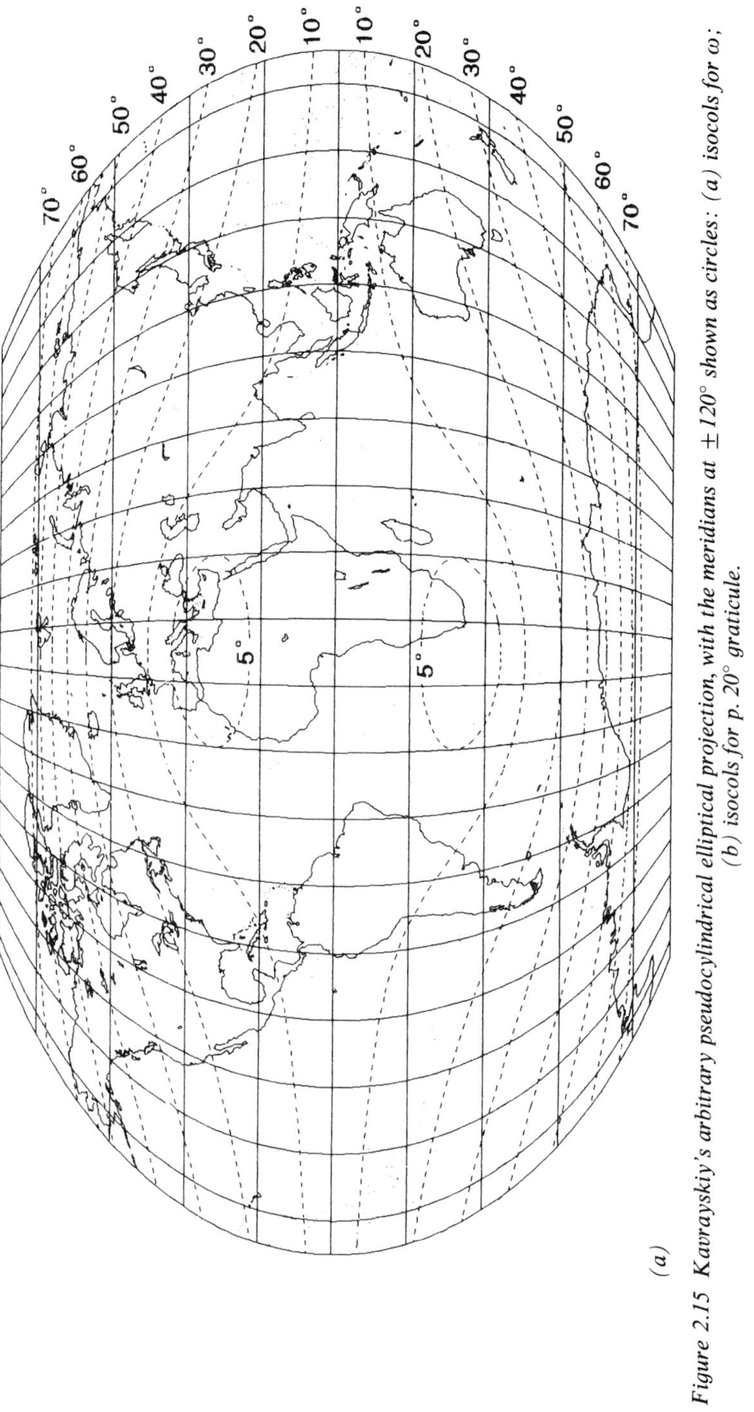

Figure 2.15 Kavrayskiy's arbitrary pseudocylindrical elliptical projection, with the meridians at $\pm 120°$ shown as circles: (*a*) isocols for ω; (*b*) isocols for *p*. 20° graticule.

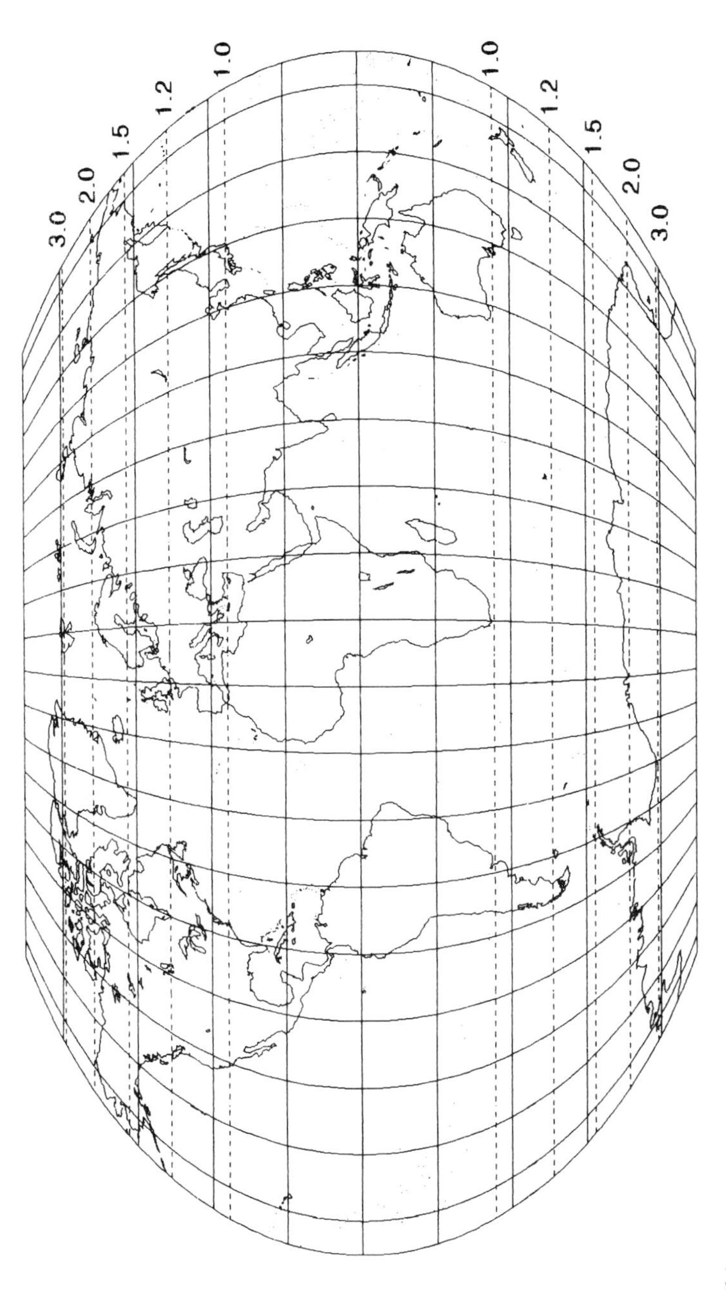

Figure 2.15 (b)

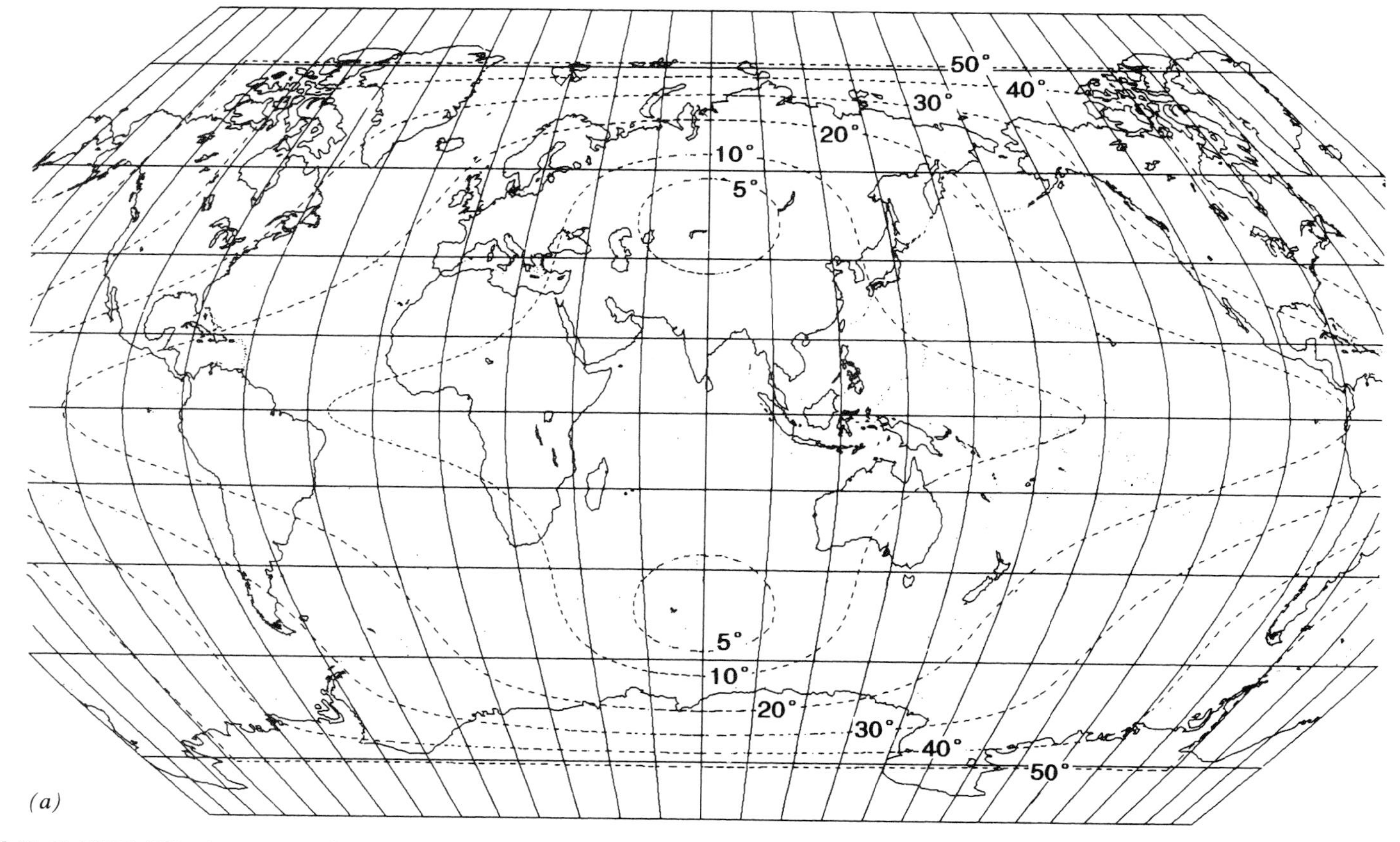

Figure 2.16 TsNIIGAiK arbitrary pseudocylindrical projection: (*a*) isocols for ω; (*b*) isocols for p. 20° graticule.

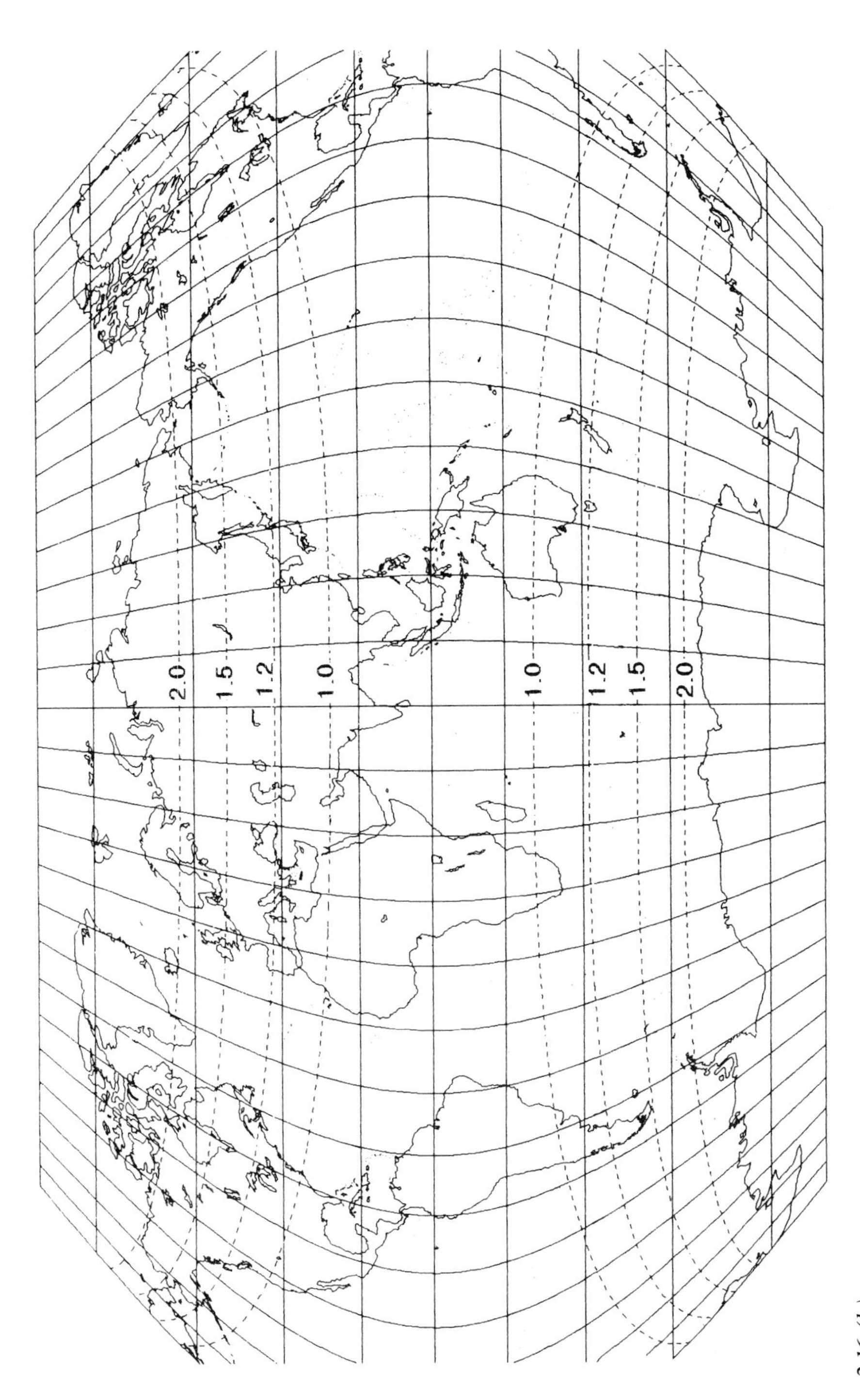

Figure 2.16 (b)

Wagner projection of 1932. (The same projection with all scales multiplied by 0.947 to produce the correct total map area was developed by Putniņš in 1934.)

The general formulas for Kavrayskiy's projection are as follows:

$$x = \frac{\lambda R}{|\lambda_1|\sqrt{3}}\sqrt{\pi^2 - 3\phi^2}, \quad y = R\phi$$

$$\tan \varepsilon = \frac{\lambda\sqrt{3}\,\phi}{|\lambda_1|\sqrt{\pi^2 - 3\phi^2}}, \quad m = \sec \varepsilon$$

$$p = n = \frac{\sqrt{\pi^2 - 3\phi^2}}{|\lambda_1|\sqrt{3}\cos\phi}, \quad \tan\frac{\omega}{2} = \frac{1}{2}\sqrt{\frac{m^2 + n^2 - 2p}{p}}$$

Isocols for this projection are of different forms. Those for constant scale along parallels and area distortion are straight lines coinciding with the parallels; the isocols characterizing the scale along meridians and maximum angular deformation are complex curves symmetrical about the central meridian and equator (Figure 2.15).

TsNIIGAiK (Ginzburg) pseudocylindrical projection

On this projection the pole line is approximately 0.6 of the length of the equator, the spaces between meridians decrease with increasing distance from the central meridian, and the spacing between parallels increases with an increase in distance from the equator.

The projection formulas take the form

$$x = R\left(1 - \frac{\phi^2}{6.16}\right)\left(0.87\lambda - \frac{\pm\lambda^4}{1049.95}\right), \quad y = R\left(\phi + \frac{\phi^3}{12}\right)$$

$$m = (1 + 0.25\phi^2)\sec\varepsilon, \quad n = [0.87 - 0.003\,81(\pm\lambda^3)]\left(1 - \frac{\phi^2}{6.16}\right)\sec\phi$$

$$\tan\varepsilon = \left(0.87 - \frac{\pm\lambda}{1049.95}\right)\frac{0.324\,68\phi}{1 + 0.25\phi^2}\lambda$$

where the $\pm$ takes the sign of λ in equations for x, n, and ε. On this projection (Figure 2.16) the regions at the edge of the map are distorted less than on the Kavrayskiy projection (angular deformation is up to 50 percent), and area distortion is greater (area scale factors are up to 4.00). This projection was used for world maps with overlapping plots in the east–west direction (Ginzburg *et al.* 1955).

Urmayev pseudocylindrical projection with arbitrary distortion

N. A. Urmayev's sinusoidal projection with low area distortion, used for ocean maps, has general formulas as follows:

$$x = 0.877\,383R\lambda\cos\phi, \quad y = 1.424\,69R(\alpha + 0.138\,175\alpha^3)$$

where $\sin\alpha = 0.8\sin\phi$.

His arbitrary projection with imposed values of constant parameters has general formulas as follows:

$$x = R a \lambda \cos \alpha, \quad y = R\left(\frac{\alpha}{ab} + \frac{k\alpha^3}{3ab}\right), \quad \sin \alpha = b \sin \phi$$

$$p = 1 + k\alpha^2, \quad n = a \cos \alpha \sec \phi, \quad m = \frac{p}{n} \sec \varepsilon$$

$$\tan \varepsilon = \frac{a^2 b \sin \alpha}{1 + k\alpha^2} \lambda$$

where the term $(k\alpha^3)/(3ab)$ influences the change of spacing between the parallels; p is the local area scale factor along particular parallels; k is a parameter which is determined by the given value of the scale p; a, b are constant parameters whose values affect the projection properties (Figure 2.17) (Ginzburg *et al.* 1955).

Robinson pseudocylindrical projection with arbitrary distortion
In 1963 A. H. Robinson (1974) of the University of Wisconsin, Madison, Wisconsin, derived a projection (Figure 2.18) that is not based on mathematical formulas. Instead, rectangular coordinates for the meridian 90° from the central meridian are listed in a table. Parallels are equidistant only between ± 38° latitude, which are the only lines true to scale. The poles are lines 0.5322 times the length of the equator, which is 1.9716 times the length of the central meridian. The dimensions and coordinates were selected to give an appearance judged suitable for a world map. The projection was adopted in 1988 by the National Geographic Society for its

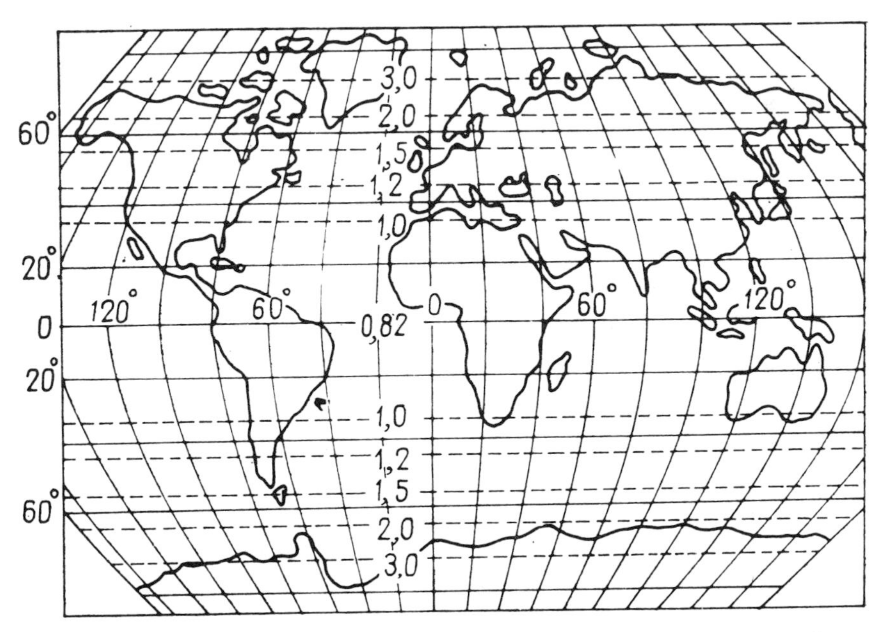

Figure 2.17 Urmayev arbitrary pseudocylindrical projection. Isocols for p. Exact design not available. 20° graticule.

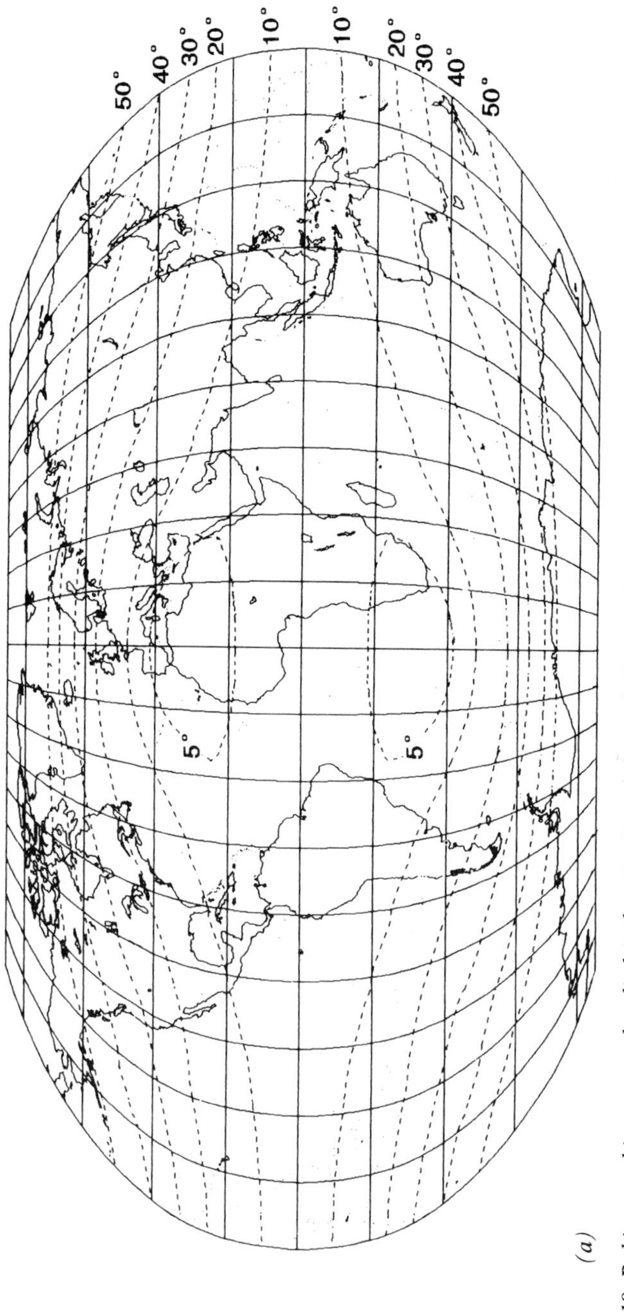

(a)

Figure 2.18 Robinson arbitrary pseudocylindrical projection: (a) isocols for ω; (b) isocols for p. 20° graticule.

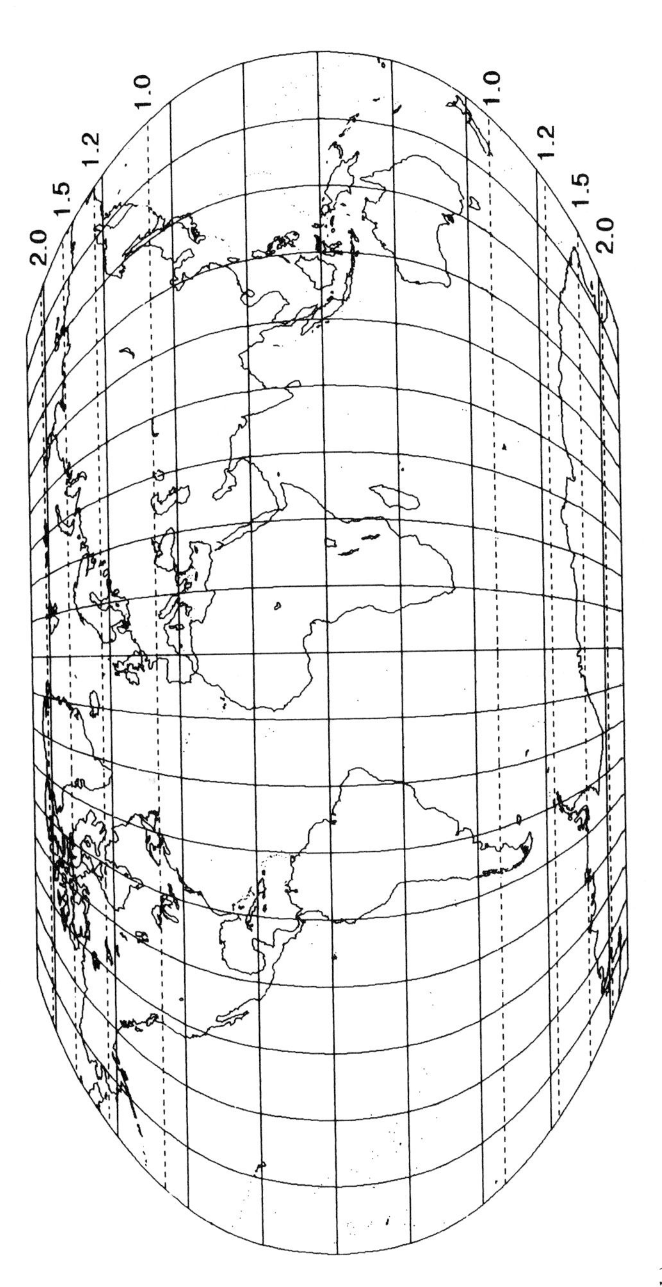

Figure 2.18 (b)

world map, although it was originally designed for Rand McNally, who has used it moderately.

Trapezoidal pseudocylindrical projection

This projection (Figure 2.19) was originally used for simply designed maps of large regions. The meridians were straight lines converging to a common point, with two of the straight parallels of latitude at true scale. This projection is now obsolete for such use, but it was used from the fifteenth to the nineteenth centuries in central Europe and the United States, and was often called the Donis projection, for Donnus Nicolaus Germanus. It was also later used for designing topographic maps of smaller regions. Its graphic analog is the Müffling projection (see section 5.1.1).

Pseudocylindrical projection of the Oxford Atlas

This projection, called a modified Gall, was adapted from the Gall perspective cylindrical projection (see section 2.1.6) with secant parallels $\phi_k = \pm 45°$. The parallels were shortened with increasing distance from the equator. The ordinates on this projection, as on Gall's, are determined by the formula

$$y = R(1 + \cos \phi_k)\tan \frac{\phi}{2}$$

No official formula for the abscissas has been published, but an empirical formula was derived from the maps by Snyder (1977):

$$x = (1 - 0.04\phi^4)R\lambda/\sqrt{2}$$

This projection is used in the *Oxford Atlas* for world maps.

Mikhaylov pseudocylindrical projection

This projection, by A. I. Mikhaylov, has arbitrary distortion. The scale increases with latitude along the straight central meridian. The other meridians are ellipses. The projection formulas are as follows:

$$x = \frac{RC\lambda}{k_1\pi}\sqrt{k_1^2\pi^2 - 4k^2\phi^2}, \quad y = 2RCk\phi,$$

where $k = k_0 + 0.003k_1\phi$, $C = \cos \phi_k$; k_0, k_1 are the given constant parameters,

$$m = 2C \sec \varepsilon(k + 0.003k_1\phi), \quad n = \frac{C}{k_1\pi \cos \phi}\sqrt{k_1^2\pi^2 - 4k^2\phi^2}$$

$$\tan \varepsilon = \frac{2k\phi\lambda}{k_1\pi\sqrt{k_1^2\pi^2 - 4k^2\phi^2}}$$

Other pseudocylindrical projections

Many other pseudocylindrical projections have been developed besides those described above. Several are briefly described in Chapter 11. Some involve modifications of earlier projections, especially the sinusoidal and Mollweide projections,

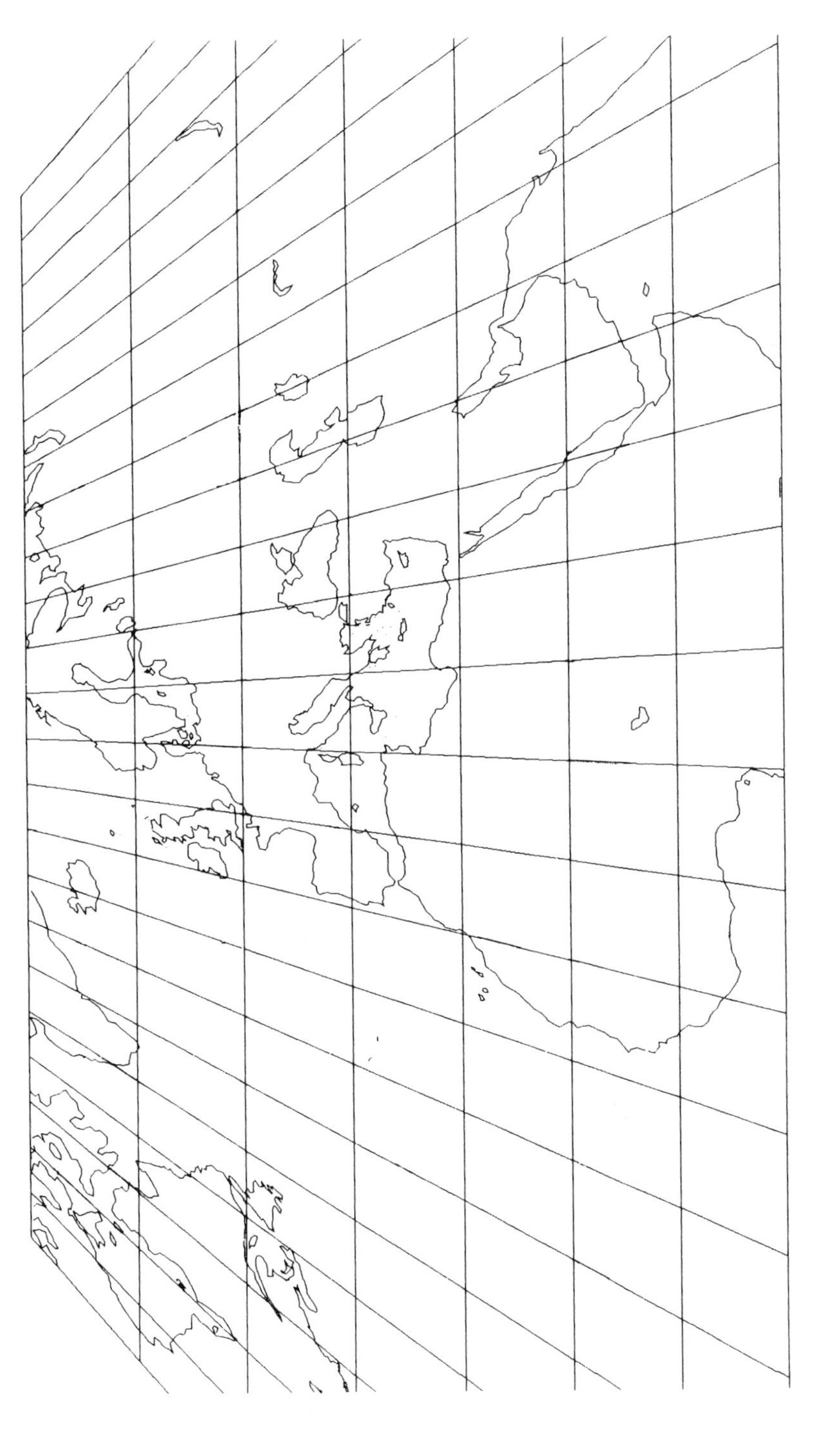

Figure 2.19 Trapezoidal pseudocylindrical projection. 10° graticule.

using interruption, averaging of coordinates, fusing, or a combination. J. P. Goode, of the University of Chicago, introduced the interruption of pseudocylindrical projections along meridians and also fusion (see section 6.3.3), principles previously applied to other types of projection. The interruption was applied by others, including Ginzburg in 1945 for the TsNIIGAiK equal-area projection, with interruptions of continents, and N. M. Volkov in 1934 for the *BSAM* variant of the Eckert VI equal-area projection, interrupting the oceans.

Fusion, best known in Goode's homolosine projection (see section 6.3.3), was later used by György Érdi-Krausz in Hungary in 1968 (see Chapter 11) and by F. W. McBryde in 1977 in the United States, each of them fitting one pseudocylindrical projection to another along an appropriate parallel of latitude. The averaging of two pseudocylindricals or of one with a cylindrical projection was discussed by Oswald Winkel in Germany in 1921, and by S. W. Boggs of the United States in 1929 (see Chapter 11).

3

Map projections with parallels in the shape of concentric circles

Conic, azimuthal, pseudoconic, and pseudoazimuthal projections are included in this group.

3.1 Conic projections

3.1.1 General formulas for conic projections

Conic projections in their normal aspects are map projections on which parallels are represented by arcs of concentric circles, and meridians by a set of straight lines drawn from the center of the circles. The angles between the meridians on the projection and on the ellipsoid (or sphere) are proportional. Consequently, a break in transformation on this type of projection arises at the pole, which may be a point (Figure 3.1) or one of the circular arcs.

From the definition, the general formulas for rectangular coordinates and local scale factors for these projections have the form

$$x = \rho \sin \delta, \quad y = \rho_s - \rho \cos \delta$$
$$\rho = f(\phi), \quad \delta = \alpha\lambda \tag{3.1}$$
$$m = -\rho_\phi/M, \quad n = \alpha\rho/r, \quad p = mn$$

where ρ_s is the polar radius of the southernmost or other chosen parallel, and r is found from equation (1.28).

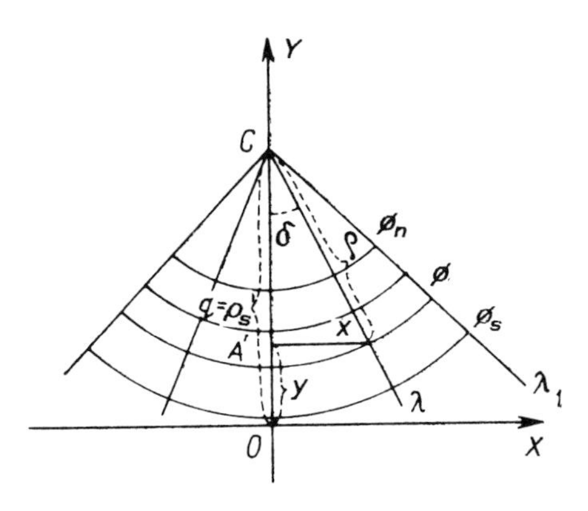

Figure 3.1 Coordinate system in a normal conic projection.

Conic projections can be conformal, equal area, or arbitrary in distortion; among the arbitrary are projections equidistant along meridians and/or along parallels.

3.1.2 Conformal conic projections

Satisfying the conditions of conformality $m = n$ and $\varepsilon = 0$, from (3.1) we can obtain

$$d\rho/\rho = -\alpha(M/r)\,d\phi$$

After integrating (see equations (1.31)–(1.32)),

$$\rho = k/U^{\alpha} \tag{3.2}$$

Formulas for local scale factors take the form

$$m = n = \alpha k/(rU^{\alpha}) \tag{3.3}$$

where α and k are constant parameters whose values affect the values and distribution of distortion.

The least scale occurs along the parallel of latitude ϕ_0. To find it, we differentiate equation (3.3) and equate it to zero:

$$m_{\phi 0} = \frac{\alpha \rho_0 M_0}{r_0^2}(\sin\phi_0 - \alpha) = 0$$

Consequently

$$\sin\phi_0 = \alpha \tag{3.4}$$

Constant parameters α (often called the cone constant) and k are determined by various methods. As a result, different conformal conic projections may be created.

Projection with a minimum local linear scale factor of unity along a given parallel
Denoting the central latitude ϕ_0, from (3.3) and (3.4) we find that

$$\alpha = \sin\phi_0, \quad k = r_0 U_0^{\alpha}/\alpha \tag{3.5}$$

Substituting (3.5) into (3.2),

$$\rho_0 = \frac{k}{U_0^{\alpha}} = \frac{r_0}{\sin\phi_0} = N_0 \cot\phi_0$$

i.e. ρ_0 is the length of the tangent to the meridian from point ϕ_0 to its intersection with the extended axis of rotation of the Earth ellipsoid.

Projection with a minimum scale factor of unity along one parallel ϕ_0 and equal distortion on the two outer parallels
Since $m_n = m_s$, the linear scale factors along the northernmost and southernmost parallels, we find that

$$\alpha = \frac{\ln r_s - \ln r_n}{\ln U_n - \ln U_s}, \quad k = \frac{r_0 U_0^{\alpha}}{\alpha}$$

Projection with equal distortion on the outer parallels, and preserving true scale along arbitrary parallel ϕ_1
On this projection

$$\alpha = \frac{\ln r_s - \ln r_n}{\ln U_n - \ln U_s}, \quad k = \frac{r_1 U_1^\alpha}{\alpha}$$

Projection preserving the scale along two standard parallels ϕ_1 and ϕ_2
Since $m_1 = m_2 = 1$, then

$$\alpha = \frac{\ln r_1 - \ln r_2}{\ln U_2 - \ln U_1}, \quad k = \frac{r_1 U_1^\alpha}{\alpha} = \frac{r_2 U_2^\alpha}{\alpha}$$

This form was developed as the first conformal conic projection (Figure 3.2) by J. H. Lambert in 1772.

The latitudes of standard parallels ϕ_1 and ϕ_2 can be determined by general rules proposed by A. R. Hinks in 1912, and later by C. H. Deetz and O. S. Adams in the United States and by V. V. Kavrayskiy in the Soviet Union, all in effect using the formulas

$$\phi_1 = \phi_s + \frac{\phi_n - \phi_s}{k}, \quad \phi_2 = \phi_n - \frac{\phi_n - \phi_s}{k}$$

where k is the value depending on the configuration of the region being mapped: Kavrayskiy suggested that when its shape is similar to a rectangle, $k = 5$; if it is somewhat circular, then $k = 4$; if it is nearly rhombic, then $k = 3$. Hinks had generally proposed that $k = 7$, and Deetz and Adams, $k = 6$, suggesting other projections if more circular.

Projection for which the absolute values of the distortion along the outer and central parallels are equal

$$\alpha = \frac{\ln r_s - \ln r_n}{\ln U_n - \ln U_s}, \quad k = \frac{1}{\alpha}\sqrt{r_n r_{\text{cen}} U_n^\alpha U_{\text{cen}}^\alpha}$$

Projection for which the mean square of linear distortion is a minimum
Assume the distortion standard to be (see equation (3.3))

$$v = \ln \mu = \ln \alpha k - \ln r - \alpha \ln U \tag{3.6}$$

Let us denote

$$\beta = \ln \alpha k, \quad a = -\ln U, \quad b = 1, \quad h = \ln r$$

Then equation (3.6) takes the form

$$a\alpha + b\beta - h = v \tag{3.7}$$

After dividing the region being mapped into elementary zones with small equal intervals of latitude $\Delta\phi$ and with intervals of longitude $\Delta\lambda = \lambda_e - \lambda_w$ (the east and

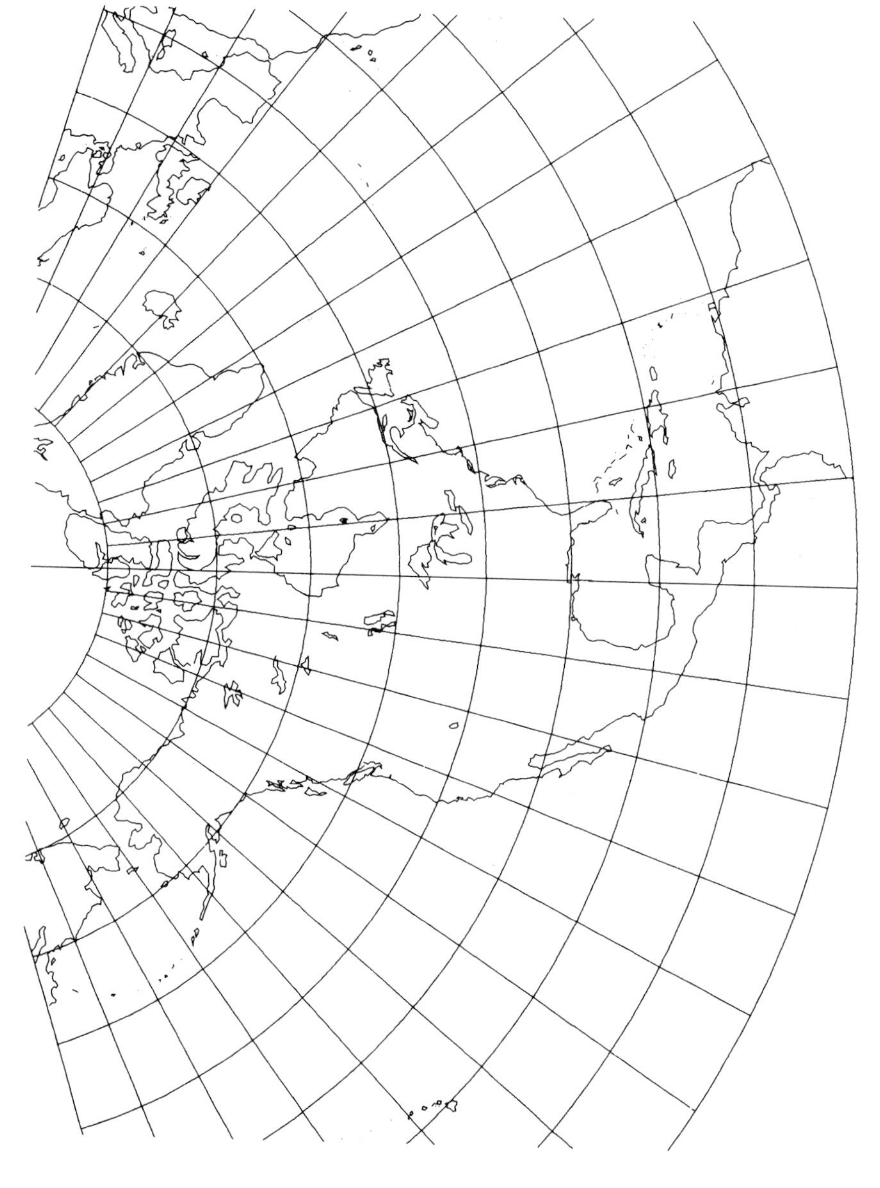

Figure 3.2 Lambert conformal conic projection, standard parallels 20° and 60° N. 10° graticule.

west limits of longitude) varying with the width of the zone at a given latitude, we can obtain a set of equations in the form of (3.7). If this set is solved by the method of least squares we can find parameters α and k. When solving the set, the area of every zone is given a weight $p = Mr\ \Delta\phi\ \Delta\lambda$. This method was first presented by N. Ya. Tsinger in Russia in 1916.

3.1.3 Equal-area conic projections

From the equal-area conditions $p = mn = 1$ and equations (3.1),

$$\alpha\rho\ d\rho = -Mr\ d\phi$$

and after integrating

$$\rho^2 = \frac{2}{\alpha}(k - S) \tag{3.8}$$

where S is the area, determined from (2.1), of a spheroidal quadrangle from the equator to the given parallel when the difference of longitude is 1 radian.

From (3.1) and (3.8),

$$n^2 = \frac{2\alpha(k - S)}{r^2}, \quad m = \frac{1}{n} = \frac{r}{\alpha\rho}, \quad \sin\frac{\omega}{2} = \frac{n - m}{n + m} \tag{3.9}$$

Depending on the method of obtaining constant parameters α and k, we can obtain different equal-area conic projections. Let us consider two of them.

Projection with a minimum scale factor of unity along a given parallel

Let us determine parallel ϕ_0 where the scale factor n_0 is minimum.

On differentiating the first equation of (3.9) and equating to zero, we find that

$$\left(\frac{d(n^2)}{d\phi}\right)_0 = \frac{2\alpha M_0}{r_0^3}[2(k - S_0)\sin\phi_0 - r_0^2] = 0$$

and

$$n_0^2 = \alpha/\sin\phi_0 = 1$$

Consequently

$$\alpha = \sin\phi_0$$

$$\rho_0 = N_0 \cot\phi_0$$

$$k = \frac{1}{2\alpha} r_0^2 + S_0$$

Projection preserving the scale along two standard parallels ϕ_1 and ϕ_2

On this projection $n_1 = n_2 = 1$, and from (3.9)

$$\alpha = \frac{1}{2}\frac{r_1^2 - r_2^2}{S_2 - S_1}$$

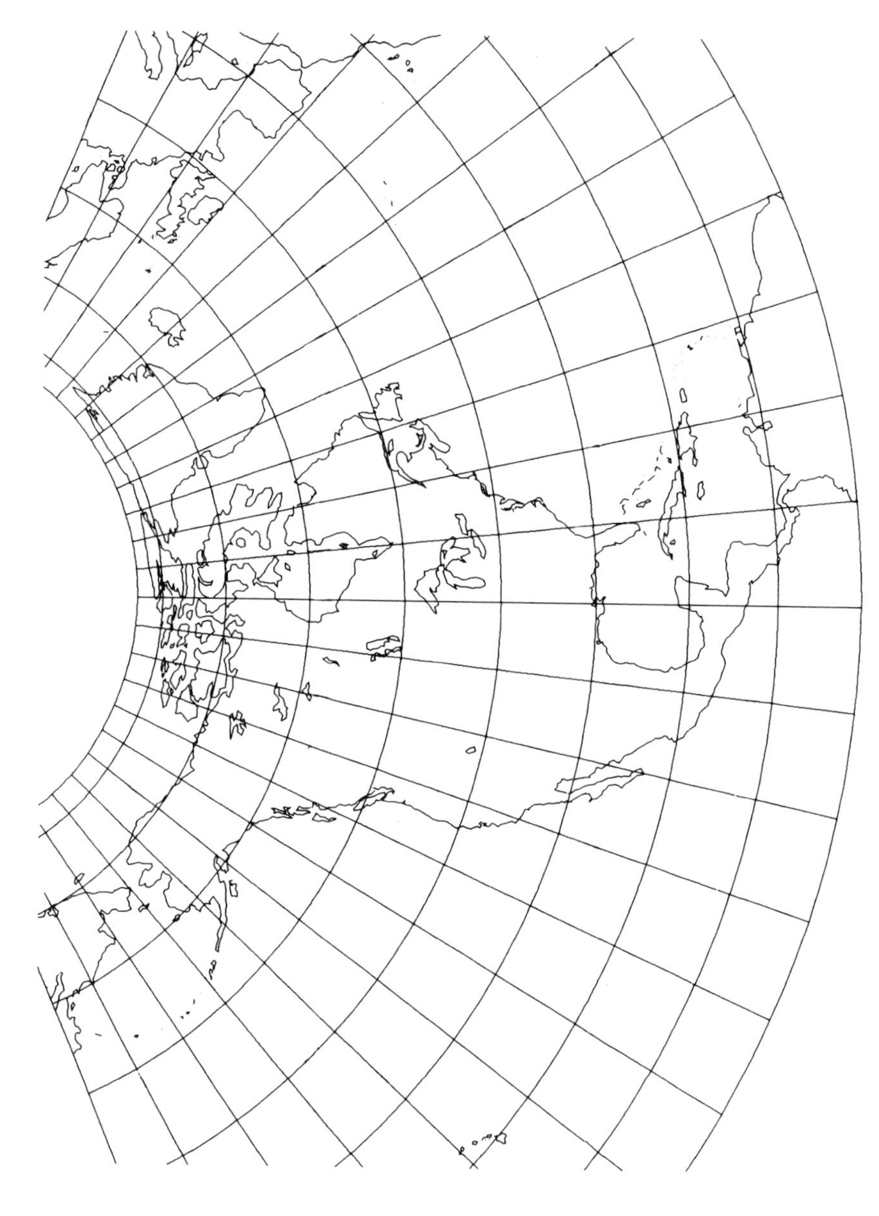

Figure 3.3 Albers equal-area conic projection, standard parallels 20° and 60° N. 10° graticule.

$$k = S_1 + r_1^2 \frac{S_2 - S_1}{r_1^2 - r_2^2} = S_2 + r_2^2 \frac{S_2 - S_1}{r_1^2 - r_2^2}$$

This form (Figure 3.3) was developed for the sphere by H. C. Albers in 1805.

Projection for which the mean square of linear distortion is a minimum
When Tsinger presented his least-squares form of the conformal conic projection in 1916, he also proposed a similar solution for an equal-area conic projection (see Snyder 1985, pp. 74–5).

3.1.4 Equidistant conic projections

Conic projections can be equidistant (at true scale) along meridians or parallels. Practically, only the former type is employed, where $m = 1$.

From equations (3.1)

$$d\rho = -M\, d\phi$$

Integration of this equation gives

$$\rho = k - s \quad \text{and} \quad n = \alpha(k - s)/r \tag{3.10}$$

where s is the arc length from the equator to the given parallel and is determined by (1.145); α and k are constant parameters. Depending on how these parameters are chosen, one can develop different projections equidistant along the meridians.

Projection with a minimum scale factor of unity preserved along a given parallel ϕ_0
Differentiating (3.10)

$$\left(\frac{dn}{d\phi}\right)_0 = \frac{\alpha M_0}{r_0^2}\left[(k - s_0)\sin\phi_0 - r_0\right] = 0$$

Hence,

$$k = s_0 + N_0 \cot\phi_0$$

From (3.10), we find that

$$n_0 = \alpha/\sin\phi_0 = 1 \quad \text{and} \quad \alpha = \sin\phi_0$$

Projection with equal distortion along the outer parallels and with a minimum scale factor of unity preserved along parallel ϕ_0
According to the conditions imposed we have $n_s = n_n$ and $n_0 = 1$. Incorporating equations (3.10),

$$k = \frac{s_n r_s - s_s r_n}{r_s - r_n}, \quad \alpha = \sin\phi_0$$

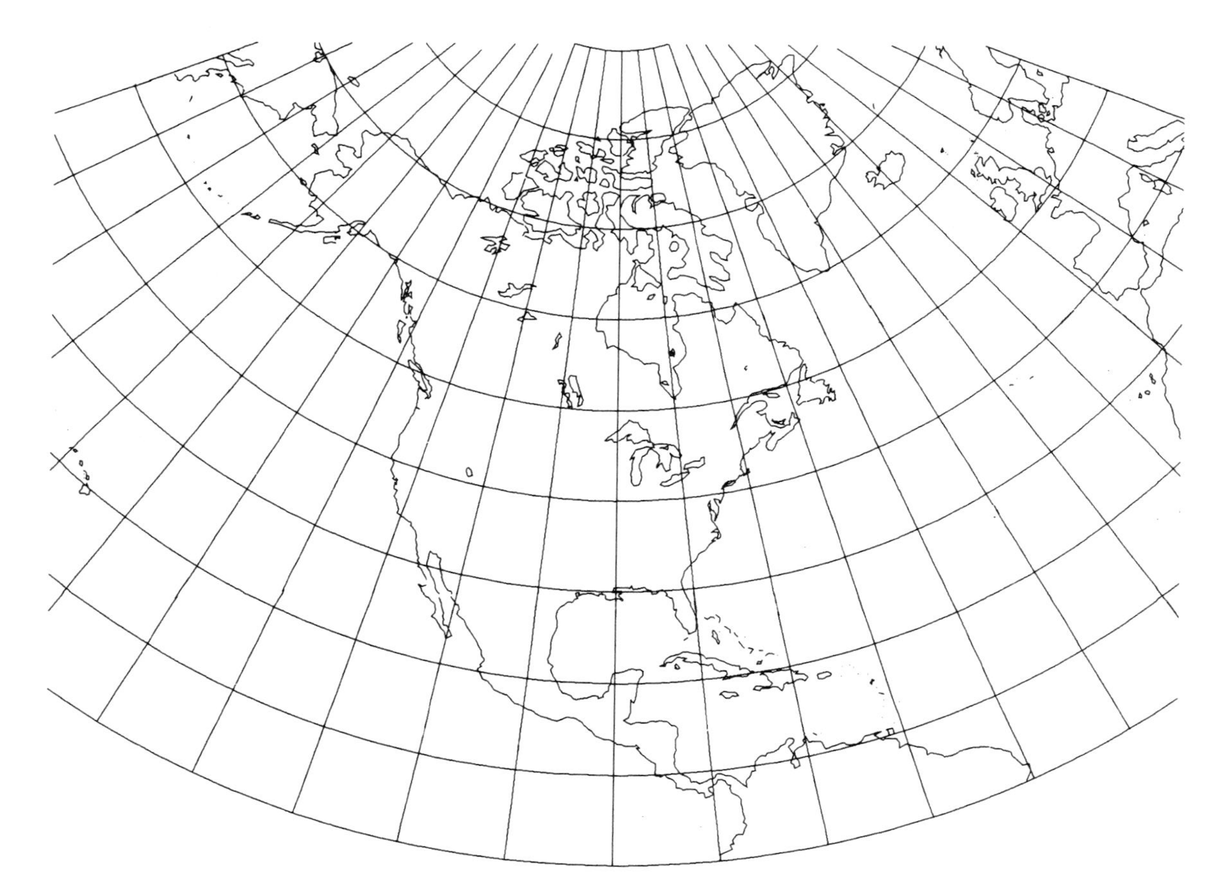

Figure 3.4 Equidistant conic projection, standard parallels 20° and 60° N. 10° graticule.

Projection preserving the scale along two standard parallels ϕ_1 and ϕ_2
For this condition we have $n_1 = n_2 = 1$; applying (3.10),

$$\alpha = \frac{r_1 - r_2}{s_2 - s_1}, \quad k = s_1 + \frac{r_1}{\alpha} = s_2 + \frac{r_2}{\alpha}$$

This was the form (Figure 3.4) originated by J. N. De l'Isle in 1745.

Projection with equal absolute values of distortion along the outer and central parallels
V. V. Vitkovskiy suggested the conditions for creating this projection. He suggested the same conditions for conformal and equal-area projections. For this projection,

$$\phi_0 = \tfrac{1}{2}(\phi_n + \phi_s), \quad |n_n - 1| = |n_s - 1| = |n_0 - 1|$$

From equations (3.10),

$$\alpha = \frac{2r_0(r_s - r_n)}{r_n(s_0 - s_s) + r_0(s_n - s_s) + r_s(s_n - s_0)}$$

$$k = s_n + r_n \frac{s_n - s_s}{r_s - r_n} = s_s + r_s \frac{s_n - s_s}{r_s - r_n}$$

where ϕ_n and ϕ_s are the outer (northern and southern) and ϕ_0 the central parallels.

Projection with equal distortion on the outer parallels of the zone to be mapped and preserving the area of the zone
Incorporating equations (3.10), we can derive the following (Urmayev 1941):

$$k = \frac{s_n r_s - s_s r_n}{r_s - r_n} = s_n + r_n \frac{s_n - s_s}{r_s - r_n} = s_s + r_s \frac{s_n - s_s}{r_s - r_n} \tag{3.11}$$

The area of the spheroidal quadrangle with longitude difference λ is equal on the surface of the ellipsoid (or sphere) to $\lambda(S_n - S_s)$, and on the map to $\frac{1}{2}\alpha\lambda(\rho_s^2 - \rho_n^2)$; hence

$$2(S_n - S_s) = \alpha(\rho_s - \rho_n)(\rho_s + \rho_n)$$

With proper consideration of (3.10) and (3.11), it is not difficult to calculate $(\rho_s - \rho_n)$ and $(\rho_s + \rho_n)$, and then

$$\alpha = \frac{2(S_n - S_s)(r_s - r_n)}{(s_n - s_s)^2(r_s + r_n)}$$

On Krasovskiy's approximately equidistant conic projection, created to represent the territory of the USSR, the area of the given zone is preserved, its range of latitude equals $2\theta = \phi_1 - \phi_2$, and linear scale on the outer parallels ($n_1 = n_2$) is preserved. The sum of the squares of linear distortion along the parallels for the whole territory is a minimum.

Parameters α and ϕ_1 are determined so that the error

$$E = \pm\sqrt{[p'v^2]/[p]} = \min$$

where p' is a weight coefficient (the distance between meridians at parallels of latitude ϕ), and $v = \ln \mu$.

For a sphere of unit radius, the formulas for the projection take the form

$$\rho = \rho_1 + m(\phi_1 - \phi) = \rho_2 + m(\phi_2 - \phi)$$

$$\phi_1 = 73°28'42'', \quad \phi_2 = 39°28'42''$$

$$2\theta = 34°, \quad \phi_0 = (\phi_1 + \phi_2)/2 = 56°28'42''$$

$$\rho_1 = \cos \phi_1/\sqrt{\alpha \cos \theta \sin \phi_0}, \quad \rho_2 = \cos \phi_2/\sqrt{\alpha \cos \theta \sin \phi_0}$$

$$m = \frac{\sin \theta}{\theta} \sqrt{\frac{\sin \phi_0}{\alpha \cos \theta}} = 0.997\,03$$

$$\alpha = 0.851\,568, \quad n = \frac{\alpha \rho}{\cos \phi}$$

Projection with the least-mean-square linear distortion

A method for creating these types of projections was first presented by Tsinger in 1916. V. V. Kavrayskiy simplified this method considerably in 1933.

The projection is created in a manner similar to (3.6)–(3.7) (see Snyder 1985, pp. 75–6). For a map of the USSR, Kavrayskiy took as standard parallels $\phi_1 = 47°$, $\phi_2 = 62°$ and determined constant parameters α and k.

From a comparative analysis of distortion on the Kavrayskiy and Krasovskiy projections, it is found that the first is advisable for a map of the territory of the former USSR when it is desired to show its continental portions. The second is used when it is necessary to include the polar territories in addition to those on the continent.

On the normal aspect of every conic projection discussed so far, the local linear scale is a function only of the latitude, and isocols coincide with the parallels.

In 1947 N. A. Urmayev developed the theory of a generalized conic projection equidistant along meridians and on which linear scale factors n along parallels depend on both the latitude and longitude, the isocols being ovals (see section 6.4.1).

The characteristic feature of all conic projections is the fact that their central lines can be made to coincide with parallels midway between the limiting parallels of a region. Consequently, conic projections are recommended for maps of regions with limiting latitudes asymmetric about the equator and with a considerable range of longitude. This is the reason that many Soviet as well as American maps have been designed using these projections.

It should be noted that on conic projections the arc lengths of meridians between parallels often change with an increase of distance from the central parallel: they increase on conformal and similar projections, they decrease on equal-area and similar projections, and they remain unchanged on projections equidistant along meridians.

Conic projections are used not only with normal but also with oblique or transverse orientations of the pole, e.g. for designing charts of aeronavigational routes.

3.1.5 Perspective conic projections

On perspective conic projections the Earth's surface is projected onto the developable surface of either a secant or a tangent cone from a point lying on the central line of the cone PO, or from lines OQ on the equatorial plane crossing the planes of the corresponding meridians (Figure 3.5). These projections are rarely used now. Information on them can be found in Snyder (1993).

3.1.6 Oblique and transverse conic projections

Oblique and transverse conic projections are for applications in which the region being mapped extends primarily along a small circle other than a parallel of latitude, i.e. along an almucantar.

On these projections or aspects the verticals are represented by a set of straight lines converging at a common point at angles proportional to the differences in azimuth. This is pole $Q(\phi_0, \lambda_0)$ on a conformal conic projection; it is beyond the pole on equal-area and equidistant projections. The almucantars are represented by arcs of concentric circles with the center at the same point.

All meridians and parallels are represented by curves, except the meridian of pole Q, which is represented by a straight line, may be taken to be the Y-axis of the projection, and is the line of symmetry of its map graticule.

On oblique and transverse conic projections, the base directions coincide with verticals and almucantars, the 'central line' on each of them being the central almucantar.

The method for creating these projections is as follows:

- For the ellipsoid, we first transform the ellipsoid onto the sphere (see section 1.7) in accordance with the method appropriate for the distortion pattern and get, as a result, spherical coordinates ϕ_{sp}, λ_{sp} from the geodetic coordinates ϕ, λ on the ellipsoid.

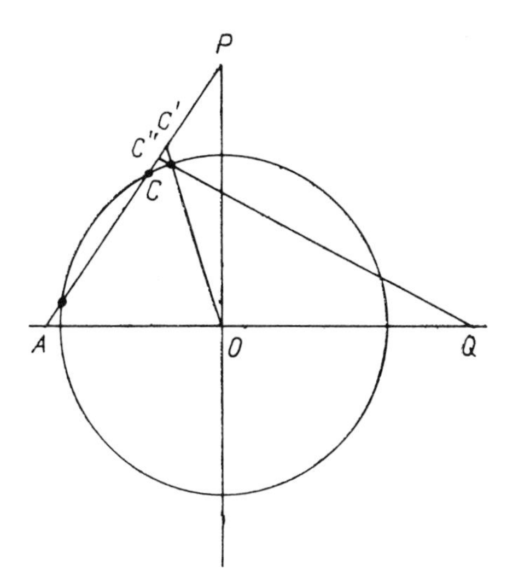

Figure 3.5 Geometry of perspective conic projections.

Table 3.1 Distortion on conformal and equal-area conic projections. Krasovskiy 1940 ellipsoid

	Conformal projection $\alpha = 0.870\,619\,2$, $k = 11\,323\,960$				Equal-area projection $\alpha = 0.852\,926\,2$, $k = 40\,929\,460$			
ϕ	n	m	p	ω	n	m	p	ω
30°	1.109	1.109	1.229	0	1.075	0.930	1.0	8°16′
40°	1.041	1.041	1.084	0	1.031	0.970	1.0	3°27′
50°	1.0	1.0	1.0	0	1.0	1.0	1.0	0°00′
60°	0.985	0.985	0.970	0	0.985	1.015	1.0	1°42′
70°	1.0	1.0	1.0	0	1.0	1.0	1.0	0°00′
80°	1.070	1.070	1.145	0	1.152	0.868	1.0	16°09′

- We then determine the pole Q (ϕ_0, λ_0) of the oblique or transverse coordinate system from section 1.8.4.
- From the given values ϕ_{sp}, λ_{sp} we calculate the polar spherical coordinates (see section 1.1.3).
- Substituting values $\phi'_{sp} = 90° - z$ and $\lambda'_{sp} = -a$ for spherical coordinates ϕ_{sp}, λ_{sp}, we calculate, in accordance with the corresponding formulas for conic projections, rectangular coordinates, local scale factors, and other characteristics.

3.1.7 Comparative characteristics of conic projections

For the purpose of analysis we list the values of local scale factors and of maximum angular deformation for some conic projections. Table 3.1 gives distortion values for typical examples of conformal and equal-area conic projections. Table 3.2 lists scale

Table 3.2 Distortion on equidistant conic projections. Krasovskiy 1940 ellipsoid

ϕ	Kavrayskiy projection $\phi_1 = 62°$ $\phi_2 = 47°$ $\alpha = 0.811\,823\,8$ $k = 10\,572\,200$ $m = 1$	Krasovskiy projection $\phi_1 = 73°29'$ – $\phi_2 = 39°29'$ – $\alpha = 0.851\,568\,0$ $k = 5\,968\,300$ $m = 0.997\,03$	Urmayev projection: $\alpha = 0.825\,527$, $k = 10\,451\,300$, $b = 0.018\,351\,8$					
			λ					
			0°	20°	40°	60°	80°	100°
30°	$\omega = 3°38'$	4°45′	3°36′	3°43′	4°06′	4°44′	5°36′	6°42′
	$n = 1.065$	1.084	1.065	1.067	1.074	1.086	1.103	1.124
40°	$\omega = 1°07'$	1°56′	0°54′	1°02′	1°25′	2°03′	2°55′	4°01′
	$n = 1.020$	1.031	1.016	1.018	1.025	1.036	1.052	1.073
50°	$\omega = 0°18'$	0°05′	0°45′	0°38′	0°30′	0°25′	1°15′	2°21′
	$n = 0.995$	0.998	0.987	0.989	0.996	1.007	1.022	1.042
60°	$\omega = 0°15'$	0°34′	1°07′	1°00′	0°37′	0°00′	0°52′	1°59′
	$n = 0.996$	0.987	0.980	0.983	0.989	1.00	1.016	1.035
70°	$\omega = 2°19'$	0°45′	0°41′	0°48′	1°11′	1°49′	2°41′	3°47′
	$n = 1.041$	1.010	1.012	1.014	1.021	1.032	1.048	1.069
80°	$\omega = 12°03'$	7°28′	8°39′	8°47′	9°10′	9°48′	10°40′	11°46′
	$n = 1.23$	1.136	1.163	1.166	1.174	1.187	1.205	1.228

factors and maximum angular deformation for the equidistant conic projections of Kavrayskiy and Krasovskiy in the first columns; the other columns give information on the Urmayev generalized equidistant conic projection (see section 6.4.1).

It follows from Tables 3.1 and 3.2 that distortion on conic projections within reasonable ranges of latitude is practically undetectable by the eye (less than 12° and 12 percent) and that each of these projections can be successfully employed for designing maps of the region of the former Soviet Union for an appropriate purpose. If angular and area distortion are equally undesirable, equidistant conic projections are advisable, and the projections of both Kavrayskiy and Krasovskiy can be used successfully. If it is necessary to minimize distortion and optimize its distribution, e.g. for carrying out cartometry, the Urmayev projection should be used for Russia. For other regions, such as the United States, predominantly east–west in extent, suitable changes in the parameters of these conic projections can be considered.

3.2 Azimuthal projections

3.2.1 General formulas for azimuthal projections

On azimuthal projections, parallels (or almucantars) are represented by concentric circles, and meridians (or verticals) are represented by straight lines passing through the center of the circles at angles equal to the difference between the corresponding longitudes of the meridians (Figure 3.6). In accordance with this definition, the general equations can be written in the form

$$x = \rho \sin a, \quad y = \rho \cos a, \quad \rho = f(z) \tag{3.12}$$

where z, a are polar spherical (or spheroidal) coordinates, determined from equations (1.10)–(1.13). The local linear scale factors along verticals (μ_1) and almucantars (μ_2) and the maximum angular deformation ω are expressed by the formulas

$$\mu_1 = \frac{d\rho}{R\,dz}, \quad \mu_2 = \frac{\rho}{R \sin z}, \quad \sin\frac{\omega}{2} = \frac{\mu_2 - \mu_1}{\mu_1 + \mu_2} \tag{3.13}$$

On these projections, there is normally no distortion at the central point ($z = 0$), and distortion increases with the increase of the distance from the center. To decrease the absolute magnitude of distortion, a reducing multiplier $k < 1$ may sometimes be introduced into the formulas, its value being determined by the fact that scale is to be preserved unchanged along the given base or standard parallel or almucantar.

Azimuthal projections are generally used for designing small-scale maps. In this case the Earth or celestial body is taken to be a sphere. The pole of the coordinate system to be used is the geographical pole for the normal or polar aspect.

3.2.2 Conformal azimuthal (stereographic) projections

Conformal azimuthal projections of the ellipsoid of revolution

The projection often used to represent polar regions is a particular case of the conformal conic projection when $\alpha = 1$.

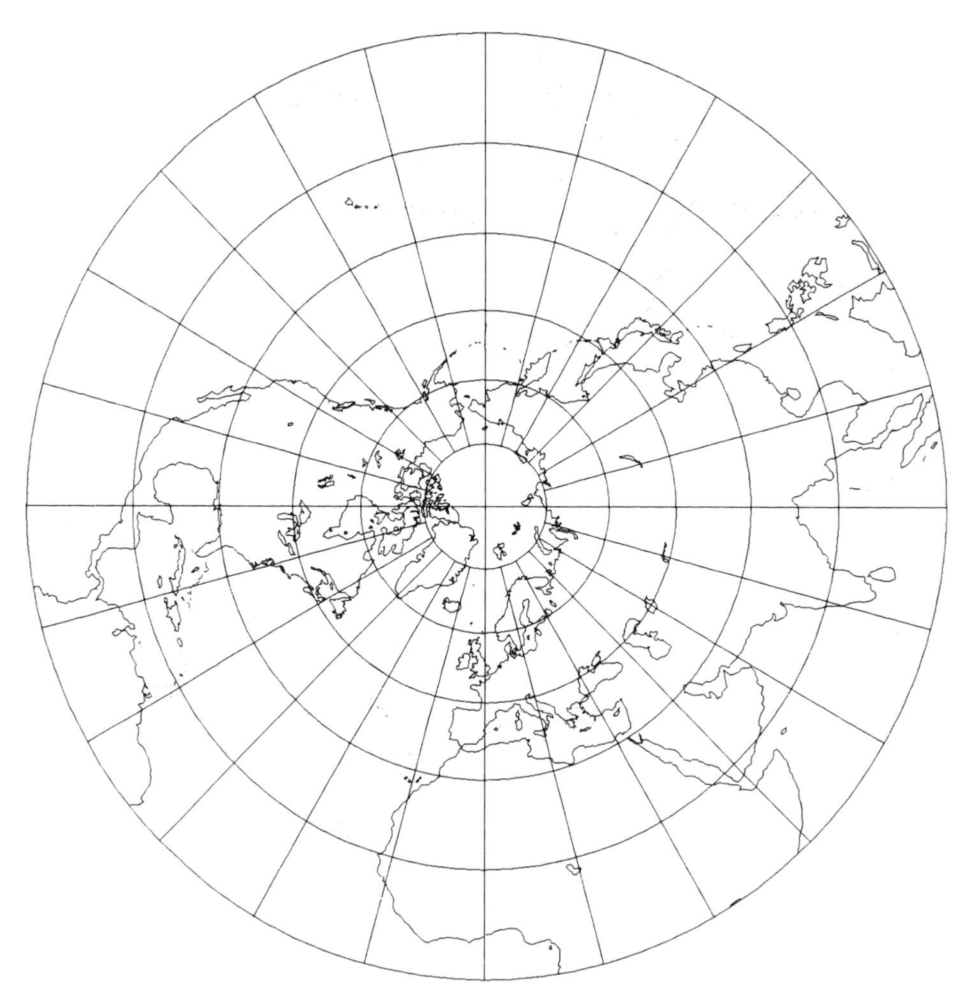

Figure 3.6 Normal or polar stereographic conformal azimuthal projection of northern hemisphere. Example of normal azimuthal projection. 15° graticule.

Equations (3.2) and (3.3) take the form

$$\rho = k/U, \quad \mu = \mu_1 = \mu_2 = k/rU, \quad p = \mu^2, \quad k = r_0 U_0$$

where U is calculated from equations (1.31)–(1.32); r_0, U_0 are determined along the latitude of the given central parallel, for which the local scale factor is equal to 1.0.

The conformal projection for representing regions with approximately circular outlines (other than polar regions) is actually a particular case of the Lagrange projection (4.3) when $\alpha = 1$:

$$x = \frac{k \cos \delta \sin \lambda}{1 + \cos \delta \cos \lambda}, \quad y = \frac{k \sin \delta}{1 + \cos \delta \cos \lambda}$$

$$\mu = \mu_1 = \mu_2 = \frac{k \cos \delta}{r(1 + \cos \delta \cos \lambda)}, \quad p = \mu^2$$

$$\delta = 2\arctan(\beta U) - \pi/2$$

Here β and k are constant parameters determined by the formulas

$$\beta = \frac{1 + \sin\phi_0}{1 - \sin\phi_0} U_0, \quad k = \mu_0 r_0 (1 + \sec\delta_0)$$

where $\delta_0 = 2\arctan(\sin\phi_0)$, ϕ_0 is the latitude of the given central parallel, and μ_0 is the given value of the local linear scale factor at the point of intersection of the central meridian and the central parallel.

The formulas for the projection created by using the oblique polar spheroidal coordinate system, to an accuracy of terms up to e^2, take the form

$$x = 2N_0 k \tan\frac{z}{2}\sin a, \quad y = 2N_0 k \tan\frac{z}{2}\cos a$$

$$\mu = k\sec^2\left(\frac{z}{2}\right)\left(1 + \frac{e^2}{2}[\sin z\cos a\cos\phi_0 + \sin\phi_0(\cos z - 1)]^2\right) + \cdots$$

where $k = \cos^2(z_k/2)$ is the reduction factor, N_0 is the radius of curvature of the first vertical section along parallel ϕ_0 determined from equation (1.2), z_k is the polar distance of the almucantar at which there is no distortion, ϕ_0, λ_0 are geographical coordinates of the new pole point, and z, a are determined by equations (1.10)–(1.12).

This projection can be employed for mapping any region with near-circular boundaries at large and medium scales. Although it is usually called stereographic, it is not quite so, nor is it quite azimuthal in the oblique aspect, except for the projection of the sphere (see below).

Conformal azimuthal projection of the sphere

This stereographic projection (Figure 3.7) can be derived from the conformal condition (1.80); it can also be considered a particular case of the previous projection variant for the ellipsoid, when $e^2 = 0$.

The projection formulas take the form

$$x = 2Rk\tan\frac{z}{2}\sin a, \quad y = 2Rk\tan\frac{z}{2}\cos a$$

$$\mu = k\sec^2\left(\frac{z}{2}\right), \quad k = \cos^2\left(\frac{z_k}{2}\right)$$

This projection is truly stereographic perspective, as well as azimuthal and conformal; that is, it may be geometrically projected from a point on the sphere (the antipode) opposite the center of the projection.

3.2.3 Equal-area azimuthal projections of the sphere

From the condition $p = \mu_1\mu_2 = 1$ for an azimuthal equal-area projection and taking into account equations (3.13), we can get the differential equation

$$\rho\, d\rho = R^2 \sin z\, dz$$

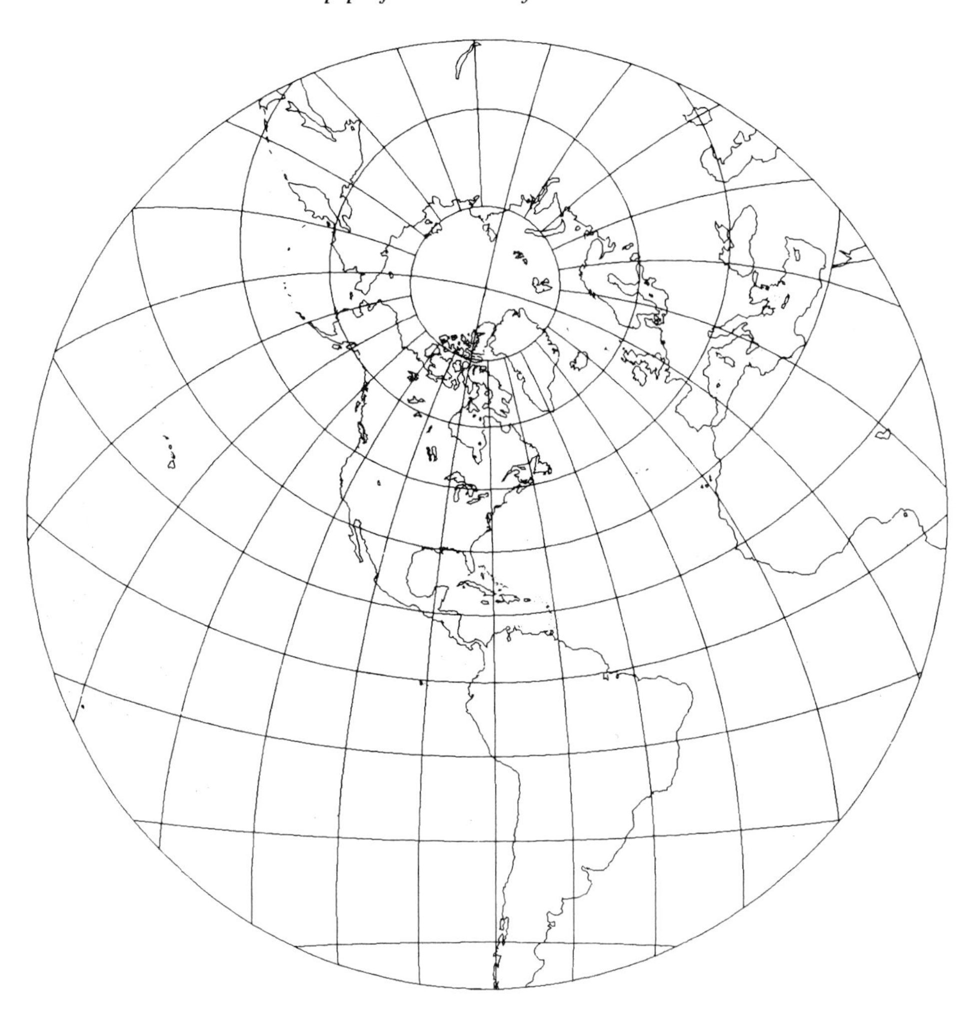

Figure 3.7 Oblique stereographic projection of one hemisphere, centered on Washington, DC. 15° graticule.

After integrating, we can obtain an equation for polar radius ρ and the general projection formulas

$$x = \rho \sin a, \quad y = \rho \cos a$$

$$\rho = 2Rk \sin(z/2)$$

$$\mu_1 = k \cos(z/2), \quad \mu_2 = k \sec(z/2)$$

$$p = k^2, \quad \tan(\pi/4 + \omega/4) = \sec(z/2)$$

When $k = 1$, there is no distortion of any type at the central projection point (at the point for the pole). If $k = \cos(z_k/2)$, linear distortion is absent along almucantar $z = z_k$, but in that case the area scale is equal not to unity, but to a constant value

$$p = \cos^2(z_k/2)$$

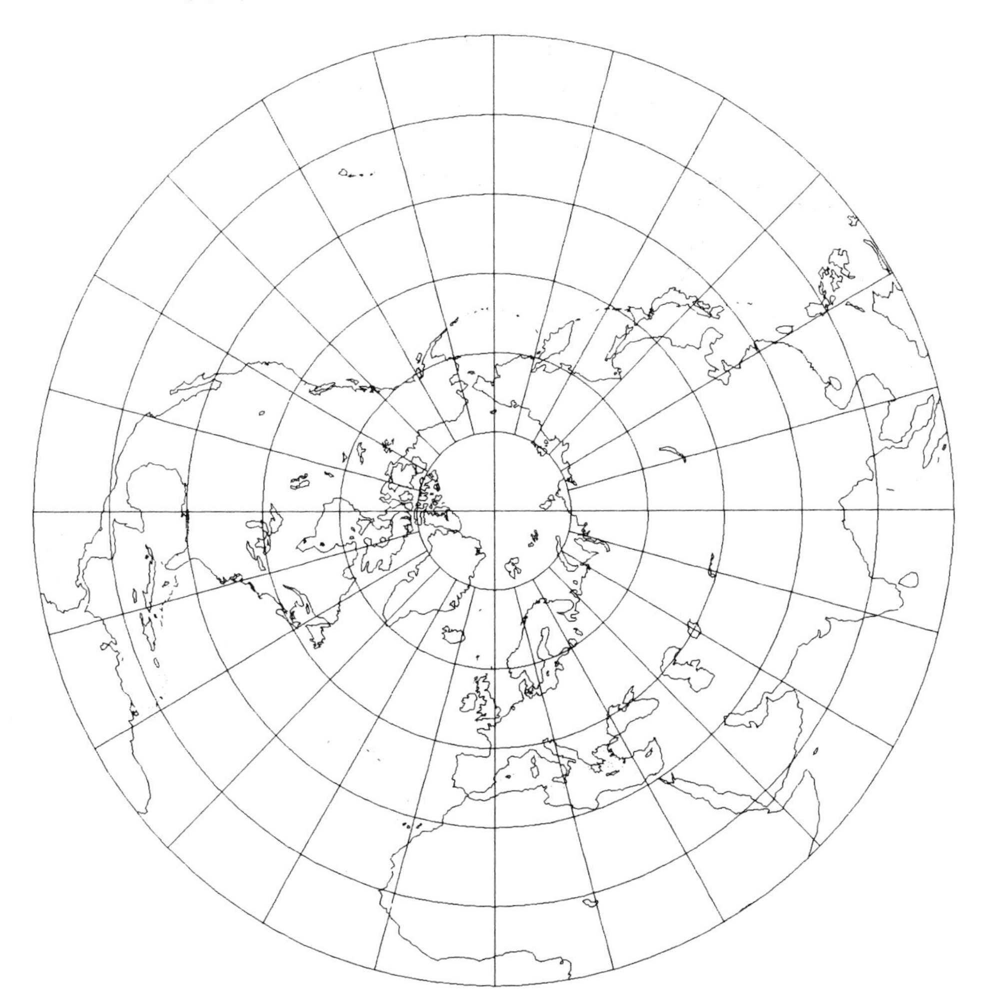

Figure 3.8 Polar azimuthal equidistant projection of northern hemisphere. 15° graticule.

The projection (with $k = 1$) was suggested by J. H. Lambert in 1772 and is used in some American atlases for equal-area maps of the polar regions. It is a limiting case of the equal-area conic projection with $\alpha = 1$, and may be computed in this manner for polar mapping of the ellipsoid. Ellipsoidal formulas are also given in Snyder (1987b, pp. 187–90).

3.2.4 Azimuthal projections equidistant along meridians (or verticals)

For this condition, $\mu_1 = 1$; applying equations (3.13),

$$d\rho = R\,dz$$

After integrating this expression and including a reducing factor k, we can get the following projection formulas:

$$x = Rkz \sin a, \quad y = Rkz \cos a$$

$$\mu_1 = k, \quad \mu_2 = kz/\sin z$$

$$p = \mu_2 k, \quad \sin\frac{\omega}{2} = \frac{z - \sin z}{z + \sin z}$$

The projection preserves scale along verticals (meridians on the normal or polar aspect): when $k = 1$, there is no distortion of any type at the central point; when $k = (\sin z_k)/z_k$, scale is preserved along the almucantar $z = z_k$, and the local linear scale factor along verticals is equal to a constant value k.

The projection (Figure 3.8) is frequently given the name of Guillaume Postel, who used it in 1581 for a polar map of the northern hemisphere. A more common name used in many countries is the azimuthal equidistant projection. It is said to have been used for star maps by early Egyptians, and appears on an existing star map of 1426 by Conrad of Dyffenbach. In its oblique aspect (Figure 3.9), it is useful centered on some city. This projection is a limiting form of the equidistant conic

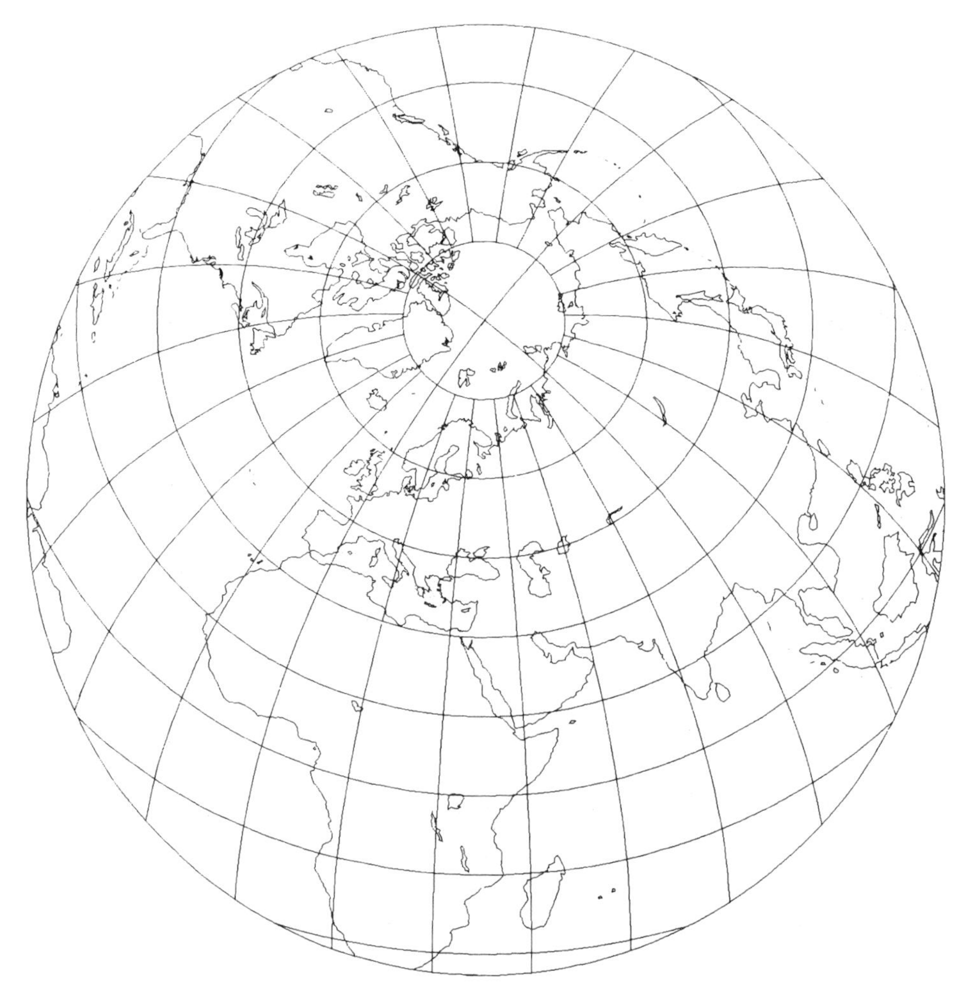

Figure 3.9 Oblique azimuthal equidistant projection of one hemisphere, centered on Moscow. 15° graticule.

projection with both the standard parallels taken as the geographic pole; thus the ellipsoidal form of the polar aspect may be derived using this relationship. Ellipsoidal formulas are also given by Snyder (1987b, pp. 197–202).

From a distortion standpoint, the azimuthal projections mentioned above are the best possible for minimizing error in representing territories with approximately circular outlines and conformal, equal-area, or equidistant distortion characteristics, respectively. On these projections the scale of arc segments along verticals (or meridians on normal aspects) increases on conformal and similar projections, decreases on equal-area and similar projections, and remains unchanged on equidistant projections, with an increase of the distance from the central point or pole.

3.2.5 Airy minimum error azimuthal projection

G. B. Airy (1861) presented an azimuthal projection with a minimum overall error (see also section 1.6.3) determined by 'balance of errors'. He minimized the error E of the expression

$$E = \int_0^{\beta} [(\mu_1 - 1)^2 + (\mu_2 - 1)^2] \sin z \, dz$$

where β is the angular range of the map from the center to the edge of the circular boundary. While Airy had made an error in his constraints, this was corrected by James and Clarke in 1862. The final formulas based on the above criterion are

$$\rho = 2R\left[\cot \frac{z}{2} \ln \sec \frac{z}{2} + \tan \frac{z}{2} \cot^2\left(\frac{\beta}{2}\right) \ln \sec \frac{\beta}{2}\right]$$

$$\mu_1 = 1 + \frac{\ln[(1 + \cos z)/2]}{1 - \cos z} - \frac{2 \ln \cos (\beta/2)}{\tan^2(\beta/2)(1 + \cos z)}$$

$$\mu_2 = -\left(\frac{\ln[(1 + \cos z)/2]}{1 - \cos z} + \frac{2 \ln \cos(\beta/2)}{\tan^2(\beta/2)(1 + \cos z)}\right)$$

At the projection center ($z = 0$), the formulas for μ_1 and μ_2 are indeterminate, but

$$\mu_{10} = \mu_{20} = \frac{1}{2} - \frac{\ln \cos(\beta/2)}{\tan^2(\beta/2)}$$

If β is 90° or less, the Airy projection resembles the azimuthal equidistant projection.

3.2.6 Generalized formulas for azimuthal projections

The formulas given above are used for creating conformal, equal-area, and equidistant azimuthal projections. To obtain the same types of projections and those with intermediate properties, a number of scientists suggested some generalized formulas. For example, Ginzburg suggested the formula

$$\rho = R\left(L_1 \sin \frac{z}{k_1} + L_2 \tan \frac{z}{k_2}\right)$$

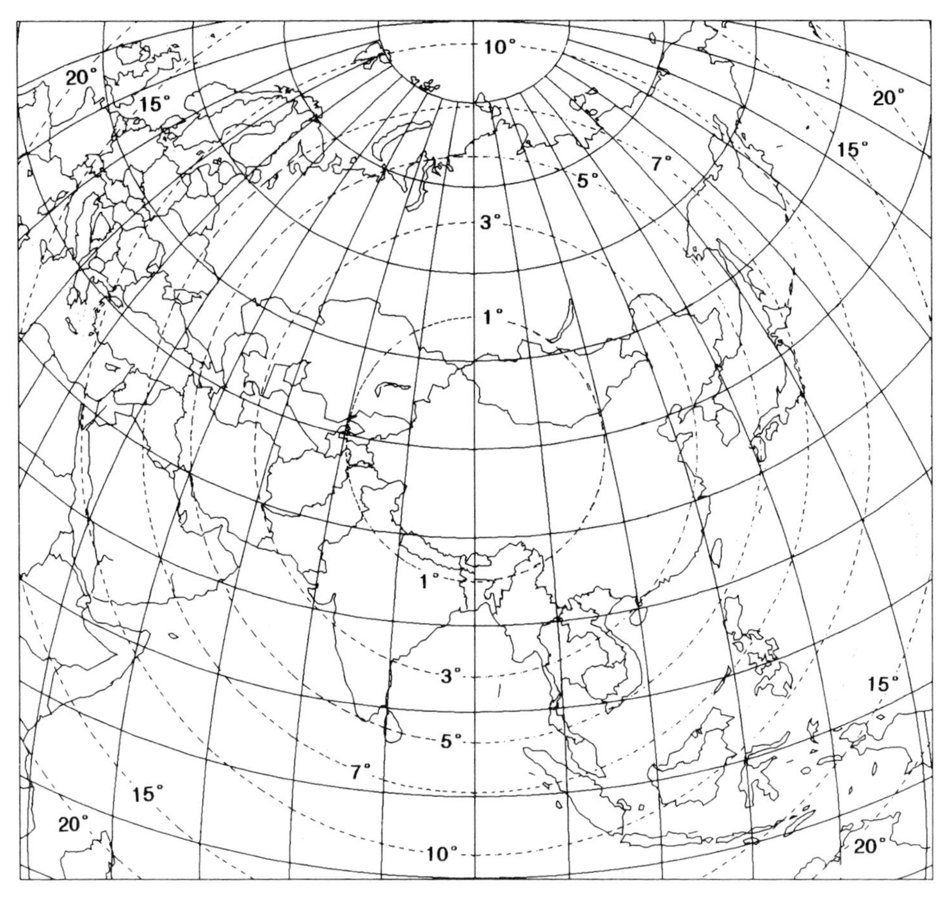

Figure 3.10 Oblique Lambert azimuthal equal-area projection of Asia centered at 40°N, 90°E. Isocols for ω. 10° graticule.

where L_1, k_1, L_2, and k_2 are constant parameters. By changing their values one can get various azimuthal projections:

stereographic projection	$L_1 = 0, \quad L_2 = k_2 = 2$
gnomonic projection	$L_1 = 0, \quad L_2 = k_2 = 1$
Lambert azimuthal equal-area projection	$L_2 = 0, \quad L_1 = k_1 = 2$
orthographic projection	$L_2 = 0, \quad L_1 = k_1 = 1$

When $L_2 = 0$ and k is in the range 3–7 we can obtain projections with low area distortion; when k is close to 1.2–1.5, we obtain projections giving a spherical appearance to the surface being mapped.

Malovichko suggested a generalized formula as follows:

$$\rho = R\left(2 \sin \frac{z}{2}\right)^k \left(2 \tan \frac{z}{2}\right)^{1-k}$$

When $k = 1/2$ we obtain the Breusing geometric projection. A further generalization of these formulas is discussed in section 6.2.

3.2.7 Oblique and transverse azimuthal projections

As a rule, when designing maps on these projections the Earth is taken to be a sphere; hence, the creation of the projections mentioned is reduced to determining the coordinates of the pole $Q(\phi_0, \lambda_0)$ of the oblique or transverse aspect, to calculating polar spherical coordinates z, a and to determining rectangular coordinates, local scale factors, and other characteristics of the corresponding azimuthal projection. When applied to azimuthal projections, transverse aspects are often called equatorial aspects, because the equator appears as a straight central line ($\phi_0 = 0$).

Oblique or equatorial azimuthal projections with various distortion patterns have been widely used for designing maps of the world and large regions. Thus, for example, the oblique or equatorial Lambert azimuthal equal-area projection (see section 3.2.3) has been used to design maps of hemispheres, of Asia (with $\phi_0 = 40°$, $\lambda_0 = 90°$ as on Figure 3.10), and of other continents. The oblique azimuthal equidistant projection has also been similarly used (see section 3.2.4), but to a lesser degree.

The equatorial orthographic projection has been used to represent hemispheres of the Moon, of the Earth as a planet, and of other celestial bodies. The TsNIIGAiK oblique azimuthal projection, developed by Ginzburg in 1946, has been used for maps of portions of the Earth's surface extending beyond one hemisphere. For this projection

$$\rho = kR \sin \frac{z}{k}, \quad \mu_1 = \cos \frac{z}{k}$$

where $k = 1.8$; the pole of the oblique system is a point with $\phi_0 = 55°$, $\lambda_0 = 50°$ (when $k = 1$ we get an orthographic projection, and when $k = 2$, an equal-area projection).

The layout of a given map with the central meridian at some angle, together with the shape of the map graticule, produce a spherical effect for the Earth's surface. Some additional data for using these projections will be given below.

3.3 Perspective azimuthal projections

Perspective azimuthal projections are subdivided into projections with negative and positive transformations. In the former the region being mapped is projected onto the pictorial plane with straight lines of sight from a point in space (the point of view) situated on the concave side of this surface, and in the latter on the convex side of the surface being mapped.

When using perspective azimuthal projections, the Earth is generally considered to be a sphere. Sometimes, however, for the design of maps of medium-sized and small regions (less than 1 000 000 km^2) and at scales larger than 1 : 10 000 000, it is necessary to take into account the ellipsoidal shape.

Suppose that we have on the surface of the ellipsoid of revolution (Figure 3.11) some point $Q_0(\phi_0, \lambda_0)$ for a new pole, a tangent plane T_0 and its normal $Q_0 O'$. At point Q_0 a spatial rectangular topocentric coordinate system $Q_0 XYZ$ is fixed, the Y-axis of which is in the direction of meridian $Q_0 P$ increasing northward, the Z-axis coincides with normal $O'Q_0$, and the X-axis completes the right-handed system.

Let us introduce symbols $S_n O' = D_n$, $S_p O' = D_p$, $O'Q_0 = N_0$, $Q_0 S_p = H$, $O'M = N'_0$, $Q_0 M' = \rho_n$, and $Q_0 M'' = \rho_p$, where S_n, S_p are the points of projection

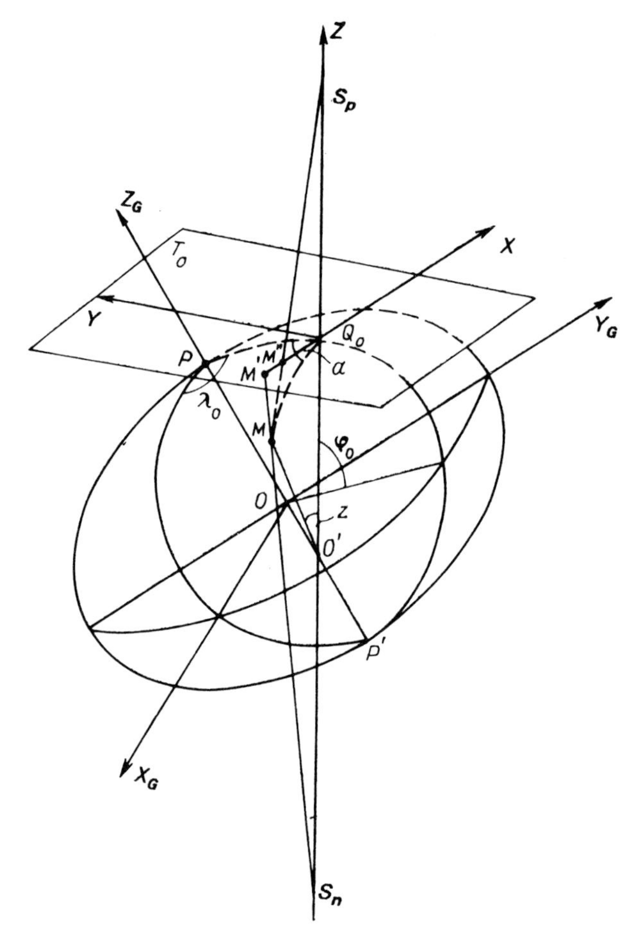

Figure 3.11 Perspective azimuthal projections when transforming onto the horizontal picture plane.

(points of view) for the negative and positive transformations, respectively, and N_0 is the radius of curvature of the first vertical section at pole point $Q_0(\phi_0, \lambda_0)$.

When preparing these projections we use equations (1.11)–(1.13) relating geodetic and polar spherical coordinates.

3.3.1 Perspective azimuthal projections with negative transformation

Among these projections, the most widely used are projections of the sphere for which

$$\rho_n = (D + R)\frac{R \sin z}{D + R \cos z} \tag{3.14}$$

where $R = N_0$, the radius of the sphere, $D = S_n O$, and $\rho_n = Q_0 M'$.

The following projections are distinguished depending on the position of the point of view: for the gnomonic projection, $D = 0$; for the stereographic projection,

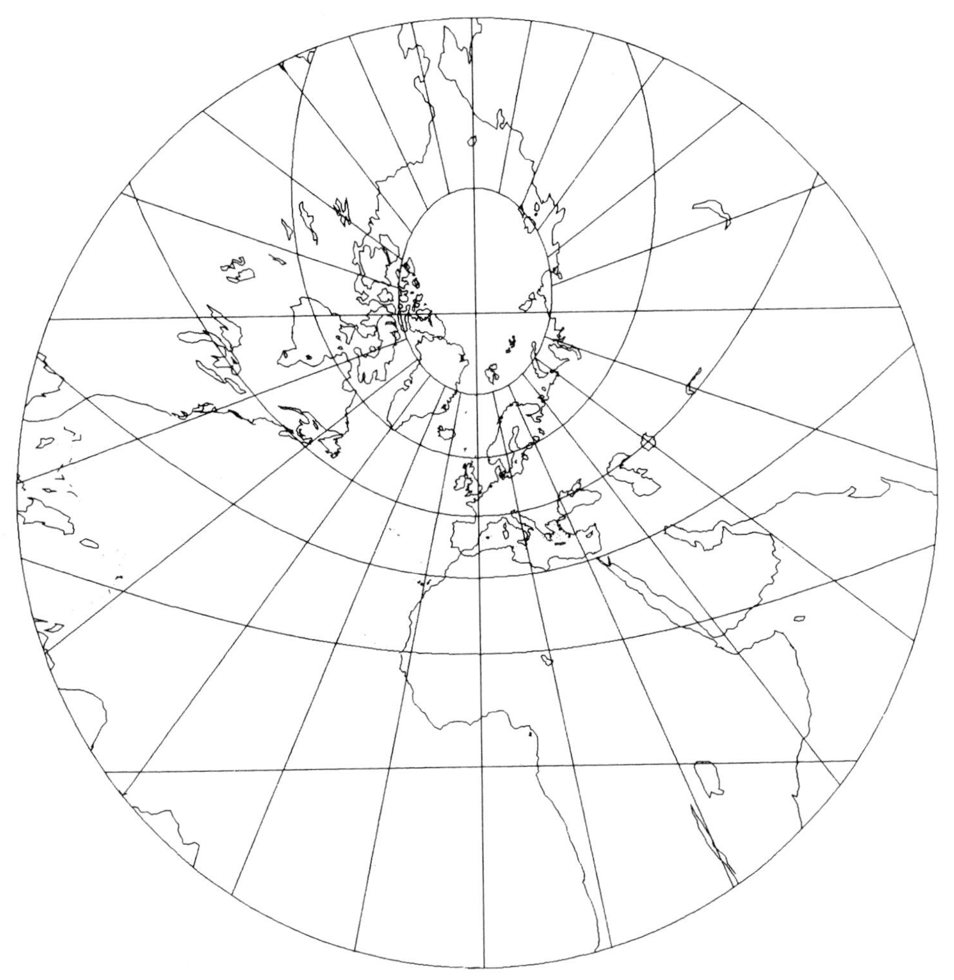

Figure 3.12 Oblique gnomonic projection centered on London, 65° range. 15° graticule.

$D = R$; for the orthographic projection, $D = \infty$; and for outer-perspective azimuthal projections, $R < D < \infty$.

Gnomonic projection

If $D = 0$, projection formulas (3.12)–(3.14) simplify to the form

$$x = R \tan z \sin a, \quad y = R \tan z \cos a$$

$$\mu_1 = \sec^2 z, \quad \mu_2 = \sec z, \quad p = \sec^3 z$$

$$\sin(\omega/2) = \frac{1 - \cos z}{1 + \cos z} = \tan^2(z/2), \quad \tan \beta = \cos z \tan \alpha$$

where β, α are azimuths of linear elements on the projection and on the sphere, respectively.

The principal advantage of the gnomonic projection (Figure 3.12) is that arcs of great circles, i.e. orthodromes or the paths of the shortest distances, are represented on it by straight lines. This fact is used to determine points along geodetic lines to assist in transferring these lines onto maps compiled on other projections.

Offsetting this feature, the projection has very great distortion. Distances between almucantars (or parallels in the polar aspect) increase rapidly with an increase in distance from the central point. Less than one hemisphere may be shown from a given center point.

Stereographic projection

Applying the condition that $D = R$, the general formulas for this projection take the form

$$x = 2R \tan(z/2)\sin a, \quad y = 2R \tan(z/2)\cos a$$

$$\mu = \sec^2(z/2), \quad p = \mu^2 = \sec^4(z/2), \quad \omega = 0$$

The projection is conformal: it has no distortion of local shape, and any circle of finite dimensions on the sphere is also represented by a circle on the projection. This property is used for the graphic solution of problems in spherical astronomy. Most of the sphere may be shown from one center point, although the far hemisphere has great areal distortion.

Orthographic projection

For this projection $D = \infty$, i.e. the projection results from a set of parallel rays.

The general formulas for the projection are

$$x = R \sin z \sin a, \quad y = R \sin z \cos a$$

$$\mu_1 = \cos z, \quad \mu_2 = 1, \quad \sin(\omega/2) = \tan^2(z/2)$$

The projection preserves true scale μ_2 along almucantars. The distance between almucantars (or parallels in the polar aspect) decreases rapidly with an increase of distance from the central point. Oblique orthographic projections (Figure 3.13) convey the effect of the spherical globe very well. Normally a full hemisphere is shown from a given center point; the far hemisphere is overlapped by the near hemisphere. The projection has been used for maps of some celestial bodies.

Outer-perspective azimuthal projections of the sphere

In considering the above perspective azimuthal projections, local linear scale factors along verticals on the near hemisphere range as follows:

- on the gnomonic projection: from 1 to ∞, increasing from the center (where $z = 0$) to the infinitely distant 'edge' (where $z = 90°$);
- on the stereographic projection: from 1 to 2, increasing from the center to the edge;
- on the orthographic projection: from 1 to 0, decreasing from the center to the edge.

Taking these ranges into account, one may develop projections intermediate between the ones mentioned, in theory even where $D < R$, but especially where $R < D < \infty$. An analysis of linear scale factors along verticals for projections with D greater than R but less than $2R$ shows that at first they increase from the center,

Figure 3.13 Oblique orthographic projection of one hemisphere centered at New Delhi. 15° graticule.

where $\mu_1 = 1$, up to some maximum, and then they continuously decrease to 0. For projections with $D \geq 2R$, this scale factor constantly decreases from 1 to 0. Among this set of projections we should note the La Hire projection, presented in 1701. In this projection the pictorial plane passes through the center of the sphere. The point of view S is located at the point of intersection of lines PP' and AH, where A, H are respectively the center points of quadrant PQ and radius CQ (Figure 3.14). Thus the parallel 45° from the pole of the projection is halfway between the pole and equator on both sphere and map, although other parallels are not uniformly spaced on the projection.

The projection equations take the form

$$\rho = \frac{DR \sin z}{D + R \cos z}, \quad D = R(1 + \sin 45^\circ)$$

$$\mu_1 = \frac{D(D \cos z + R)}{(D + R \cos z)^2}, \quad \mu_2 = \frac{D}{D + R \cos z}$$

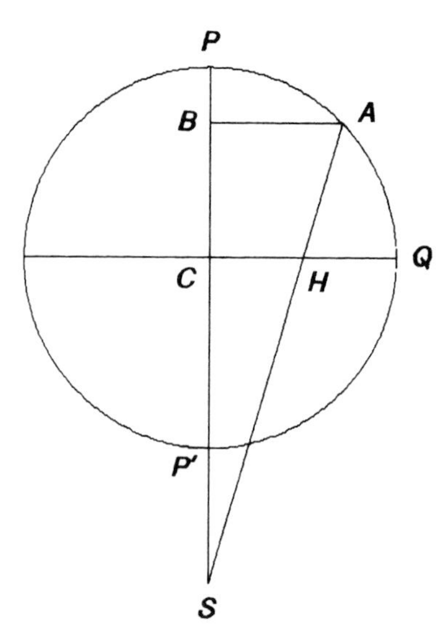

Figure 3.14 Construction of the La Hire projection.

The maximum scale factor $\mu_1 = 1.030$ is reached when $z = 57°37'$; on almucantar $z = 78°03'$, μ_1 is equal to unity for the second time. The maximum area scale factor $p = 1.474$ is reached when $z = 91°26'$, and on almucantar $z = 114°21'$ the area scale factor $p = 1$ for the second time.

On the La Hire projection, areas are not preserved and the projection is not conformal, but shapes are less distorted, and the angular and area distortion is somewhat equalized. The projection can be used for representing the northern and southern hemispheres.

A series of projections by Clarke of England in 1862 is also of interest. On these the location of the point of view is determined by minimizing the sum of squares of local scale errors throughout the desired range of the projection (normally not to the horizon), applying the Airy criteria (see sections 1.6.3 and 3.2.5) to a perspective projection. For example, Clarke chose a minimum error range of 40° for Africa or South America, and 54° for Asia, each appropriately centered and with the horizon of the projection extended further.

A number of arbitrary perspective azimuthal projections have been proposed for which the point of view is determined using other conditions. These include projections by Antoine Parent, John Lowry, Philipp Fischer, H. F. Gretschel, and Ernst Hammer.

Perspective projections by Solov'ev with multiple transformations can also be classified as azimuthal projections in this category. To design these, the Earth's surface is projected sequentially onto a number of auxiliary spherical surfaces of which the radii are multiples of the original radius ($2R$, $4R$, and so on). Only at the final stage is the projection cast onto a plane (Ginzburg and Salmanova 1964, p. 117).

In this case the projection onto spherical surfaces can be carried out according to the principles of one of the perspective projections (e.g. gnomonic), and the final projection can be made in accordance with some other perspective projection (e.g.

stereographic). This method makes it possible to create a great variety of perspective azimuthal projections.

Perspective azimuthal projections of the ellipsoid with negative transformation
The value of the polar radius (see Figure 3.11) is equal to

$$\rho = (N_0 + D_n)\frac{N'_0 \sin z}{D_n + N'_0 \cos z} \tag{3.15}$$

where to an accuracy of its terms up to e^2 (Bugayevskiy and Portnov 1984)

$$N'_0 = N_0\left(1 - \frac{e^2}{2}[\sin z \cos a \cos \phi_0 + \sin \phi_0(\cos z - 1)]^2\right) + \cdots \tag{3.16}$$

N_0 is the radius of curvature of the first normal section at the pole point $Q_0(\phi_0, \lambda_0)$. For other symbols see section 1.7.5.

The general equations take the form

$$X_n = \frac{N_0(N_0 + D_n)}{D_n + N_0 t_5}\left\{t_4\left[1 + \frac{e^2}{2}\tau\left(2 \sin \phi - \frac{\tau D_n + t_6}{D_n + N_0 t_5}\right)\right] + \cdots\right\}$$

$$Y_n = \frac{N_0(N_0 + D_n)}{D_n + N_0 t_5}\left\{t_1 + \frac{e^2}{2}\tau\left[2(t_1 \sin \phi - \cos \phi_0) - (\tau D_n + t_6)\frac{t_1}{D_n + N_0 t_5}\right] + \cdots\right\}$$

$$\mu_1 = (N_0 + D_n)\frac{D_n \cos z + N_0}{(D_n + N_0 \cos z)^2}$$

$$\times\left[1 - \frac{e^2}{2}\left(t_z \frac{\sin z(D_n + N_0 \cos z)}{D_n \cos z + N_0} - \tau^2 \frac{N_0 \cos z}{D_n + N_0 \cos z}\right)\right]$$

$$\mu_2 = \frac{N_0 + D_n}{D_n + N_0 \cos z}\left(1 + \frac{e^2}{2}\tau^2 \frac{N_0 \cos z}{D + N_0 \cos z}\right)$$

$$p = \mu_1\mu_2, \quad \sin\frac{\omega}{2} = \frac{\mu_2 - \mu_1}{\mu_1 + \mu_2}$$

where

$$t_z = \frac{2\tau}{D_n + N_0 \cos z}\left[D_n(\cos z \cos a \cos \phi_0 - \sin z \sin \phi_0) + \frac{\tau D_n N_0 \sin z}{2(D_n + N_0 \cos z)}\right]$$

$$t_6 = 2N_0(t_5 \sin \phi - \sin \phi_0)$$

where z, a, t_1, t_4, t_5, τ are determined from equations (1.11)–(1.12).

The formulas given make it possible to create a set of perspective azimuthal projections of the ellipsoid with negative transformation, depending on the position of the projection point S_n. Among these variants there are projections corresponding to gnomonic, orthographic, and stereographic projections of the sphere.

Thus, for the case of the central perspective (corresponding to the gnomonic) $D = 0$, from equation (3.15) we can get

$$x = N_0\left\{t_4\left[1 + \frac{e^2}{2}\tau\left(2 \sin \phi - \frac{t_6}{N_0 t_5}\right)\right]\right\}\left(\frac{1}{t_5}\right)$$

$$y = N_0\left\{t_1 + \frac{e^2}{2}\tau\left[2(t_1 \sin \phi - \cos \phi_0) - \frac{t_1 t_6}{N_0 t_5}\right]\right\}\left(\frac{1}{t_5}\right)$$

$$\mu_1 = \sec^2 z(1 + e^2\tau^2/2), \quad \mu_2 = \sec z(1 + \tau^2 e^2/2)$$

$$p = \sec^3 z(1 + e^2\tau^2), \quad \sin(\omega/2) = \tan^2(z/2)$$

On this projection the line for the shortest distance (or geodesic on the ellipsoid) is represented with a slight curvature.

For the orthographic projection of the ellipsoid $D \to \infty$. From the same formulas one can get

$$x = N_0\left\{t_4\left[1 + \frac{e^2}{2}\tau(2 \sin\phi - \tau)\right]\right\}$$

$$y = N_0\left\{t_1 + \frac{e^2}{2}\tau[2(t_1 \sin\phi + \cos\phi_0) - \tau t_1]\right\}$$

$$\mu_1 = \cos z[1 - e^2\tau(\sin z \cos a \cos\phi_0 - \sin z \tan z \sin\phi_0)]$$

$$\mu_2 = 1, \quad p = \mu_1\mu_2$$

$$\sin(\omega/2) = \frac{\{1 - \cos z[1 - e^2\tau(\sin z \cos a \cos\phi_0 - \sin z \tan z \sin\phi_0)]\}}{\{1 + \cos z[1 - e^2\tau(\sin z \cos a \cos\phi_0 - \sin z \tan z \sin\phi_0)]\}}$$

This projection is recommended not only for solving cartographic problems, but also for creating orthophotos at medium scale, representing large regions.

It is clear that when eccentricity $e = 0$ all the formulas given for transforming the ellipsoid take the form of the formulas for the corresponding projections of the sphere, considered above.

3.3.2 Perspective azimuthal projections with positive transformation

Projections with positive transformation onto a horizontal pictorial plane
For projections of the ellipsoid

$$\rho = \frac{HN_0' \sin z}{D_p - N_0' \cos z} \tag{3.17}$$

The projection formulas take the form

$$\begin{aligned} X_p &= HN \cos\phi \sin(\lambda - \lambda_0)/\{N[\sin\phi \sin\phi_0 + \cos\phi \cos\phi_0 \cos(\lambda - \lambda_0)] \\ &\quad - (N_0 + H) + e^2(N_0 \sin\phi_0 - N \sin\phi)\sin\phi_0\} \\ Y_p &= H\{N[\sin\phi \cos\phi_0 - \cos\phi \sin\phi_0 \cos(\lambda - \lambda_0)] \\ &\quad + e^2(N_0 \sin\phi_0 - N \sin\phi)\cos\phi_0\}/\{N[\sin\phi \sin\phi_0 \\ &\quad + \cos\phi \cos\phi_0 \cos(\lambda - \lambda_0)] \\ &\quad + e^2(N_0 \sin\phi_0 - N \sin\phi)\sin\phi_0 - (N_0 + H)\} \end{aligned} \tag{3.18}$$

$$\mu_{1H} = \frac{H(D_p \cos z - N_0)}{(D_p - N_0 \cos z)^2}\left[1 - \frac{e^2}{2}\left(p_z \frac{\sin z(D_p - N_0 \cos z)}{D_p \cos z - N_0} + \tau^2 \frac{N_0 \cos z}{D_p - N_0 \cos z}\right)\right]$$

$$\mu_{2H} = \frac{H}{D_p - N_0 \cos z}\left(1 - \frac{e^2}{2}\tau^2 \frac{N_0 \cos z}{D_p - N_0 \cos z}\right) \tag{3.19}$$

Figure 3.15 Oblique vertical perspective azimuthal projection, centered on Washington, DC, from a height of 5000 km. 10° graticule.

where

$$p_z = \frac{2\tau}{D_p - N_0 \cos z}\left(D_p(\cos z \cos a \cos \phi_0 - \sin z \sin \phi_0) - \frac{\tau D_p N_0 \sin z}{2(D_p - N_0 \cos z)}\right)$$

Note again that when eccentricity $e = 0$, the formulas given reduce to the perspective azimuthal projection (Figure 3.15) of the sphere with positive transformation onto a horizontal pictorial plane.

Perspective azimuthal projection with positive transformation onto an inclined pictorial plane (the mathematical model for ideal aerial and space photographs)

Suppose that we have previously calculated rectangular coordinates X, Y for a perspective azimuthal projection with positive transformation onto a horizontal pictorial plane from equations (3.12) and (3.16)–(3.18), and that we know the elements of inner and outer orientation for the corresponding oblique photograph (Figure 3.16).

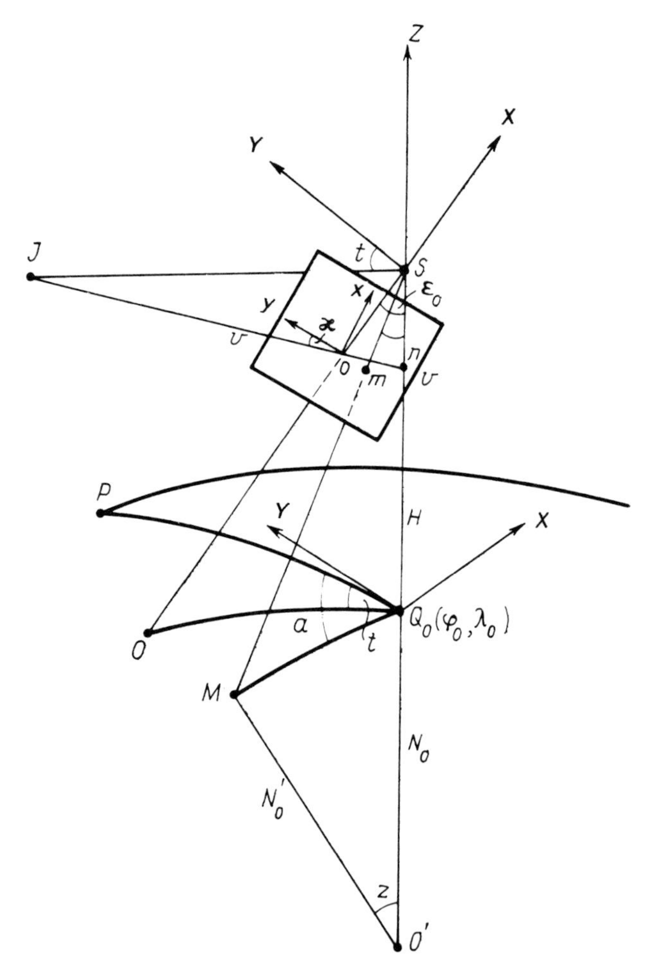

Figure 3.16 Construction of perspective azimuthal projection with positive transformation of an included pictorial plane.

Among the elements of inner orientation are the focal length f of the camera lens and the rectangular coordinates x_0, y_0 of the optical center of the photograph.

The elements of outer orientation include the geodetic coordinates ϕ_0, λ_0 of the photographic nadir Q_0 (the pole point), the height H above Q_0 (a linear element), and the azimuth t of the direction of 'the plane of the base vertical' or plane of photography. The other elements of outer orientation are the angle ε_0 between the principal optical beam and the normal to the ellipsoid from the point of view S in the principal vertical plane, and the angle χ between the Y-axis of the photograph and the principal vertical line, i.e. the plane of intersection of the principal vertical and that of the photograph.

The problem of determining the rectangular coordinates x, y of this photograph (and therefore of this projection onto an inclined pictorial plane) can be reduced to the following (Bugayevskiy 1980).

Let us move the coordinate system Q_0XYZ in a manner parallel to the original axes, so that its origin is at point S, and then turn the coordinate system at angles t, ε_0, χ:

$$x - x_0 = -f\frac{a_2 Y_n + b_2 X_n - c_2 H}{a_3 Y_n + b_3 X_n - c_3 H}$$

$$y - y_0 = -f\frac{a_1 Y_n + b_1 X_n - c_1 H}{a_3 Y_n + b_3 X_n - c_3 H}$$

and

$$X_n = -H\frac{b_1(y - y_0) + b_2(x - x_0) - b_3 f}{c_1(y - y_0) + c_2(x - x_0) - c_3 f}$$

$$Y_n = -H\frac{a_1(y - y_0) + a_2(x - x_0) - a_3 f}{c_1(y - y_0) + c_2(x - x_0) - c_3 f}$$

where a_i, b_i, c_i are direction vectors found thus:

$$a_1 = \cos t \cos \varepsilon_0 \cos \chi - \sin t \sin \chi$$

$$a_2 = -\cos t \cos \varepsilon_0 \sin \chi - \sin t \cos \chi$$

$$a_3 = -\cos t \sin \varepsilon_0$$

$$b_1 = \sin t \cos \varepsilon_0 \cos \chi + \cos t \sin \chi$$

$$b_2 = -\sin t \cos \varepsilon_0 \sin \chi + \cos t \cos \chi$$

$$b_3 = -\sin t \sin \varepsilon_0$$

$$c_1 = \sin \varepsilon_0 \cos \chi$$

$$c_2 = -\sin \varepsilon_0 \sin \chi$$

$$c_3 = \cos \varepsilon_0$$

(When working with space photographs, one can also use other sets of elements of external orientation by changing the expressions used for the definition of direction vectors.)

If we suppose that, when determining coordinates X, Y from formulas (3.12), (3.16), and (3.17), values $a' = a - t$ are used instead of spheroidal values a, and that we have calculated the abscissas and ordinates (X, Y) rotated an angle t in azimuth, then the coordinate transformation matrix, taking into account the other angles ε_0, χ, will take the form

$$A = \begin{pmatrix} \cos \varepsilon_0 \cos \chi & -\cos \varepsilon_0 \sin \chi & -\sin \varepsilon_0 \\ \sin \chi & \cos \chi & 0 \\ \sin \varepsilon_0 \cos \chi & -\sin \varepsilon_0 \sin \chi & \cos \varepsilon_0 \end{pmatrix} = \begin{pmatrix} a_1 & a_2 & a_3 \\ b_1 & b_2 & b_3 \\ c_1 & c_2 & c_3 \end{pmatrix}$$

The formulas relating coordinates x, y of points on the inclined pictorial plane, i.e. on a photograph (with the origin of the coordinates at the optical center), and coordinates X, Y of points on a horizontal pictorial plane (with the origin of the coordinates at the pole Q_0, i.e the photographic nadir) take the form

$$x = f\frac{-(Y \cos \varepsilon_0 - H \sin \varepsilon_0)\sin \chi + X \cos \chi}{Y \sin \varepsilon_0 + H \cos \varepsilon_0}$$

$$y = f\frac{(Y \cos \varepsilon_0 - H \sin \varepsilon_0)\cos \chi + X \sin \chi}{Y \sin \varepsilon_0 + H \cos \varepsilon_0}$$

$$X = H \frac{y \sin \chi + x \cos \chi}{-(\sin \varepsilon_0 \cos \chi)y + (\sin \varepsilon_0 \sin \chi)x + f \cos \varepsilon_0}$$

$$Y = H \frac{(\cos \varepsilon_0 \cos \chi)y - (\cos \varepsilon_0 \sin \chi)x + f \sin \varepsilon_0}{-(\sin \varepsilon_0 \cos \chi)y + (\sin \varepsilon_0 \sin \chi)x + f \cos \varepsilon_0} \tag{3.20}$$

Let us shift the initial coordinate system xoy of the inclined pictorial plane T to the photographic nadir n, and impose the condition that the local scale factor μ_2 along the almucantar at this point be equal to unity. Then let us express the value of coordinates in meters, introduce the symbol $\tan \beta = Y/H$, and assume the Earth (or any celestial body) not to be an ellipsoid of revolution, but a sphere. From equations (3.20) we can then find that

$$x = X \frac{\cos \beta \cos \varepsilon_0}{\cos(\beta - \varepsilon_0)}, \quad y = Y \frac{\cos \beta}{\cos(\beta - \varepsilon_0)}$$

Formulas for this particular case were suggested by Volkov.

Now, based on the general theory of map projections, we can get (to an accuracy of terms up to e^2) formulas for local scale factors: along verticals,

$$\mu_1 = \mu_{1H} k^2 (1 - \sin^2 a \sin^2 \varepsilon_0)^{1/2} \tag{3.21}$$

along almucantars,

$$\mu_2 = \mu_{2H} k^2 \left[\sin^2 a + \left(\cos a \cos \varepsilon_0 + \frac{\rho}{H} \sin \varepsilon_0 \right)^2 + 2 \frac{\rho_a}{\rho} \sin a \sin \varepsilon_0 \left(\frac{\rho}{H} \cos \varepsilon_0 - \cos a \sin \varepsilon_0 \right) \right]^{1/2} \tag{3.22}$$

where

$$k = H/(H \cos \varepsilon_0 + Y \sin \varepsilon_0)$$

$$\frac{\rho_a}{\rho} = e^2 \tau \frac{D_p \sin z \sin a \cos \phi_0}{D_p - N_0 \cos z}, \quad \rho = \sqrt{X^2 + Y^2}$$

and μ_{1H}, μ_{2H} are determined from equations (3.19).

From equations (3.21)–(3.22), it follows that for the point of the principal vertical

$$\mu_1 = \mu_{1H} k^2, \quad \mu_2 = \mu_{2H} k$$

In particular, at the optical center, at the point of zero distortion, and at the photographic nadir n, we find that

$$\mu_{1_o} = \mu_{1H} \cos^2 \varepsilon_0, \quad \mu_{2_o} = \mu_{2H} \cos \varepsilon_0$$

$$\mu_{1_c} = \mu_{1H}, \quad \mu_{2_c} = \mu_{2H}$$

$$\mu_{1_n} = \sec^2 \varepsilon_0, \quad \mu_{2_n} = \sec \varepsilon_0$$

Consequently, distortion at the point of the photographic nadir is due only to the inclination of the pictorial plane at an angle ε_0. At the normal point of zero distortion, the distortion is due only to the spheroidal or spherical shape of the surface being mapped. At the optical center as well as at every other point of the projection

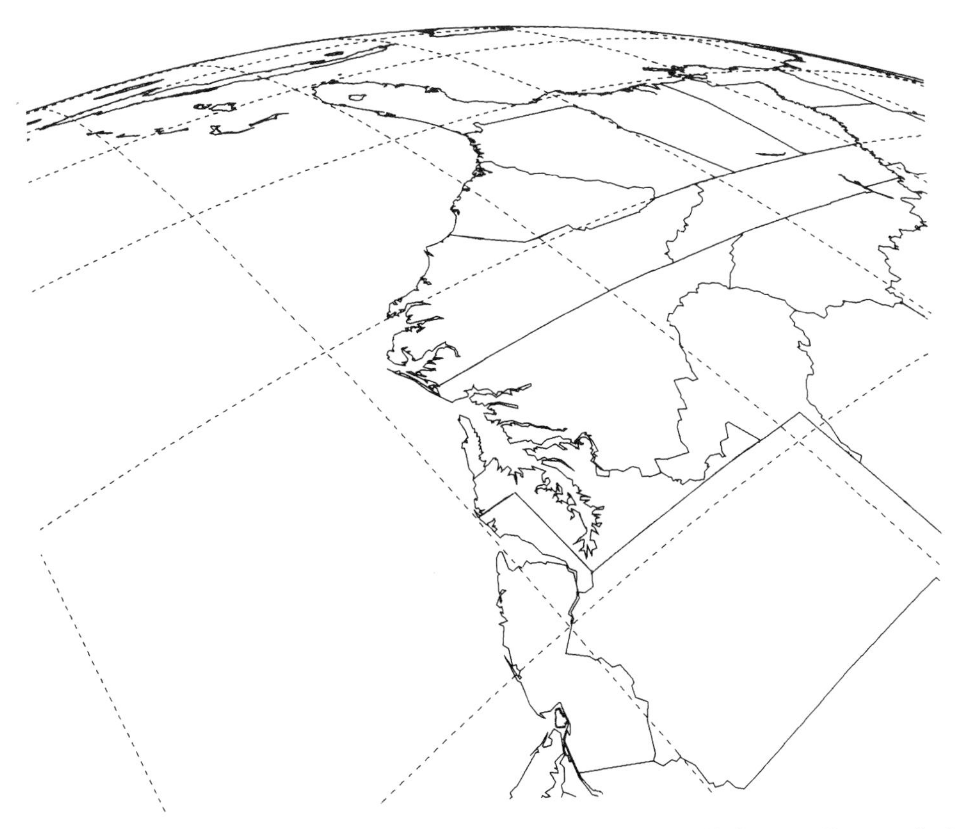

Figure 3.17 Tilted-perspective azimuthal projection, viewed from 1000 km over Boston, tilted 50° with direction of view at azimuth 220° clockwise from north. 5° graticule.

(or photograph), distortion is due to the combined effect of both factors. Figure 3.17 is an example of this tilted-perspective projection.

Local area scale factors and maximum angular deformation at the point of projection are determined to an accuracy of their terms up to e^2 by established formulas:

$$p = \mu_1\mu_2, \quad \sin(\omega/2) = |\mu_2 - \mu_1|/(\mu_2 + \mu_1)$$

3.3.3 Characteristics of azimuthal projections

Conformal, equal-area, and equidistant azimuthal projections are best for mapping regions with near-circular boundaries, because they provide the least distortion.

For the design of some maps, it is occasionally desirable to use azimuthal projections with a prescribed intermediate distortion pattern. Oblique azimuthal projections both with negative transformation (gnomonic, orthographic, stereographic, etc.) and positive transformation possess certain specific properties, and they can therefore be applied to designing various maps and solving a variety of problems.

For example, as stated previously, on a gnomonic projection of the sphere, geodetic lines (great-circle arcs) are represented by straight lines. On a stereographic projection, finite-sized circles on the surface of the sphere are represented on the projection by circles as well, and there is no local shape distortion. Outer-perspective oblique azimuthal projections with positive transformation are mathematical models of ideal aerospace photographs.

Table 3.3 lists the local scale factors and maximum angular deformation for some azimuthal projections (Ginzburg and Salmanova 1964, p. 120). Table 3.4 lists the same parameters, but for two particular cases using projections for ideal vertical aerospace photographs. The scale factors and angular deformation were calculated

Table 3.3 Distortion on some azimuthal projections

Characteristic of the projection	z, polar distance			
	0°	30°	60°	90°
1	2	3	4	5
	Central (gnomonic) projection			
μ_1	1	1.333	4.000	∞
μ_2	1	1.155	2.000	∞
p	1	1.540	8.000	∞
ω	0°	8°14′	38°57′	180°
	Conformal (stereographic) projection			
μ_1	1	1.072	1.333	2.000
μ_2	1	1.072	1.333	2.000
p	1	1.149	1.778	4.000
ω	0°	0°	0°	0°
	Lambert azimuthal equal-area projection			
μ_1	1	0.996	0.866	0.707
μ_2	1	1.035	1.155	1.414
p	1	1	1	1
ω	0°	3°58′	16°26′	38°57′
	Ginzburg projection (with low area distortion)			
μ_1	1	0.966	0.870	0.853
μ_2	1	1.035	1.155	1.437
p	1	1.000	1.005	1.225
ω	0°	3°58′	16°12′	29°34′
	Ginzburg projection (showing sphericity)			
μ_1	1	0.940	0.766	0.500
μ_2	1	1.026	1.113	1.299
p	1	0.964	0.853	0.650
ω	0°	5°02′	21°18′	52°44′
	Solov'ev projection (with multiple perspectives)			
μ_1	1	1.017	1.072	1.172
μ_2	1	1.053	1.238	1.657
p	1	1.071	1.326	1.941
ω	0°	1°59′	8°14′	19°45′
	Orthographic projection			
μ_1	1	0.866	0.500	0.000
μ_2	1	1.000	1.000	1.000
p	1	0.866	0.500	0.000
ω	0°	8°14′	38°57′	180°

Table 3.4(a) Variant 1
Perspective azimuthal projection, $\phi_0 = 55°$, $H = 300\,000$ m

ϕ	Characteristic of the projection	λ 0°	1°	5°	10°	15°
55°	μ_2	1.000	0.9989	0.9740	0.8037	0.8072
	μ_1	1.000	0.9968	0.9222	0.7257	0.4886
	p	1.000	0.9957	0.8983	0.6558	0.3944
	$\omega°$	0.000	0.1252	3.1310	12.5471	28.4647
56°	μ_2	0.9968	0.9957	0.9716	0.9033	0.8090
	μ_1	0.9902	0.9870	0.9150	0.7244	0.4927
	p	0.9870	0.9828	0.8890	0.6544	0.3985
	$\omega°$	0.3806	0.5028	3.4337	12.6152	28.1290
60°	μ_2	0.9250	0.9242	0.9055	0.8518	0.7756
	μ_1	0.7830	0.7808	0.7304	0.5935	0.4193
	p	0.7243	0.7216	0.6614	0.5056	0.3252
	$\omega°$	9.5389	9.6489	12.2907	20.5925	34.6905
70°	μ_2	0.5793	0.5791	0.5741	0.5588	0.5352
	μ_1	0.0805	0.0802	0.0735	0.0541	0.0260
	p	0.0466	0.0465	0.0422	0.0302	0.0139
	$\omega°$	98.2246	98.3428	101.2210	110.8464	130.2933

using formulas given above as well as the following formulas by Ginzburg

$$\rho = R\left(2 \sin \frac{z}{2} + 0.000\,25 z^{10}\right), \quad \rho = 1.5R \sin \frac{z}{1.5}$$

to obtain, respectively, the projections with low area distortion and showing the sphericity effect. A formula by Solov'ev

$$\rho = 2^c R \tan \frac{z}{2^c} = 4R \tan \frac{z}{4}$$

was used for his projection with multiple perspectives (for 'stereo–stereo', $C = 2$).

From an analysis of Tables 3.3 and 3.4 it follows that the greatest distortion occurs on the gnomonic projection and on the projection of a vertical aerospace photograph. When using oblique aerospace photographs (outer-perspective oblique azimuthal projections with a positive transformation onto an inclined pictorial plane), distortion due to the tilted perspective is added to the distortion resulting from the curvature of the surface being mapped.

3.4 Pseudoconic projections

3.4.1 General formulas for pseudoconic projections

On pseudoconic projections, parallels are arcs of concentric circles, and meridians are curves symmetrical about the straight central meridian, along which the center of the circles for the parallels is situated (Figure 3.18).

Table 3.4(b) Variant 2
Perspective azimuthal projection, $\phi_0 = 40°$, $H = 5\,000\,000$ m

ϕ	Characteristic of the projection	λ: 0°	1°	5°	10°	20°	30°	40°
1	2	3	4	5	6	7	8	9
40°	μ_2	1.0000	0.9999	0.9972	0.9887	0.9567	0.9087	0.8508
	μ_1	1.0000	0.9996	0.9893	0.9578	0.8416	0.6779	0.4975
	p	1.0000	0.9994	0.9864	0.9470	0.8052	0.6161	0.4233
	$\omega°$	0.0000	0.0182	0.4553	1.8240	7.3430	16.7275	30.3785
41°	μ_2	0.9998	0.9997	0.9970	0.9887	0.9573	0.9098	0.8525
	μ_1	0.9993	0.9988	0.9887	0.9577	0.8431	0.6814	0.5027
	p	0.9991	0.9985	0.9857	0.9469	0.8070	0.6200	0.4285
	$\omega°$	0.0310	0.0490	0.4797	1.8282	7.2644	16.5022	29.9220
45°	μ_2	0.9952	0.9951	0.9926	0.9849	0.9555	0.9111	0.8571
	μ_1	0.9818	0.9814	0.9720	0.9434	0.8371	0.6857	0.5162
	p	0.9771	0.9765	0.9648	0.9291	0.7998	0.6248	0.4424
	$\omega°$	0.7765	0.7934	1.1984	2.4662	7.5721	16.2279	28.7434
50°	μ_2	0.9810	0.9809	0.9787	0.9718	0.9458	0.9061	0.8572
	μ_1	0.9290	0.9286	0.9206	0.8957	0.8030	0.6692	0.5166
	p	0.9113	0.9109	0.9009	0.8705	0.7595	0.6063	0.4428
	$\omega°$	3.1184	3.1339	3.5064	4.6722	9.3629	17.2956	28.7097
60°	μ_2	0.9285	0.9284	0.9269	0.9221	0.9037	0.8752	0.8393
	μ_1	0.7437	0.7434	0.7382	0.7222	0.6616	0.5713	0.4639
	p	0.6905	0.6902	0.6843	0.6660	0.5979	0.4950	0.3893
	$\omega°$	12.6910	12.7038	13.0089	13.9635	17.7987	24.2596	33.4835
70°	μ_2	0.8539	0.8535	0.8529	0.8502	0.8394	0.8224	0.8004
	μ_1	0.5066	0.5065	0.5039	0.4957	0.4643	0.4159	0.3557
	p	0.4326	0.4325	0.4298	0.4214	0.3897	0.3420	0.2847
	$\omega°$	29.5739	29.5836	29.8154	30.5403	33.4481	38.3277	45.2419
80°	μ_2	0.7699	0.7699	0.7695	0.7684	0.7639	0.7566	0.7470
	μ_1	0.2769	0.2769	0.2760	0.2731	0.2619	0.2442	0.2212
	p	0.2132	0.2132	0.2124	0.2098	0.2001	0.1848	0.1652
	$\omega°$	56.1887	56.1946	56.3386	56.7885	58.5884	61.5894	65.7951

The general equations for these projections take the form

$$x = \rho \sin \delta, \quad y = \rho_s - \rho \cos \delta,$$
$$\rho = f_1(\phi), \quad \delta = f_2(\phi, \lambda) \tag{3.23}$$

where ρ_s is a constant, the polar distance of the southernmost or other convenient parallel on the projection.

After differentiating these equations with respect to ϕ and λ and introducing the values of derivatives into the formulas of the general theory of map projections

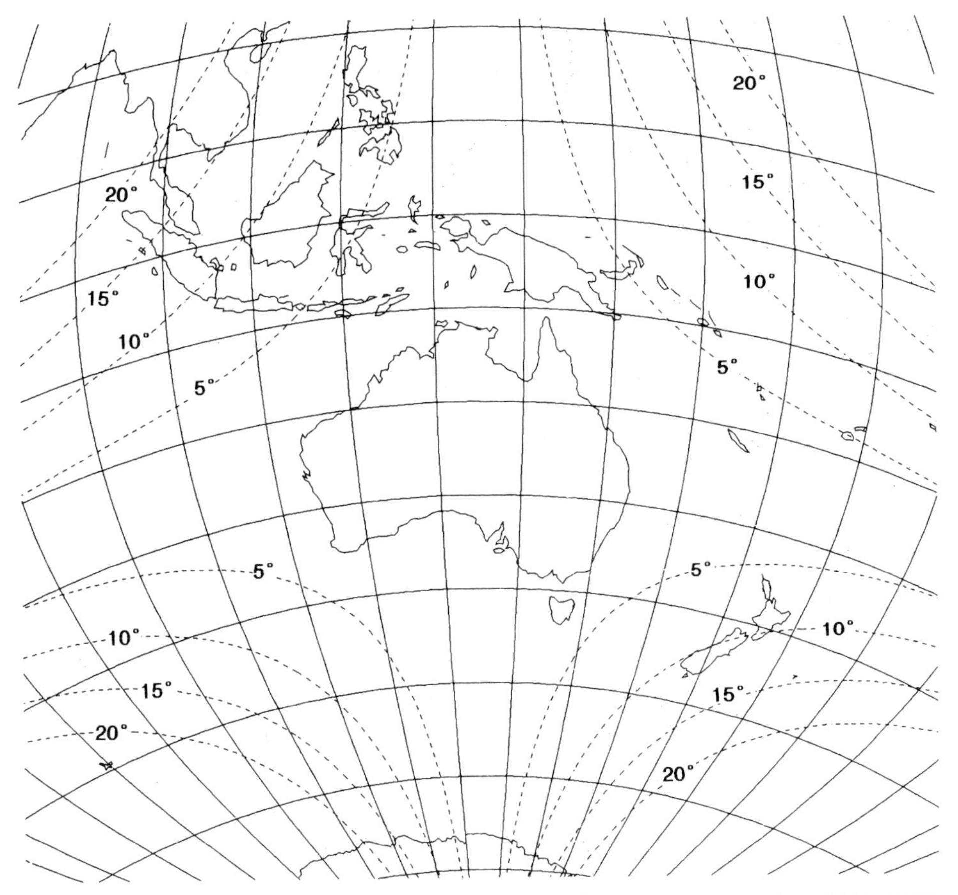

Figure 3.18 Bonne equal-area pseudoconic projection of Australia, centered at 25°S, 135°E, with isocols for ω. 10° graticule.

(see section 1.3), we have

$$f = \rho^2 \delta_\phi \delta_\lambda, \quad h = -\rho \rho_\phi \delta_\lambda$$
$$\tan \varepsilon = \rho \delta_\phi / \rho_\phi \tag{3.24}$$

$$n = \rho \delta_\lambda / r, \quad p = -\rho \rho_\phi \delta_\lambda / Mr$$
$$m = (p/n) \sec \varepsilon \tag{3.25}$$

where p, m, and n are local area and linear scale factors.

From the definition and formulas for these projections, it follows that the map graticules are not orthogonal, and arc lengths of meridians are functions of the latitude and longitude. Consequently, these projections cannot be conformal and do not preserve lengths along meridians other than frequently the central meridian. They can be equal area or arbitrary in the distribution of distortion.

In the particular cases where $\delta = \alpha\lambda$ or $\delta = \lambda$, the meridians are straight lines and the projections will be, respectively, conic or azimuthal.

The Bonne equal-area projection has been used more often than any other pseudoconic projection. Rigobert Bonne of France used it extensively in 1752, but it had been used by others beginning in the sixteenth century.

3.4.2 Equal-area pseudoconic projections

Bonne projection

The condition for equivalency, applying (3.24) and (3.25), takes the form

$$h = -\rho\rho_\phi\delta_\lambda = -\rho_\phi nr = Mr$$

Hence, the polar distance

$$\rho = C - \int \frac{M}{n}\, d\phi \tag{3.26}$$

where C is a constant of integration.

Imposing the condition of symmetry for the projection about the central meridian, we obtain from equations (3.25)

$$\delta = \frac{nr}{\rho}\lambda$$

Defining the form of function $n = f(\phi)$, we can develop from (3.26) and general formulas (3.23)–(3.25) various equal-area pseudoconic projections.

To obtain the Bonne projection, the condition is imposed that local linear scale factors along parallels and along the central meridian be unity:

$$n = 1, \quad m_0 = 1$$

The projection formulas take the form

$$\rho = C - s, \quad \delta = \frac{r}{\rho}\lambda, \quad \tan\varepsilon = \lambda\left(\sin\phi - \frac{r}{\rho}\right)$$

$$p = 1, \quad m = \sec\varepsilon, \quad \tan(\omega/2) = \tfrac{1}{2}\tan\varepsilon$$

$$C = s_0 + N_0 \cot\phi_0, \quad \rho = (s_0 - s) + N_0 \cot\phi_0$$

where s is the arc length of the meridian from the equator to the given parallel and is determined by equation (1.145).

From the formulas given, it follows that all types of distortion (ε, ω, v_m) are equal to zero along the central meridian $\lambda = \lambda_0 = 0$ and along the given parallel $\phi = \phi_0$ for which

$$\sin\phi_0 - r_0/\rho_0 = 0$$

Isocols on the Bonne projection that are close to the central meridian and the central parallel resemble equilateral hyperbolas symmetrical about these lines.

In 1937–8 Solov'ev suggested modified formulas for this projection so that, through the introduction of three constant parameters, the curvature of the plotted parallels was decreased, an improvement of great importance for the design of some maps, e.g. school maps (Figure 3.19):

$$\rho = C_0 + C_1(s_0 - s), \quad \delta = C_2\frac{r}{\rho}\lambda$$

$$\tan\varepsilon = C_2\lambda\left(\frac{\sin\phi}{C_1} - \frac{r}{\rho}\right), \quad \tan\frac{\omega}{2} = \frac{1}{2}\sqrt{\frac{(C_1 - C_2)^2 + C_1^2\tan^2\varepsilon}{C_1C_2}} \tag{3.27}$$

$$m = C_1\sec\varepsilon, \quad n = C_2, \quad p = C_1C_2$$

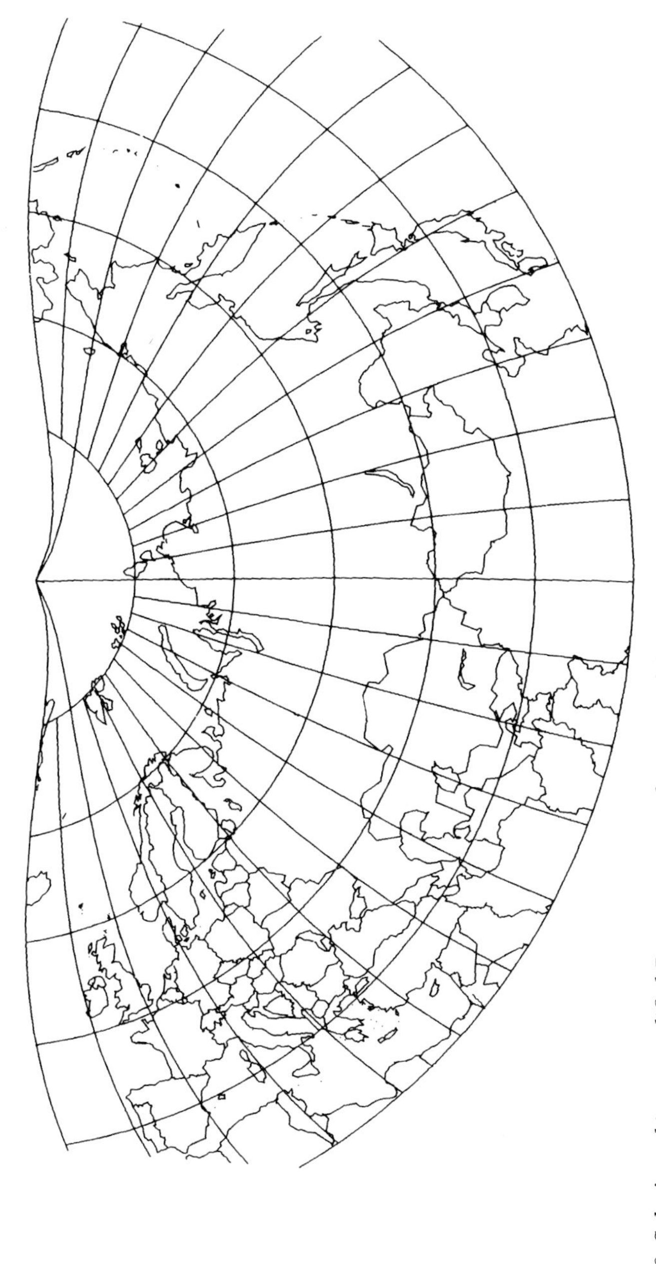

Figure 3.19 Solov'ev arbitrary modified Bonne projection for a map of the former USSR, replotted with current boundaries. 10° graticule.

Constant parameters C_0, C_1, C_2 for the modified Bonne projection were arbitrarily selected. They can also be determined analytically from the given conditions, starting with given curvatures for the parallels.

Taking into account (3.27) and the fact that on a pseudoconic projection the curvature of a parallel is equal to $k_P = -1/\rho$, we can get

$$-1/k_P = C_0 + C_1(s_0 - s)$$

After supplying curvatures k_P for two parallels, it is easy to find coefficients C_0, C_1. Coefficient C_2 may be readily found from the given local scale factor n or p; consequently $C_2 = n_k$ or $C_2 = p_k/C_1$. Solov'ev chose $C_0 = \cot 52°18'$, $s_0 = \text{arc } 52°18'$, $C_1 = 1.02$, $C_2 = 0.95$.

If we impose the condition that $p = 1$, then $C_2 = 1/C_1$. This projection has equally divided parallels (see section 3.4.3). The linear scale along parallels and along the central meridian as well as the local area scale are constant values.

3.4.3 Map projections with equally divided parallels

In 1953 N. A. Urmayev, assuming the Earth to be a sphere with unit radius, developed a theory of map projections with equally divided parallels, a series of pseudoconic and pseudocylindrical projections being a particular case (see sections 2.2 and 3.4.2).

He proceeded from conditions specifying that the local linear and area scale factors along parallels are equal and functions only of the latitude, i.e.

$$n = p = f(\phi) \tag{3.28}$$

Hence, $m = \sec \varepsilon$.

On the basis of the formulas

$$m^2 = x_\phi^2 + y_\phi^2, \quad v^2 = n^2 \cos^2 \phi = x_\lambda^2 + y_\lambda^2$$

the values of partial derivatives were rewritten in the form

$$x_\phi = -\sec \varepsilon \sin(\varepsilon + \tau), \quad x_\lambda = v \cos \tau$$

$$y_\phi = \sec \varepsilon \cos(\varepsilon + \tau), \quad y_\lambda = v \sin \tau$$

where τ is the angle between the normal to the parallel and the Y-axis.

After differentiating these expressions with respect to ϕ and λ, and introducing their derivatives into the conditions of their integration, we have

$$-\tau_\lambda \tan \varepsilon = v\tau_\phi \tag{3.29}$$

$$\tau_\lambda + (\tan \varepsilon)_\lambda = -v_\phi = -dv/d\phi \tag{3.30}$$

As $v = n \cos \phi$ is a function only of latitude, and the projections under consideration are symmetrical about the central meridian, the integration of (3.30) gives

$$\tau + \tan \varepsilon = -\lambda v_\phi \tag{3.31}$$

Equation (3.30) takes the form

$$v\tau_\phi - (\tau + \lambda v_\phi)\tau_\lambda = 0 \tag{3.32}$$

As the values of n and v are given, we can obtain by integrating (3.32)

$$\tau\phi + \lambda v = f(\tau) \tag{3.33}$$

where $f(\tau)$ is an arbitrary function.

Equations (3.31) and (3.33) provide the basis of the theory of map projections with equally divided parallels, where condition (3.28) holds. In particular cases, if $\tau = 0$ and $f(\tau) = 0$ or $f(\tau) = c\tau$, from equations (3.31) and (3.33), respectively, we can obtain the sinusoidal (Sanson–Flamsteed) equal-area pseudocylindrical projection or the Bonne equal-area pseudoconic projection.

3.4.4 Stab–Werner heart-shaped pseudoconic projection

The equations for this equal-area projection, suggested in the early sixteenth century, take the form

$$x = \rho \sin \delta, \quad y = \rho_s - \rho \cos \delta$$

$$\rho = R\left(\frac{\pi}{2} - \phi\right), \quad \delta = \frac{\lambda \cos \phi}{(\pi/2) - \phi}$$

$$\tan \varepsilon = \delta - \lambda \sin \phi, \quad p = n = 1, \quad m = \sec \varepsilon$$

On this projection there is no distortion along the central meridian, including the North Pole but not the South Pole. With increasing distance from it distortion increases greatly, becoming maximum on the parallels close to the south polar region. The Stab–Werner projection is a limiting form of the Bonne projection on which $\phi_0 = 90°$.

3.4.5 Pseudoconic projections with arbitrary distortion

Using general formulas (3.23)–(3.25), we can derive a set containing various pseudoconic projections with arbitrary distortion depending on the initial conditions, given in pairs:

$$\rho = f_1(\phi) \quad \text{and} \quad \delta = f_2(\phi, \lambda), \quad \rho = f_1(\phi) \quad \text{and} \quad n = f_3(\phi)$$

$$\rho = f_1(\phi) \quad \text{and} \quad p = f_4(\phi, \lambda), \quad \rho = f_1(\phi) \quad \text{and} \quad \delta = f_5(\phi, \lambda)$$

$$n = f_3(\phi) \quad \text{and} \quad \delta = f_2(\phi, \lambda), \quad n = f_3(\phi) \quad \text{and} \quad \rho = f_4(\phi, \lambda)$$

$$p = f_4(\phi, \lambda) \quad \text{and} \quad \delta = f_2(\phi, \lambda), \quad \text{etc.}$$

3.5 Pseudoazimuthal projections

Projections of this type were developed by Ginzburg in 1952 to project the surface of a sphere onto a plane for those cases where it is desirable to show the sphericity of the Earth. As a rule, they are used in an oblique aspect.

On polar aspects of pseudoazimuthal projections, parallels are represented by concentric circular arcs, and meridians are shown as curves or straight lines, converging in the center of the parallels, the meridians of longitudes 0° and 360° being

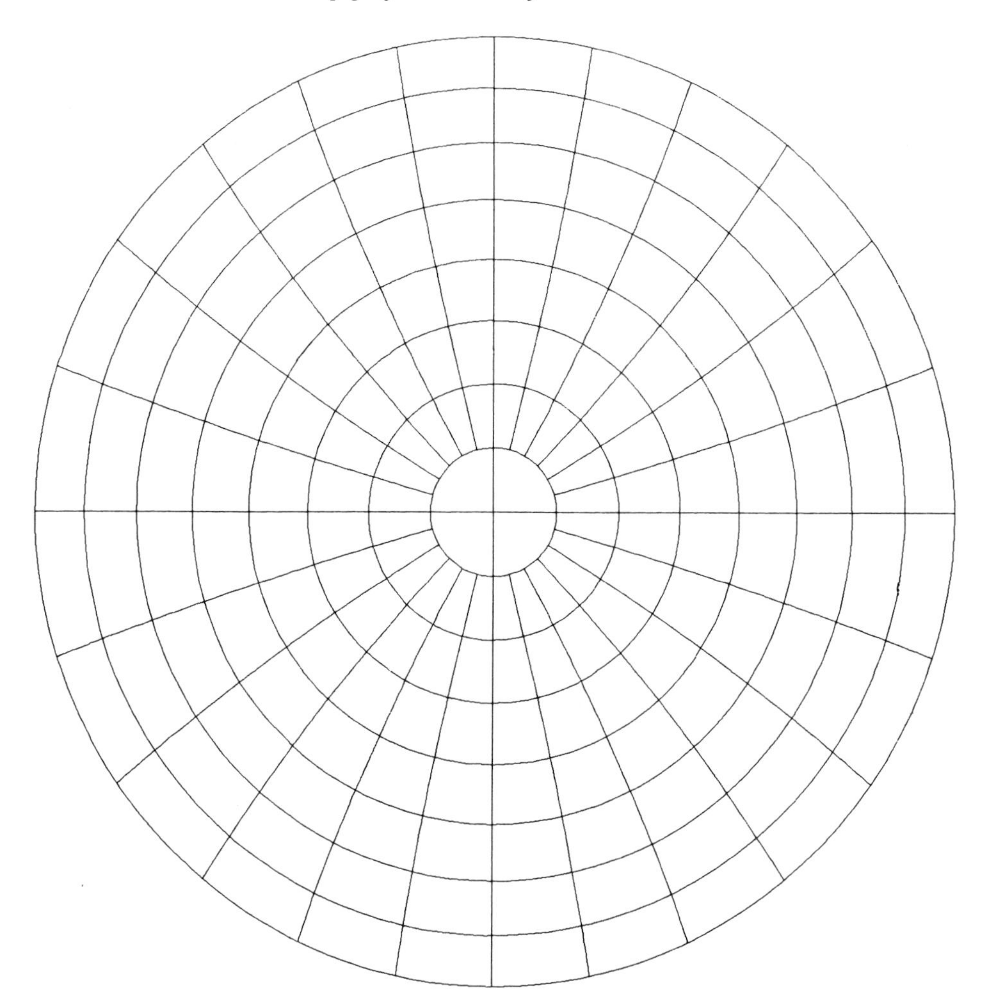

Figure 3.20 Polar aspect of the Ginzburg TsNIIGAiK pseudoazimuthal projection, with meridians curved except at each 90°. 15° graticule.

represented by straight lines on most of them, including the Ginzburg versions (Figure 3.20). The lines of constant maximum angular distortion appear as ovals.

From the definition, the general equations for these projections have the form

$$x = \rho \sin \delta, \quad y = \rho \cos \delta$$

$$\rho = f_1(z), \quad \delta = a + f_2(z) \sin ka$$

where z, a are polar spherical coordinates and k is a constant which affects the shape of the meridians. When $k = 1$, meridians for longitudes 0° and 180° are shown as straight lines. When $k = 2$, meridians for longitudes 0°, 90°, 180°, and 270° are straight lines. When parameter k is a fraction, the projections become pseudoconic rather than pseudoazimuthal.

To define polar angle δ between meridians on this projection, Ginzburg suggested the following formula for the case where the major oval axis is in the direction of the central meridian:

$$\delta = a - c(z/z_{max})^q \sin ka$$

where q is a constant. If the major oval axis is orthogonal to the central meridian, the formula is

$$\delta = (90^\circ + a) - c(z/z_{max})^q \sin[k(90^\circ + a)]$$

The projection was used for designing a combined map of the Atlantic and Arctic Oceans (Figure 3.21) used in the Russian *Atlas Mira* (*World Atlas*) beginning in 1954, where it was assumed that $\rho = 3R \sin(z/3)$, and that $c = 0.1$, $q = 1$, $k = 2$, and $z_{max} = 120^\circ$ in the middle equation for δ. In England, Guy Bomford designed a pseudoazimuthal projection based on the azimuthal equidistant projection, and curving the meridians on the polar aspect to give approximately equal scale error along a bounding oval. The equatorial aspect was used for maps of oceans in the *Oxford Atlas*, beginning in 1951.

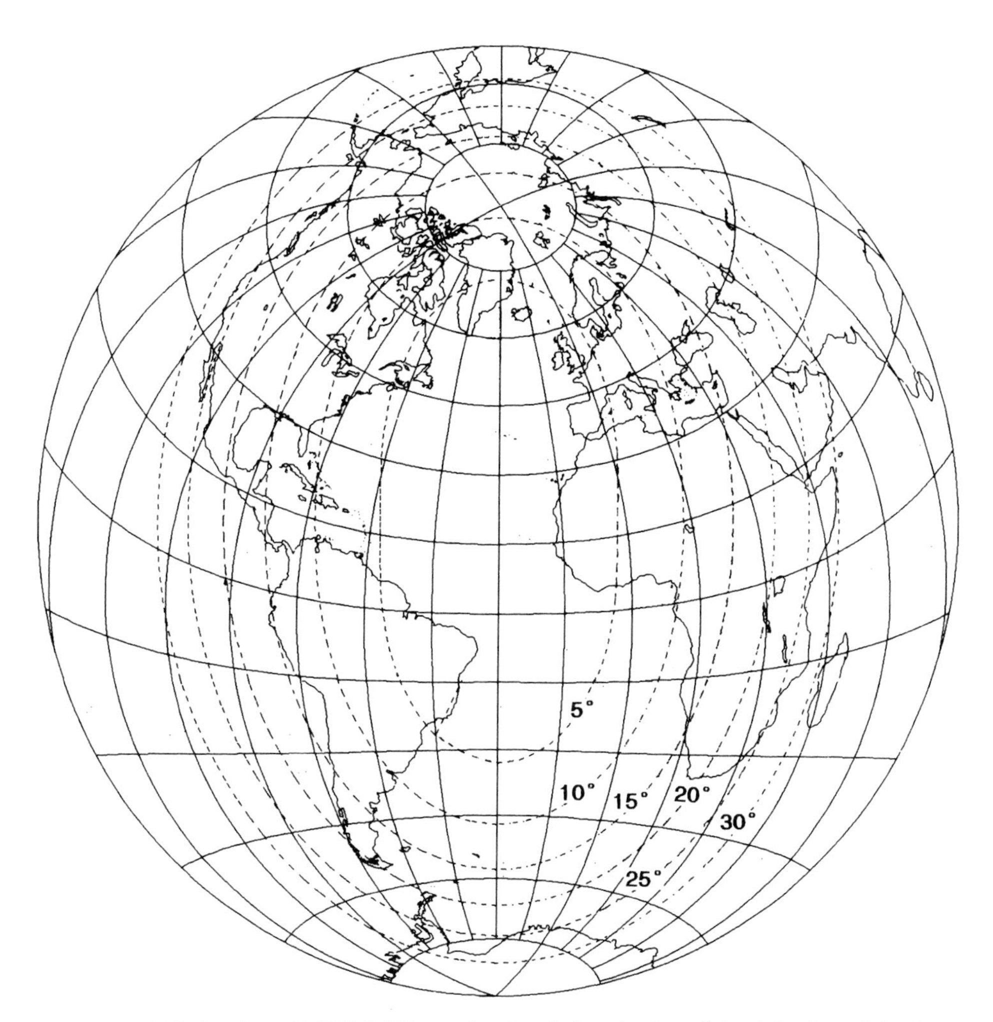

Figure 3.21 Ginzburg TsNIIGAiK pseudoazimuthal projection of the Atlantic and Arctic Oceans, centered at 20°N, 30°W. 15° graticule.

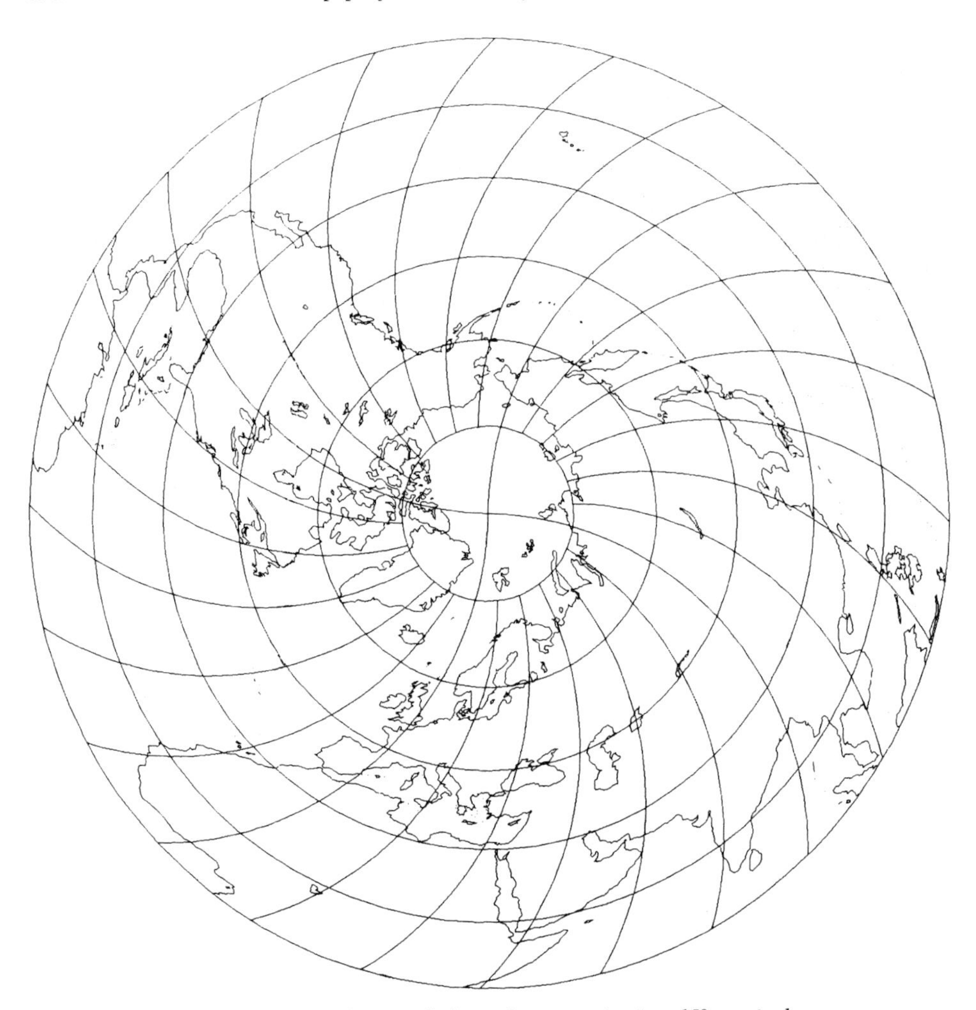

Figure 3.22 Wiechel polar pseudoazimuthal equal-area projection. 15° graticule.

The first pseudoazimuthal projection (Figure 3.22) was equal-area, presented by H. Wiechel (1879) in Germany. It is only of interest in the polar aspect, on which parallels are spaced as on the Lambert azimuthal equal-area projection, but all meridians are identical circular arcs at true scale, radiating like pinwheels from the central pole. The above general formulas apply to the Wiechel projection, except that $\delta = \alpha + f(z)$. Specifically, the formulas for the Wiechel projection may be written as

$$x = R[\sin \lambda \cos \phi - (1 - \sin \phi)\cos \lambda]$$

$$y = -R[\cos \lambda \cos \phi + (1 - \sin \phi)\sin \lambda]$$

3.6 Retroazimuthal projections

In a sense the opposite of azimuthal projections, on retroazimuthal projections the azimuth *from* every point on the map *to* the center point is shown correctly, as the angle between the vertical line through the point and the vector joining the point to the center. They are useful, for example, for showing the direction in which to point antennas to receive radio signals.

The first such projection was developed by J. J. Littrow in 1833 (see section 7.3.7); it showed correct azimuths to any point along the central meridian. J. I. Craig of the Survey of Egypt designed a retroazimuthal 'Mecca' projection in 1909 with Makkah (Mecca) as the center so that Islamic worshippers could determine the direction to face for prayers (Figure 3.23). In 1910 Hammer presented a retro-azimuthal projection showing both correct azimuths and distances from all points to the center; this version was extended to a world map centered on Rugby, England, the location of a powerful radio station, by Hinks (1929) and E. A. Reeves.

Figure 3.23 Craig Mecca retroazimuthal projection. 10° graticule.

4

Map projections with parallels in the shape of non-concentric circles

4.1 General formulas for polyconic projections

On polyconic projections, parallels are arcs of non-concentric circles, with the centers located along the straight central meridian, and meridians are curves symmetrical about the central meridian (Figure 4.1).

The general equations for these projections have the form

$$x = \rho \sin \delta, \quad y = q - \rho \cos \delta \tag{4.1}$$

where

$$q = f_1(\phi), \quad \rho = f_2(\phi), \quad \delta = f_3(\phi, \lambda)$$

After differentiating these equations with respect to ϕ and λ we can find in accordance with the general theory of map projections that

$$\tan \varepsilon = -\frac{f}{h} = -\frac{q_\phi \sin \delta + \rho\delta_\phi}{q_\phi \cos \delta - \rho_\phi}$$

$$n = \rho\delta_\lambda / r, \quad p = \frac{h}{Mr} = \rho\delta_\lambda \frac{q_\phi \cos \delta - \rho_\phi}{Mr}$$

$$m = \frac{p}{n} \sec \varepsilon = \frac{q_\phi \cos \delta - \rho_\phi}{M} \sec \varepsilon \tag{4.2}$$

$$\tan \frac{\omega}{2} = \frac{1}{2}\sqrt{\frac{m^2 + n^2 - 2p}{p}}$$

$$a = (A + B)/2, \quad b = (A - B)/2$$

where

$$A = \sqrt{m^2 + n^2 + 2mn \cos \varepsilon}, \quad B = \sqrt{m^2 + n^2 - 2mn \cos \varepsilon}$$

These projections may be conformal, equal area (see section 4.2.2), or arbitrary in distortion distribution.

4.2 Polyconic projections in a general sense

4.2.1 Polyconic projections with circular meridians and parallels

Lagrange projection

Suppose we have in Figure 4.2 a circle of radius k, a circular parallel BAB_1 of radius C_2B_1 perpendicular to radius OB_1, a circular meridian PAP' of radius C_1P, and an

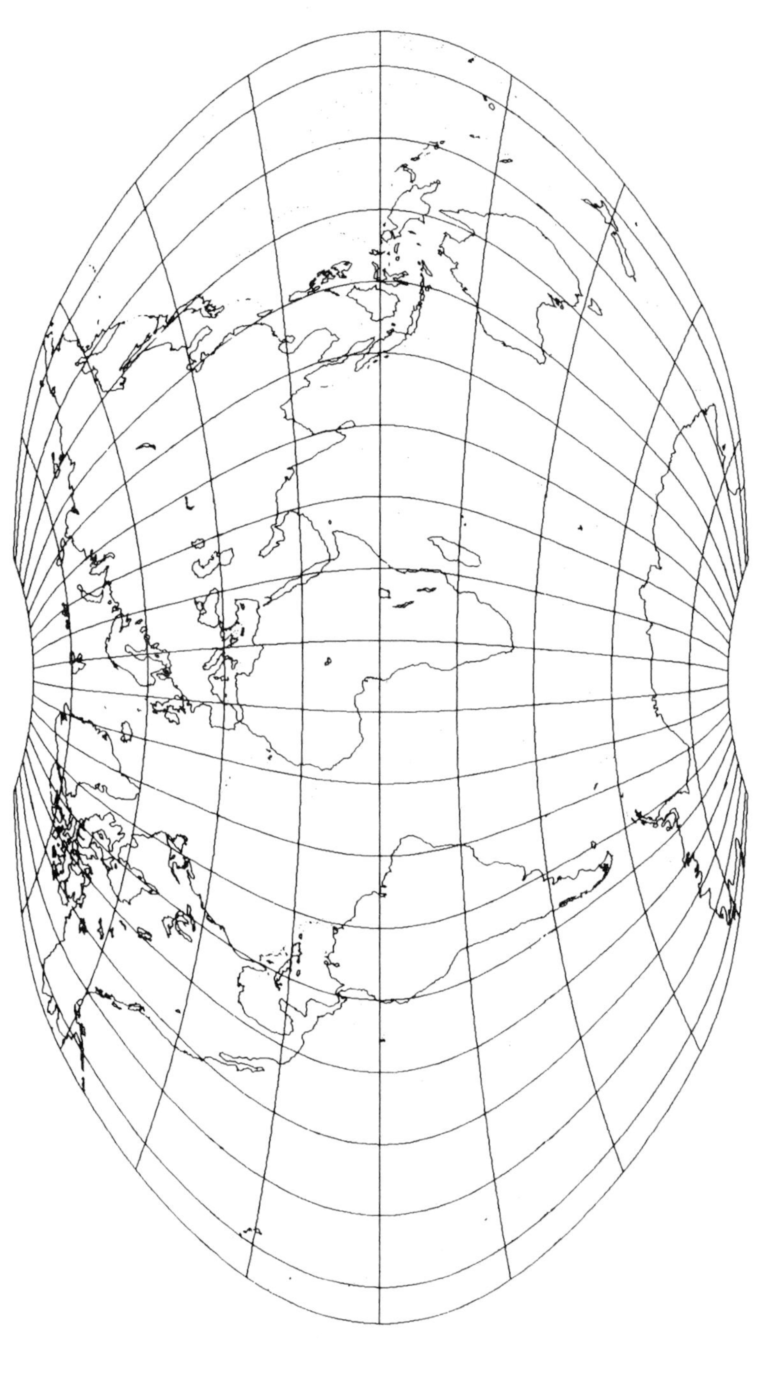

Figure 4.1 TsNIIGAiK *modified polyconic projection (1939–49 variant). 20° graticule.*

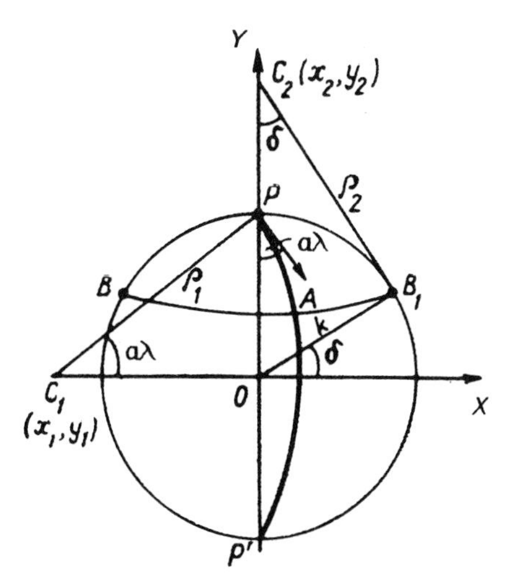

Figure 4.2 Coordinate system for the Lagrange projection.

angle $\alpha\lambda$ tangent to the meridian at point P. Then for the meridian with center C_1 and parallel with center C_2, intersecting at point A, we have

$$x_1 = -k \cot \alpha\lambda, \quad y_1 = 0, \quad \rho_1 = k \csc \alpha\lambda$$

$$x_2 = 0, \quad y_2 = k \csc \delta, \quad \rho_2 = k \cot \delta$$

Using these values and forming an equation for circles we can solve the equations simultaneously:

$$x = \frac{k \cos \delta \sin \alpha\lambda}{1 + \cos \delta \cos \alpha\lambda}, \quad y = \frac{k \sin \delta}{1 + \cos \delta \cos \alpha\lambda} \tag{4.3}$$

These formulas hold for all circular polyconic projections. For example, Bulgarian scientist Andreyev (1983, 1984) derived a number of polyconic projections arbitrary in distortion distribution using these formulas.

The Lagrange projection is conformal. Owing to the function δ in equations (4.3), it may be written using the Cauchy–Riemann conditions in the form

$$\tan(\pi/4 + \delta/2) = \beta U^{\alpha}$$

where U is determined from equations (1.31)–(1.33).

Local linear scale factors are equal to

$$m = n = \frac{\alpha k \cos \delta}{r(1 + \cos \delta \cos \alpha\lambda)}$$

In the projection there are three constant parameters α, β, k. Parameter α is found by analyzing the shapes of isocols near a central point $O(\phi_0, \lambda_0)$. In practice,

$$\alpha = \sqrt{1 + \frac{1 - \eta^2}{1 + \eta^2} \cos^2 \phi_0}$$

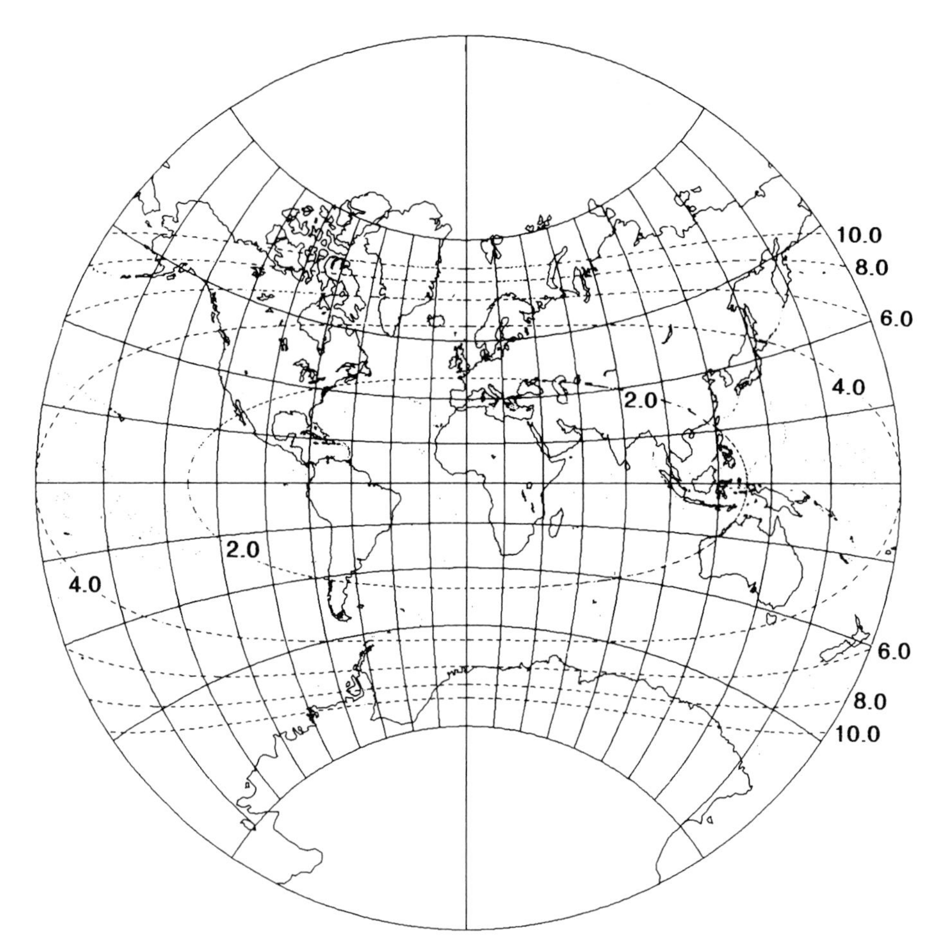

Figure 4.3 Lagrange conformal projection of the world in a circle, with isocols for p. 20° graticule.

where $\eta = b/a$, ϕ_0 is the parallel shown straight along C_1O, and b, a are the semi-axes of the isocol in the direction of the parallel and meridian, respectively, approximating an outline of the region being mapped.

Parameters β and k are determined from the following formulas:

$$\beta = \tan\left(\frac{\pi}{4} + \frac{\delta_0}{2}\right)U_0^{-\alpha}, \quad \frac{\delta_0}{2} = \arctan\left(\frac{\sin \phi_0}{\alpha}\right)$$

$$k = \frac{m_0 r_0}{\alpha}(1 + \sec \delta_0)$$

where m_0 is the given value of the local linear scale factor at the central point.

The presence of these three parameters in the projection formulas makes it possible to approach closely the outline of the region being mapped through the shape of the outer isocol of the projection.

Parameter α affects the shape of the isocols: when $\alpha < 1$ the isocols are ovals extended along parallels; if $\alpha > 1$ they are ovals extended along meridians; if $\alpha = 1$, they are circles and the projection becomes stereographic; if $\alpha = 0$, they become parallel lines and the projection becomes conformal cylindrical or the Mercator.

The Lagrange projection can be successfully used for representing any region on maps, except for polar regions, where distortion becomes excessive. It was actually originated as a conformal projection of the world in a circle (Figure 4.3, $\alpha = 0.5$) by Lambert in 1772, but generalized by J. L. Lagrange in 1779.

Van der Grinten projection

This projection is arbitrary in distortion distribution and is intermediate between conformal and equal-area projections. Continental shapes are represented somewhat better than on some other projections.

On this projection (Figure 4.4), presented by A. J. van der Grinten (1904) of Chicago, the primary circle is used for the creation of world maps, one of its diameters PP' (Figure 4.2) being the Y-axis and the other EE' (coinciding with C_1O) being the X-axis. Equator EE' is divided into equal parts in accordance with the graticule interval adopted; through the points of division obtained and the two poles P and P' circular arcs, i.e. meridians, are drawn.

In that case the radius ρ of any meridian λ and distance Q of its center from the projection center O is determined by the formulas

$$\rho = k \csc \lambda_1, \quad Q = k \cot \lambda_1$$

where $\lambda_1 = 2 \arctan(\lambda/\pi)$ and $k = \pi R$.

Parallels are drawn through three points, the points of intersection of this parallel with the central meridian and with the primary circle. Distance C from the equator to the point of intersection of the parallel at latitude ϕ with the central meridian is equal to

$$C = k(\pi - \sqrt{\pi^2 - 4\phi^2})/(2\phi)$$

Distance d_0 from the equator to the points of intersection of the parallels with the primary circle is determined by the formula

$$d_0 = k\phi/(\pi - \phi)$$

Here the radius ρ_1 of any parallel on the projection and the distance q from its center to the equator are equal to

$$\rho_1 = \frac{k^3 - C^3}{2C^2}, \quad q = \rho_1 + C = \frac{k^3 + C^3}{2C^2}$$

On this projection the graticule is not orthogonal:

$$\sin \varepsilon = \phi/(\pi + \phi)$$

Local linear scale factors m and n along the equator equal unity, and at the pole m and n approach infinity; at other points they are determined by complex formulas. They vary considerably; for example, along the parallel $\phi = 60°$, m varies from 1.537 along the central meridian to 2.598 along the meridian $\lambda = 180°$, and n correspondingly changes from 1.708 to 1.789.

Note that the al-Bīrūnī (or Nicolosi) globular projection belongs to the general group of all-circular polyconic projections. It will be discussed in section 6.3.6.

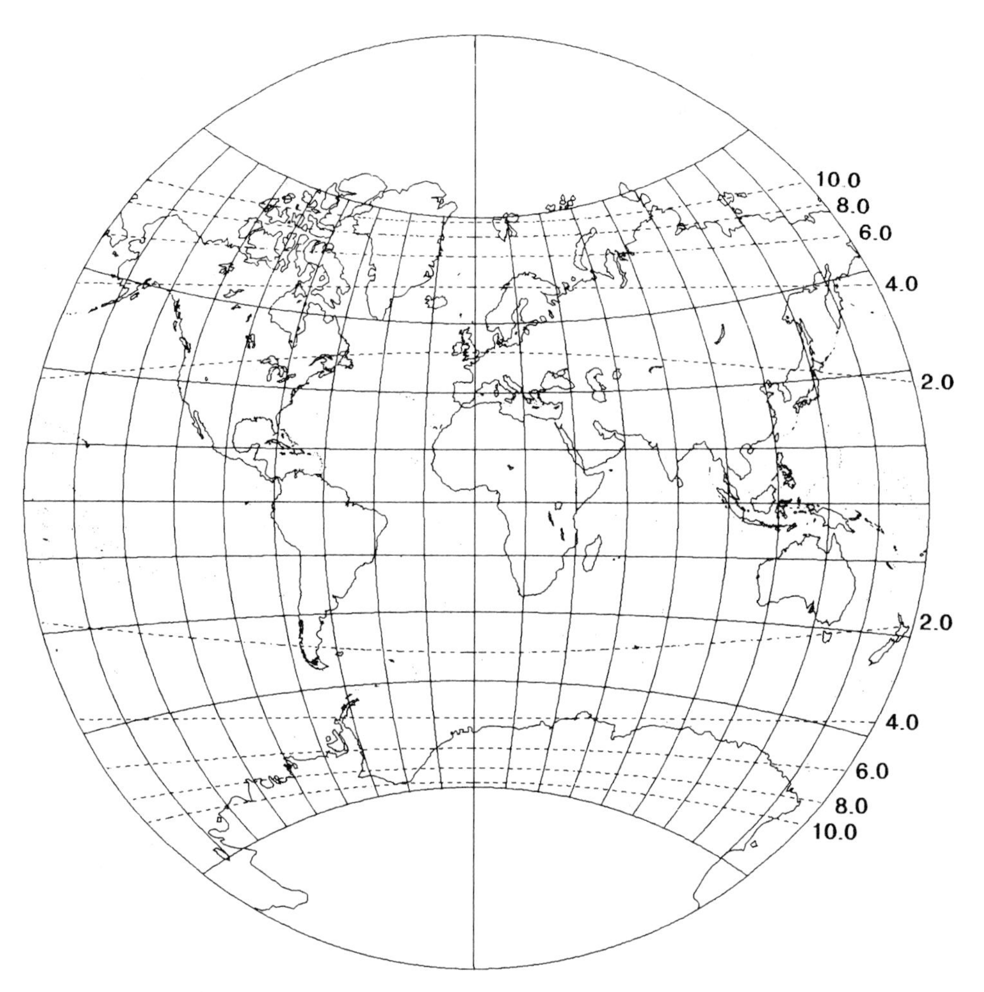

Figure 4.4 Van der Grinten (I) arbitrary projection of the world in a circle, with isocols for p. 20° graticule.

4.2.2 Equal-area polyconic projections of the ellipsoid

Applying the general formulas for local linear scale factors along parallels and the condition for equivalency of a projection we obtain

$$\begin{aligned} q_\phi \rho \cos \delta\delta_\lambda - \rho\rho_\phi \delta_\lambda &= Mr \\ \delta_\lambda &= nr/\rho \end{aligned} \tag{4.4}$$

where

$$q_\phi = dq/d\phi, \quad \rho_\phi = d\rho/d\phi, \quad \delta_\lambda = \partial\delta/\partial\lambda$$

n is the local linear scale factor along a parallel; $r = N \cos \phi$ is the radius of curvature of a parallel; M, N are radii of curvature of the meridional and first vertical sections, respectively; and ϕ, λ are geodetic coordinates of the point on the surface of the ellipsoid. The set (4.4) of two simultaneous equations includes four functions.

By considering various methods of redefining this set, we can develop a large number of different equal-area polyconic projections (Bugayevskiy 1986b).

As an example, we specify the following redefined equations:

$$q = f_1(\phi), \quad \rho = f_2(\phi) \tag{4.5}$$

In this case the problem being considered is reduced to defining polar angle δ. Integrating equations (4.4) under the condition requiring that the projections be symmetrical about the central meridian, we will have an expression similar to Kepler's equation:

$$\delta = c + b \sin \delta \tag{4.6}$$

where

$$c = -\lambda \frac{Mr}{\rho\rho_\phi}, \quad b = \frac{q_\phi}{\rho_\phi} \tag{4.7}$$

Now, using as a first approximation

$$\delta^{(1)} = c/(1 - b)$$

it is not difficult using (4.6) and (4.7) to compute the values of polar angles δ by iteration and, consequently, a group of various projections depending on the given functions (4.5).

In particular, let us assume that

$$\rho = N \cot \phi, \quad q = s + N \cot \phi \tag{4.8}$$

where s is the length of the arc of the meridian from the equator to the given parallel.

The variables in equations (4.7) are determined as follows:

$$\begin{aligned} c &= \lambda \sin^3 \phi/(1 + e'^2 \cos^4 \phi) \\ b &= \cos^2 \phi(1 + e'^2 \cos^2 \phi)/(1 + e'^2 \cos^4 \phi) \end{aligned} \tag{4.9}$$

where e' is the second eccentricity of the ellipsoid, defined as

$$e' = \sqrt{a^2 - b^2}/b = e/\sqrt{1 - e^2}$$

Introducing the values of polar angles from (4.9) and (4.6) and values of q and ρ from (4.8), all into (4.1), the rectangular coordinates for the projection may be calculated.

To determine the positions of meridians along the equator, taking into account (4.1) and (4.6)–(4.9) and evaluating their intermediate form, we obtain

$$x = a\lambda(1 - \tfrac{1}{6}\lambda^2 + \tfrac{1}{12}\lambda^4 - \cdots)$$

The formulas for local linear scale factors on the projection take the form:

1. along parallels:

$$n = c_1/(1 - b \cos \delta)$$

where

$$c_1 = \sin^2 \phi/(1 + e'^2 \cos^4 \phi)$$

2. along meridians:

$$m = (1/n)\sec \varepsilon$$

Angle ε is the deviation from a right angle of the angle of intersection between a given meridian and parallel and is determined from a formula which is general for all polyconic projections:

$$\tan \varepsilon = \frac{q_\phi \sin \delta + \rho \delta_\phi}{\rho_\phi - q_\phi \cos \delta}$$

where

$$\delta_\phi = \frac{c_\phi + b_\phi \sin \delta}{1 - b \cos \delta}$$

$$c_\phi = \frac{\lambda}{t} \sin^2 \phi \cos \phi[3 + e'^2 \cos^2 \phi(4 - \cos^2 \phi)]$$

$$b_\phi = -\frac{1}{t} \sin 2\phi[1 + e'^2 \cos^2 \phi(2 - \cos^2 \phi)]$$

$$t = (1 + e'^2 \cos^4 \phi)^2$$

The maximum angular deformation is determined from the following formula:

$$\tan(\omega/2) = \tfrac{1}{2}\sqrt{m^2 + n^2 - 2}$$

On this projection there is no distortion of any kind along the central meridian. These projections can be used especially when designing maps of regions considerably extended in latitude and less extended in longitude. The projection is symmetrical about the central meridian and the equator.

4.2.3 Rectangular polyconic projections of the ellipsoid (with orthogonal map graticule)

The condition for orthogonality of a map graticule can be expressed by the formula (see also equation (1.50))

$$f = x_\phi x_\lambda + y_\phi y_\lambda = 0 \qquad (4.10)$$

On differentiating (4.1) and introducing the various terms into (4.10), we obtain

$$\frac{d\delta}{\sin \delta} = \frac{-q_\phi}{\rho}\, d\phi \qquad (4.11)$$

The integration of (4.11) gives

$$\ln \tan \frac{\delta}{2} = -\int \frac{q_\phi}{\rho}\, d\phi + \ln c(\lambda) \qquad (4.12)$$

where $c(\lambda)$ is the integration function.

Given the functions q and ρ, we can develop many different polyconic projections with orthogonal map graticules from (4.12). These are often called rectangular polyconic projections.

For a particular case, if we let $q = s + \rho$ and $\rho = N \cot \phi$, and take into account $q_\phi = -N \cot^2 \phi$, we obtain

$$\ln \tan \frac{\delta}{2} = \ln \sin \phi + \ln c(\lambda) \quad \text{and} \quad \tan \frac{\delta}{2} = c(\lambda) \sin \phi \tag{4.13}$$

The integration function may be determined from a requirement to preserve linear scale along the equator (although another parallel may have scale preserved instead); then

$$x = a\lambda = \rho \sin \delta = a(1 + \cos \delta)c(\lambda)\cos \phi \tag{4.14}$$

From (4.13) and (4.14), we find that

$$c(\lambda) = \frac{\lambda}{2} \quad \text{and} \quad \tan \frac{\delta}{2} = \frac{\lambda}{2} \sin \phi$$

Local linear scale factors m along meridians and n along parallels take the form

$$m = \frac{4 + \lambda^2(1 + \cos^2 \phi + 2e'^2 \cos^4 \phi)}{4 + \lambda^2 \sin^2 \phi}, \quad n = \frac{4}{4 + \lambda^2 \sin^2 \phi}$$

where e' is the second eccentricity of the ellipsoid.

Local area scale factors and the maximum angular deformation can be easily determined from the general formulas for map projections:

$$p = mn, \quad \sin(\omega/2) = (m - n)/(m + n)$$

The rectangular polyconic projection was apparently developed by the US Coast Survey by 1853 and later used as the War Office projection in Great Britain. We should note that N. A. Urmayev considered a somewhat different method of producing this projection, and Vitkovskiy (1907) suggested a corresponding geometric projection for transforming the surface of a sphere.

On this ellipsoidal projection (and of course its spherical limit) there is no distortion of any type along the central meridian, distance distortion along parallels is absent along the equator, and there is no distance distortion along meridians at the poles.

This projection is especially recommended for mapping regions extending primarily along parallels of latitude and comparatively less along meridians of longitude.

4.2.4 Arbitrary polyconic projections for world maps developed from sketches of the map graticule

The theory of developing projections of this kind using numerical analysis, the theory of approximations, and function interpolation was developed by N. A. Urmayev.

This projection-making process can be divided into two stages: drawing a sketch of the graticule, and its mathematical treatment. In the first stage, using various maps of the region being mapped, one or two variants of original projections are taken to be the base, and all necessary changes are introduced onto them. On this base a sketch of the graticule of meridians and parallels is drawn at small scale on graph paper.

From that sketch local scale and angular distortion are determined approximately by graphical means or by numerical analysis (see section 8.2). If the values

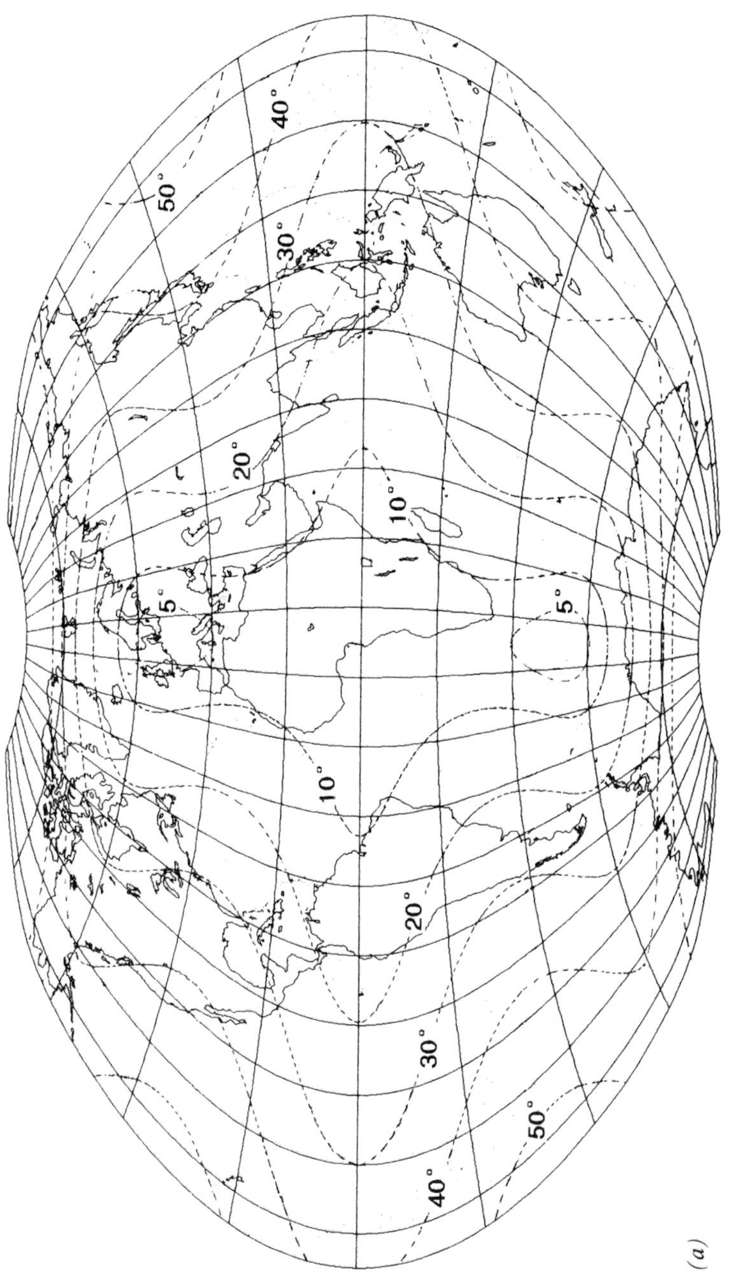

Figure 4.5 TsNIIGAiK modified polyconic projection (1939–49 variant): (a) isocols for ω; (b) isocols for p. 20° graticule.

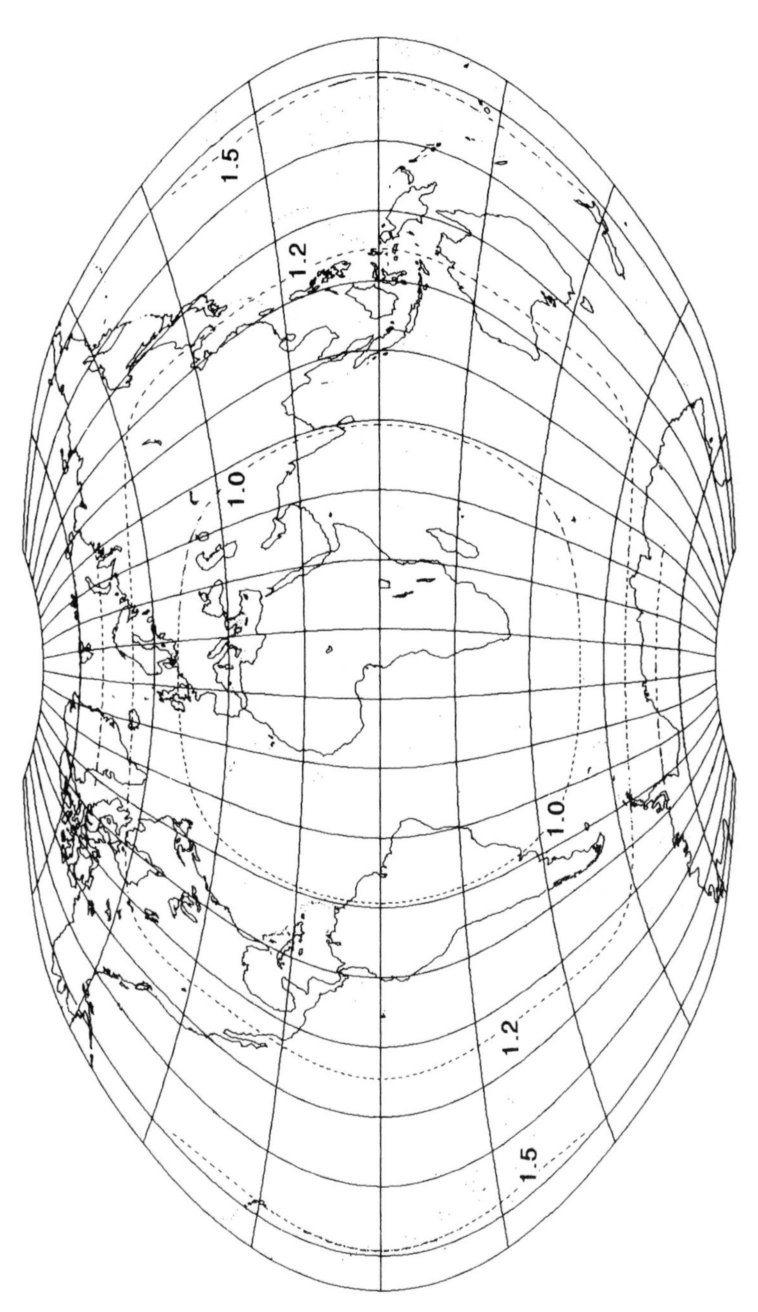

Figure 4.5 (b)

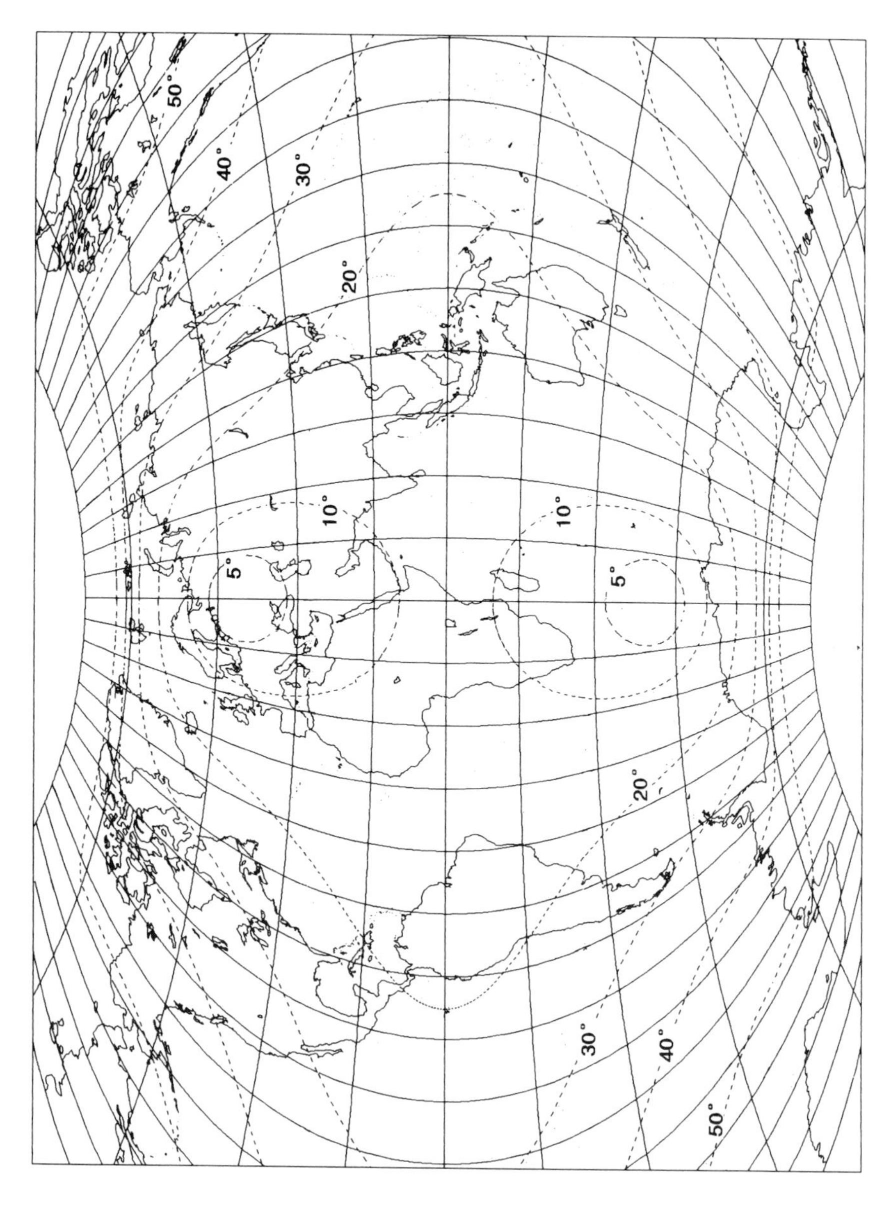

(a)

Figure 4.6 TsNIIGAiK *modified polyconic projection (BSE variant): (a) isocols for* ω; *(b) isocols for p.* 20° *graticule.*

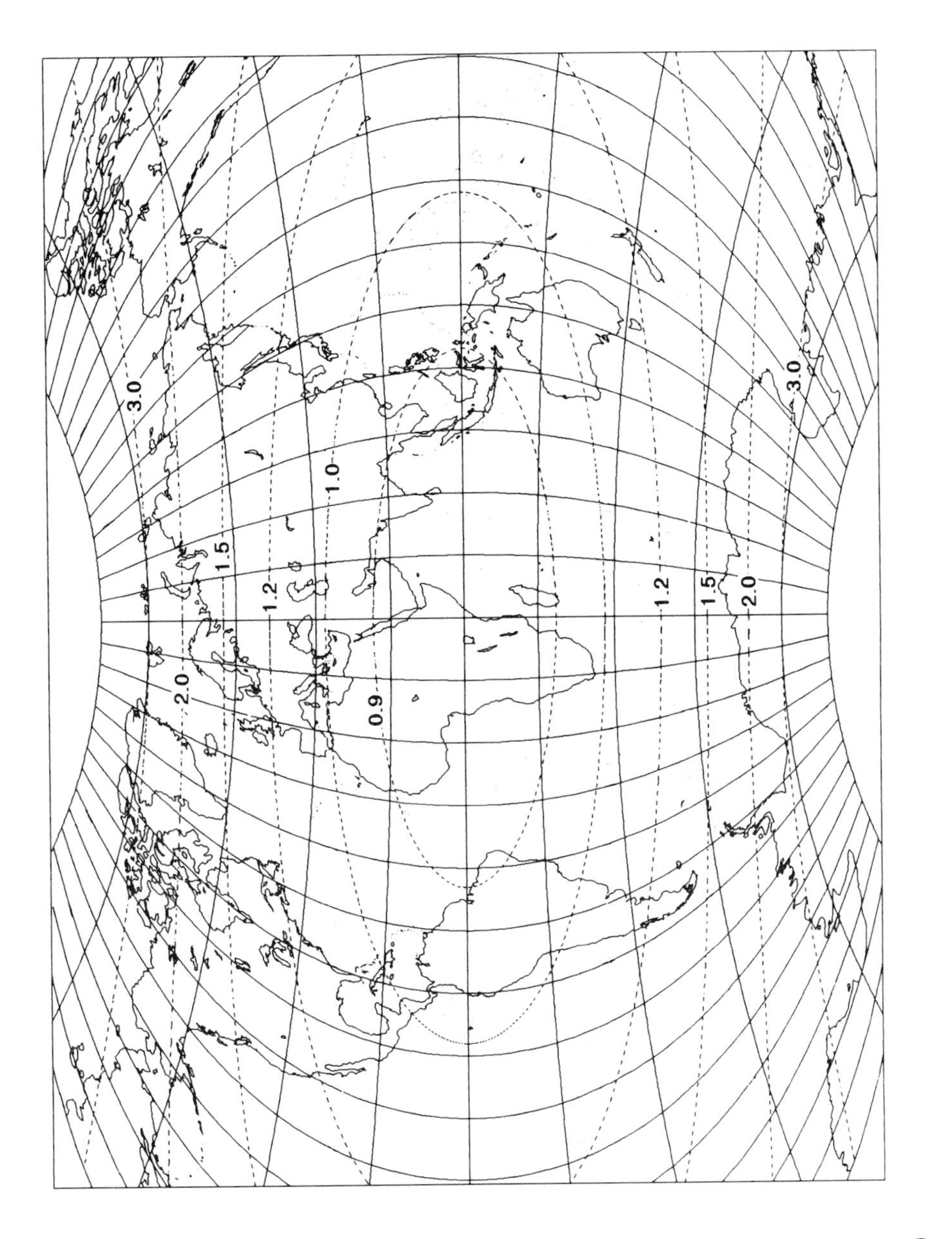

Figure 4.6 (b)

and pattern of distortion do not satisfy the condition imposed, some corrections are introduced onto the sketch.

In the second stage, the sketch is treated mathematically, as a result of which rectangular coordinates of the nodal (intersection) points of the graticule and distortion values are determined.

Note that the methods and peculiarities of preparing polyconic and other projections in this way are partially touched upon below. Details are provided in Ginzburg and Salmanova (1962). The method described was used for a number of modified polyconic projections by TsNIIGAiK, especially for world maps. The first was developed by Ginzburg and used in 1939–49 (Figure 4.5).

It was assumed that the central meridian and parallels were equally divided. The coordinates of seven main points, taken from the sketch, were used as the initial points; two of them were on the central meridian at latitudes of 0° and 80°, and the other five were on the meridian $\lambda = 180°$ at intervals of 20° from the equator.

The coordinates of these points were adjusted by least-squares approximation, and the values of the coordinates of graticule intersections (at intervals of 10° in longitude and latitude) were determined by interpolation.

On this variant of the projection, the degrees of maximum angular deformation and percentage of area distortion are roughly equal, not exceeding 50° and 80 percent, respectively (except in the polar regions, where they are greater). The projection was used for designing a series of world maps in the *Geograficheskiy Atlas* (*Geographical Atlas*) for secondary-school teachers.

The TsNIIGAiK 1950 polyconic projection was also developed by Ginzburg for the design of school world maps. For this development a scheme of arranging nodal points and also the method of drawing sketches and making calculations as in the previous variant were used. On the second variant, however, parallels are represented as arcs of circles of less curvature, and areas in an east–west direction from the center are less distorted, but the angular distortion is up to 60°.

A third variant of polyconic projections was developed by Ginzburg for the maps of the *Bol'shoy Sovetskiy Entsiklopediya* (*BSE* or *Great Soviet Encyclopedia*) in 1950. Here it was assumed that the central meridian is not equally divided for latitude. Ten nodal points (five on the central meridian and five on the outer one) were used, and their coordinates were determined from a sketch (Figure 4.6).

On this variant the curvature of a given parallel has values intermediate between those on the previous two projections; angular and area distortion are approximately of the same order.

A modified polyconic projection was developed by T. D. Salmanova for a series of maps of the Soviet Union for higher schools in 1949–50. She used numerical analysis. Parallels on this projection are unequally divided; their curvature is less than on conic projections, which creates a more correct visual perception of the geographical position of a region when reading the map. The spacing of meridians along the parallels decreases with an increase of distance from the also unequally divided central meridian. On the projection the isocol $\omega = 10°$ is close to the general outline of the Soviet Union, angular distortion reaches the greatest value of $\omega = 20°$ in the polar regions, and area distortion varies from -2 percent at the center of the region to 30 percent near the polar regions.

A TsNIIGAiK polyconic projection with a compound graticule was made by joining two parts of a projection along the central meridian, each being designed to provide a better representation of the western and eastern halves of the territory of the USSR. Some other polyconic projections were also developed at the TsNIIGAiK, e.g. a dome-shaped asymmetric projection for the world map.

Further development of theoretical proposals and methods for making polyconic and other projections with arbitrary distortion distribution from sketches of the map graticule is given in works by Boginskiy.

4.3 Polyconic projections in a narrow sense

On these projections two additional conditions are imposed: the radii of parallels on the projection are $\rho = N \cot \phi$, fitting a cone tangent to the ellipsoid (or sphere) at these parallels; and distances along the central meridian are preserved, i.e $m_0 = 1$.

The ordinary polyconic and modified (IMW) polyconic projections are the most widely used of this type of projection.

4.3.1 Ordinary polyconic projection

In developing this projection (Figure 4.7) an additional condition is imposed: the lengths of all parallels on the projection are represented without distortion ($n = 1$).

The projection formulas take the form

$$\begin{gathered} x = N \cot \phi \sin \delta, \quad y = s + N \cot \phi (1 - \cos \delta) \\ \delta = \lambda \sin \phi, \quad \tan \varepsilon = \frac{\delta - \sin \delta}{\cos \delta - [1 + (M/N)\tan^2 \phi]} \\ p = 1 + 2 \frac{N}{M} \cot^2 \phi \sin^2\left(\frac{\delta}{2}\right), \quad m = p \sec \varepsilon \\ \tan \frac{\omega}{2} = \frac{1}{2} \sqrt{\frac{m^2 + n^2 - 2p}{p}} \end{gathered} \tag{4.15}$$

and s is the meridian distance of ϕ from the latitude of the origin of rectangular coordinates. If $\phi = 0$, $x = a\lambda$.

The distortion for this projection depends on both latitude and longitude. The isocols are in the shape of curves symmetrical about the central meridian. The distortion of distance along meridians and angular and area distortion increase considerably with an increase of distance from the central meridian; parallels (especially at high latitudes) are represented with considerable and correct curvature. Regions extending along meridians are represented with little distortion.

The projection found considerable application in the United States for designing maps of narrow and occasionally wide regions (east to west). It was especially used as the basis for topographic quadrangles produced by the US Geological Survey. It was first developed by F. R. Hassler, the first director of the US Survey of the Coast, in about 1820, and it was used for many coastal charts. Since the 1950s it has been almost abandoned for new maps.

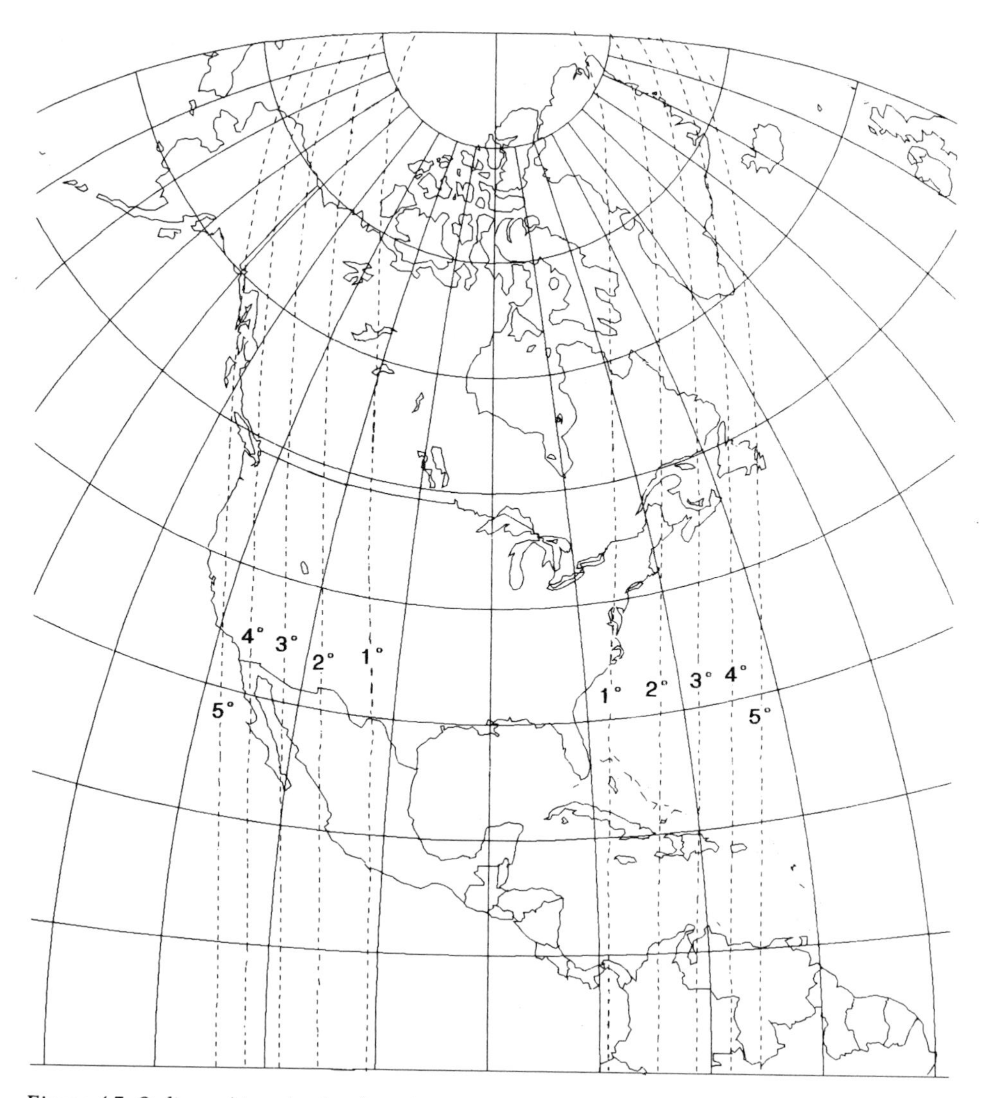

Figure 4.7 Ordinary (American) polyconic projection; isocols for ω. 10° graticule.

For narrow zones the projection formulas (4.15) may be given in the following form:

$$x = \lambda N \cos \phi - \frac{\lambda^3}{6} N \cos \phi \sin^2 \phi + \cdots$$

$$y = s + \frac{\lambda^2}{2} N \cos \phi \sin \phi + \cdots \tag{4.16}$$

When $\lambda \leq 3°$ angle $\varepsilon \leq 3''$; therefore, in practice,

$$p = m = 1 + \frac{\lambda^2}{2} \cos^2 \phi = 1 + 0.000\,152\,3(\lambda°)^2 \cos^2 \phi$$

$$\omega' = 0.52'(\lambda°)^2 \cos^2 \phi \tag{4.17}$$

where ω' is ω in minutes of arc with λ in degrees. The greatest distortion is at the points of intersection of the outer meridians ($\pm 3°$) with the equator and reaches up to $v_p = v_m = 0.14$ percent, $\omega = 4.7'$.

4.3.2 Modified polyconic projection for the International Map of the World

In 1891 Albrecht Penck of Germany advanced a proposal for the design of a 1 : 1 000 000 scale International Map of the World (IMW). In 1909 a modified polyconic projection designed by Charles Lallemand of France was approved for this map at the International Geographical Congress in London, and the division of the map into sheets and the nomenclature on the sheets were determined as well.

The following modifications to the ordinary polyconic projection were adopted. The projection is applied to numerous quadrangle sheets. All meridians are represented by straight lines; all parallels on each sheet are circular arcs of radii $\rho = N \cot \phi$ drawn from centers lying along the central meridian, and the linear scale along the extreme northern and southern parallels of the quadrangle is preserved. At latitudes less than 60°, the scale along two standard meridians at a distance from the central meridian of $\pm 2°$ is also preserved; other parallels are drawn through points marked along the two standard meridians in accordance with the correct scale between the extreme parallels of the quadrangle.

The dimensions of most quadrangles are $\Delta\phi = 4°$ and $\Delta\lambda = 6°$. At latitudes from 60° to 76° the sheets are doubled in longitude range ($\Delta\lambda = 12°$) and the standard meridians are $\pm 4°$ from the central meridian; higher than latitudes of 76° they are increased four-fold ($\Delta\lambda = 24°$) and standard meridians are $\pm 8°$ from the center.

As each sheet has its own central meridian, distortion within sheets is small, but when forming blocks of sheets (mosaicking four or more sheets together in all directions) there are angular and linear gaps:

$$\varepsilon' = \frac{\rho'}{\rho^{\circ 2}} \Delta\phi^\circ \, \Delta\lambda^\circ \cos \phi_{\text{cen}}$$

where ϕ and λ are expressed in degrees, ε in minutes, and ρ', ρ° are conversion factors given in Appendix 7.

As a result of the various conditions, the scale factors for local area and distance along meridians, using (4.17), are equal to, with sufficient accuracy,

$$p = m = \frac{1 + 0.000\,152\,3\lambda^{\circ 2} \cos^2 \phi}{1 + 0.000\,152\,3 \times 4° \cos^2 \phi} = 1 + 0.000\,152\,3(\lambda^{\circ 2} - 4°)\cos^2 \phi$$

Hence, along the central meridian ($\lambda^\circ = 0$),

$$m_0 = 1 - 0.000\,609\,2 \cos^2 \phi$$

i.e. the central meridian is shortened by

$$\Delta y_0 = 0.000\,609\,2 \Delta s \cos^2 \phi$$

Assuming with considerable accuracy that the length of a segment of the meridian with $\Delta\phi \approx 4°$ is equal to $\Delta s \approx 444$ km at a scale of 1 : 1 000 000, we obtain in millimeters

$$\Delta y_0 = 0.271 \cos^2 \phi$$

Table 4.1 Distortion characteristics of various polyconic projections

ϕ	Characteristics of the projection	λ 0°	30°	60°	90°	120°	150°	180°
		Lagrange projection[1]						
0°	$m = n$	1.000	1.017	1.072	1.172	1.333	1.589	2.000
	p	1.000	1.035	1.149	1.373	1.777	2.524	4.000
30°	$m = n$	1.132	1.152	1.212	1.323	1.501	1.780	2.224
	p	1.283	1.327	1.469	1.749	2.252	3.169	4.947
60°	$m = n$	1.795	1.832	1.910	2.068	2.316	2.693	3.263
	p	3.222	3.323	3.649	4.275	5.364	7.253	10.649
		Rectangular polyconic projection						
0°	m	1.000	1.137	1.548	2.234	3.193	4.427	5.935
	n	1.000	1.000	1.000	1.000	1.000	1.000	1.000
	p	1.000	1.137	1.548	2.234	3.193	4.427	5.935
	$\omega°$	0	7.35	24.84	44.86	63.07	78.32	90.73
30°	m	1.000	1.101	1.385	1.802	2.239	2.799	3.289
	n	1.000	0.983	0.936	0.866	0.785	0.700	0.618
	p	1.000	1.082	1.296	1.561	1.758	1.959	2.033
	$\omega°$	0	6.49	22.31	41.08	57.48	73.72	86.26
60°	m	1.000	1.033	1.114	1.211	1.301	1.375	1.433
	n	1.000	0.951	0.829	0.684	0.549	0.438	0.351
	p	1.000	0.982	0.924	0.828	0.714	0.602	0.503
	$\omega°$	0	4.74	16.87	32.29	47.97	62.24	74.67
90°	m	1.000	1.000	1.000	1.000	1.000	1.000	1.000
	n	1.000	0.936	0.785	0.618	0.477	0.369	0.288
	p	1.000	0.936	0.785	0.618	0.477	0.369	0.288
	$\omega°$	0	3.79	13.84	27.31	41.48	54.89	67.12
		Equal-area polyconic projection[2]						
0°	m	1.000	1.138	1.552	2.242	3.208	4.450	5.968
	n	1.000	0.879	0.644	0.446	0.312	0.225	0.168
	$\omega°$	0	14.77	48.82	83.85	110.75	129.34	141.95
30°	m	1.000	1.105	1.364	1.688	2.026	2.358	2.667
	n	1.000	0.912	0.752	0.619	0.522	0.454	0.403
	$\omega°$	0	13.09	36.18	58.07	75.27	88.27	98.20
60°	m	1.000	1.034	1.124	1.254	1.406	1.568	1.737
	n	1.000	0.968	0.894	0.812	0.742	0.688	0.649
	$\omega°$	0	4.18	14.30	27.08	39.91	51.58	61.80
90°	m	1.000	1.000	1.000	1.000	1.000	1.000	1.000
	n	1.000	1.000	1.000	1.000	1.000	1.000	1.000
	$\omega°$	0	0	0	0	0	0	0

Formulas for the rectangular coordinates of this projection take the form

$$x = \left(\lambda r_s - \frac{\lambda^3}{6} r_s \sin^2 \phi_s\right) + \left(\lambda(r_n - r_s) - \frac{\lambda^3}{6}(r_n \sin^2 \phi_n - r_s \sin^2 \phi_s)\right) \frac{\phi - \phi_s}{4}$$

$$y = \frac{\lambda^2}{2} r_s \sin \phi_s + \left((s_n - s_s - \Delta y_0) + \frac{\lambda^2}{2}(r_n \sin \phi_n - r_s \sin \phi_s)\right) \frac{\phi - \phi_s}{4}$$

where r_n, r_s are found from (1.28) for the limiting parallels ϕ_n, ϕ_s. On writing

Table 4.1 Continued

ϕ	Characteristics of the projection	λ 0°	30°	60°	90°	120°	150°	180°
				Ordinary polyconic projection				
0°	m	1.000	1.137	1.548	2.234			
	n	1.000	1.000	1.000	1.000			
	p	1.000	1.138	1.552	2.242			
	ω	0°00′	7°24′	24°54′	44°54′			
30°	m	1.000	1.102	1.404	1.894			
	n	1.000	1.000	1.000	1.000			
	p	1.000	1.103	1.404	1.883			
	ω	0°00′	5°36′	19°36′	36°42′			
60°	m	1.000	1.034	1.129	1.270			
	n	1.000	1.000	1.000	1.000			
	p	1.000	1.034	1.128	1.264			
	ω	0°00′	1°54′	7°12′	14°06′			
90°	m	1.000	1.000	1.000	1.000			
	n	1.000	1.000	1.000	1.000			
	p	1.000	1.000	1.000	1.000			
	ω	0°00′	0°00′	0°00′	0°00′			

[1] There is no angular distortion on Lagrange projections.
[2] There is no area distortion on equal-area polyconic projections.

derivatives

$$x_\lambda = r_s - \frac{\lambda^2}{2} r_s \sin^2 \phi_s + \left((r_n - r_s) - \frac{\lambda^2}{2} (r_n \sin^2 \phi_n - r_s \sin^2 \phi_s) \right) \frac{\phi - \phi_s}{4}$$

$$y_\lambda = \lambda \left(r_s \sin \phi_s + (r_n \sin \phi_n - r_s \sin \phi_s) \frac{\phi - \phi_s}{4} \right)$$

It is now feasible, from the formula for the general theory $n = (1/r)(x_\lambda^2 + y_\lambda^2)^{1/2}$, to find the values of local linear scale factors along parallels on this projection. Formulas derived from a different approach are given in Snyder (1987b, pp. 131–7).

4.4 Characteristics of polyconic projections

Polyconic projections possess more generalized properties than other projections considered above. As a rule, local scale is a function of latitude and longitude, isocols are sometimes ovals, the magnitude of distortion is less, and the distortion distribution pattern is better for these projections than for many others.

Polyconic projections have found wide application in the creation of world maps. To characterize the advantages of these projections apart from the models with isocols mentioned above, we list the values of local scale factors and maximum angular deformation for some of these projections (Table 4.1).

A distortion analysis has shown that polyconic projections can be satisfactorily used for world maps as well as for maps of large regions, especially those extending along the central meridian.

5

Projections for topographic and named-quadrangle maps; projections used in geodesy

5.1 Topographic map projections

Various countries use the following projections for designing topographic maps: pseudocylindrical trapezoidal; Bonne equal area; pseudoconic; sinusoidal; conformal azimuthal (stereographic); azimuthal equidistant; conformal conic; ordinary polyconic; transverse cylindrical; Laborde; Gauss–Krüger; and UTM (Universal Transverse Mercator).

We will outline the development of three projections that were not considered in the previous chapters.

5.1.1 Trapezoidal pseudocylindrical projection

The projection equations are of the form

$$
\begin{gathered}
x = \alpha(a - ks)\lambda, \quad y = ks \\
\tan \varepsilon = \alpha\lambda, \quad m = k \sec \varepsilon, \quad n = \alpha(a - ks)/r
\end{gathered}
\tag{5.1}
$$

where s is the arc length of the meridian from the equator to the given parallel; k, α, and a are constant parameters determined from the condition that the scales along parallels of latitude ϕ_1 and ϕ_2 and along meridians of longitude $\pm\lambda_0$ from the central meridian are to be preserved.

Then, incorporating (5.1) we obtain

$$
\begin{aligned}
k &= \left(1 - \frac{\lambda_0^2(r_1 - r_2)^2}{(s_2 - s_1)^2}\right)^{1/2} \approx 1 - \frac{\lambda_0^2}{2} \sin^2 \phi_m \\
\phi_m &= \tfrac{1}{2}(\phi_1 + \phi_2) \\
\alpha &= \frac{1}{k} \frac{r_1 - r_2}{s_2 - s_1} \approx \frac{1}{k} \sin \phi_m \\
a &= ks_1 + \frac{r_1}{\alpha} = ks_2 + \frac{r_2}{\alpha}
\end{aligned}
$$

The projection was treated as if it were projected onto the faces of a polyhedron, and it was constructed graphically from the straightened meridian and parallel arcs for 1 : 200 000 and larger-scale map sheets. It was called the Müffling projection and was used in Germany and later in Russia and the Soviet Union until 1928 (see also section 6.3.6).

Within the limits of each map sheet, distortion was small (not more than $(\lambda_0^2/2)\sin^2\phi_m$), but when mosaicking sheets into blocks, angular (ε', in minutes) and linear gaps occurred:

$$\varepsilon' = \frac{\rho'}{\rho^{\circ 2}}(\phi_2 - \phi_1)^\circ(\lambda_2 - \lambda_1)^\circ \cos\phi_{\text{cen}}$$

5.1.2 Transverse cylindrical projections

Let PO be an initial meridian (Figure 5.1). Let us mark off arc $OQ = 90°$ on the equator and join poles P and Q with the arc of a great circle. Then the position of any point A is defined by the geographical coordinates ϕ,λ of the normal coordinate system and the corresponding spherical polar coordinates ϕ', λ' of the transverse system.

From spherical triangles PQA and AQA_2,

$$\cos\phi'\cos\lambda' = \cos\phi\cos\lambda, \quad \cos\phi'\sin\lambda' = \sin\phi$$

Hence,

$$\tan\lambda' = \tan\phi\sec\lambda, \quad \sin\phi' = \cos\phi\sin\lambda \tag{5.2}$$

where

$$\phi' = 90° - z, \quad \lambda' = 90° - a$$

and z, a are polar spherical coordinates (see section 1.1.3).

Assuming that distances along the central meridian are preserved, the equations for all transverse cylindrical projections for the sphere can be rewritten in the form

$$x = f(\phi'), \quad y = R\lambda'$$

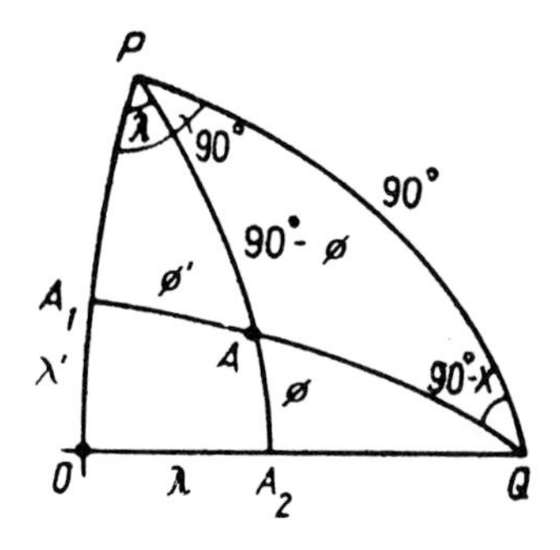

Figure 5.1 Relationship of coordinates for a transverse cylindrical projection.

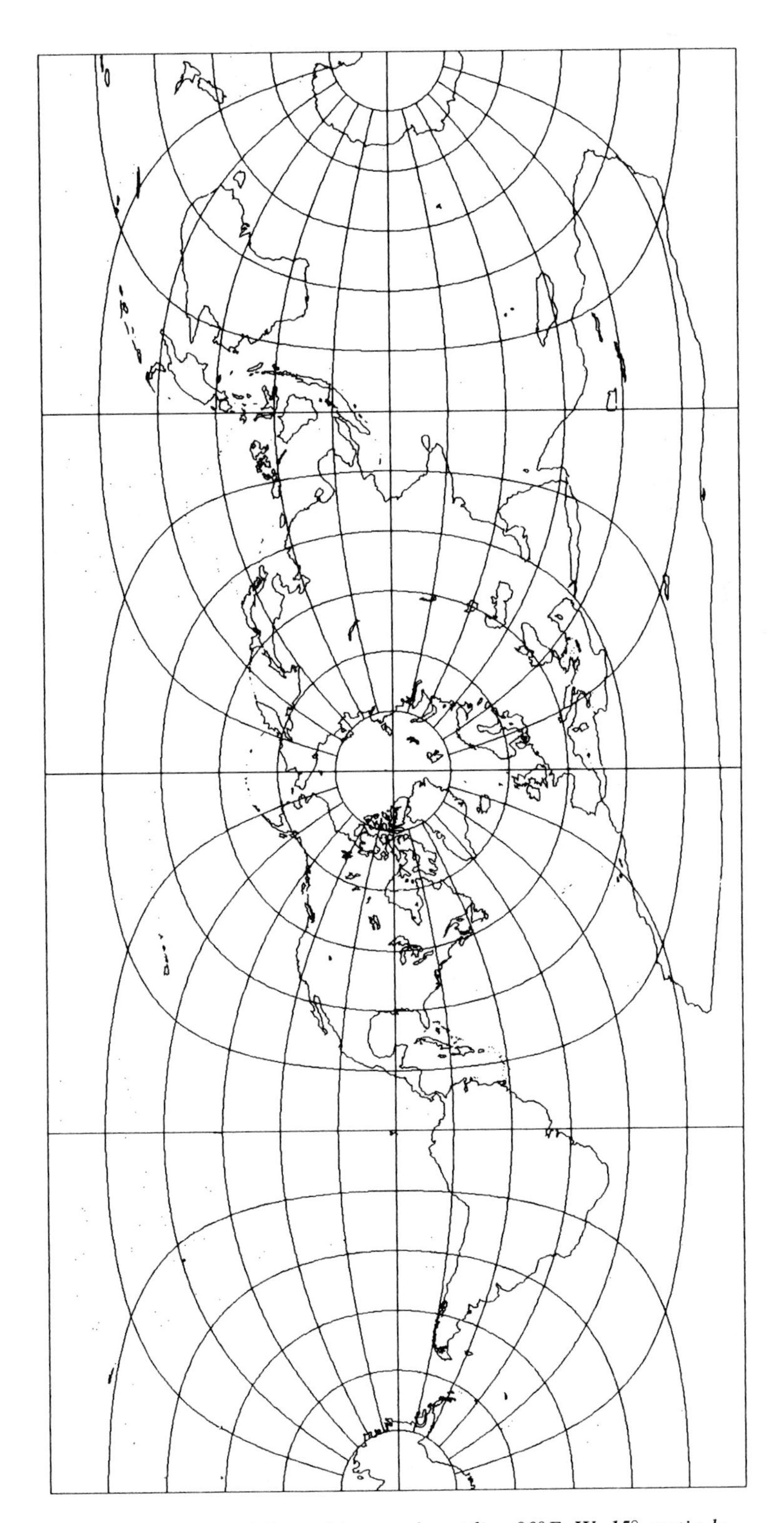

Figure 5.2 Cassini projection of the world, central meridian 90° E–W. 15° graticule.

Cassini projection

The normal analog to this projection (Figure 5.2) is the plate carrée or equidistant cylindrical projection (Figure 2.3)

$$x = R\lambda, \quad y = R\phi$$

Taking into account the fact that on transverse projections x and y exchange places, and that ϕ' corresponds to the value of ϕ and λ' to that of λ on normal projections, we have

$$x = R\phi', \quad y = R\lambda'$$

Applying (5.2), we obtain

$$x = R \arcsin(\cos \phi \sin \lambda), \quad y = R \arctan(\tan \phi \sec \lambda)$$

Formulas for local linear scale factors take the form

$$\mu_x = 1, \quad \mu_y = \sec \phi' \approx 1 + \frac{x^2}{2R^2} + \cdots$$

This is the spherical form. For topographic mapping, ellipsoidal forms developed by C. F. Cassini in the eighteenth century and J. G. von Soldner in the nineteenth century (the Cassini–Soldner projection) were used (see Snyder 1987b, p. 95).

Transverse Mercator or Gauss–Lambert projection

This is a transverse cylindrical projection (Figure 5.3), analogous to the normal Mercator projection (Figure 2.1). Projection formulas take the form (for the sphere)

$$x = R \ln \tan\left(\frac{\pi}{4} + \frac{\phi'}{2}\right), \quad y = R\lambda'$$

Since

$$\tan^2\left(\frac{\pi}{4} + \frac{\phi'}{2}\right) = \frac{1 + \sin \phi'}{1 - \sin \phi'}$$

the formula for x may take the form

$$x = \frac{R}{2} \ln \frac{1 + \sin \phi'}{1 - \sin \phi'}$$

Applying (5.2),

$$x = \frac{R}{2} \ln \frac{1 + \cos \phi \sin \lambda}{1 - \cos \phi \sin \lambda}, \quad y = R \arctan(\tan \phi \sec \lambda)$$

Local linear scale factors can be determined by the formula

$$\mu = \sec \phi' \approx 1 + \frac{x^2}{2R^2} + \cdots$$

The formulas are of closed form and can be used for producing a projection of the sphere within the zone $\lambda = \pm 90^\circ$ except for the point $\phi_0 = 0$, $\lambda = \pm 90^\circ$ and its vicinity.

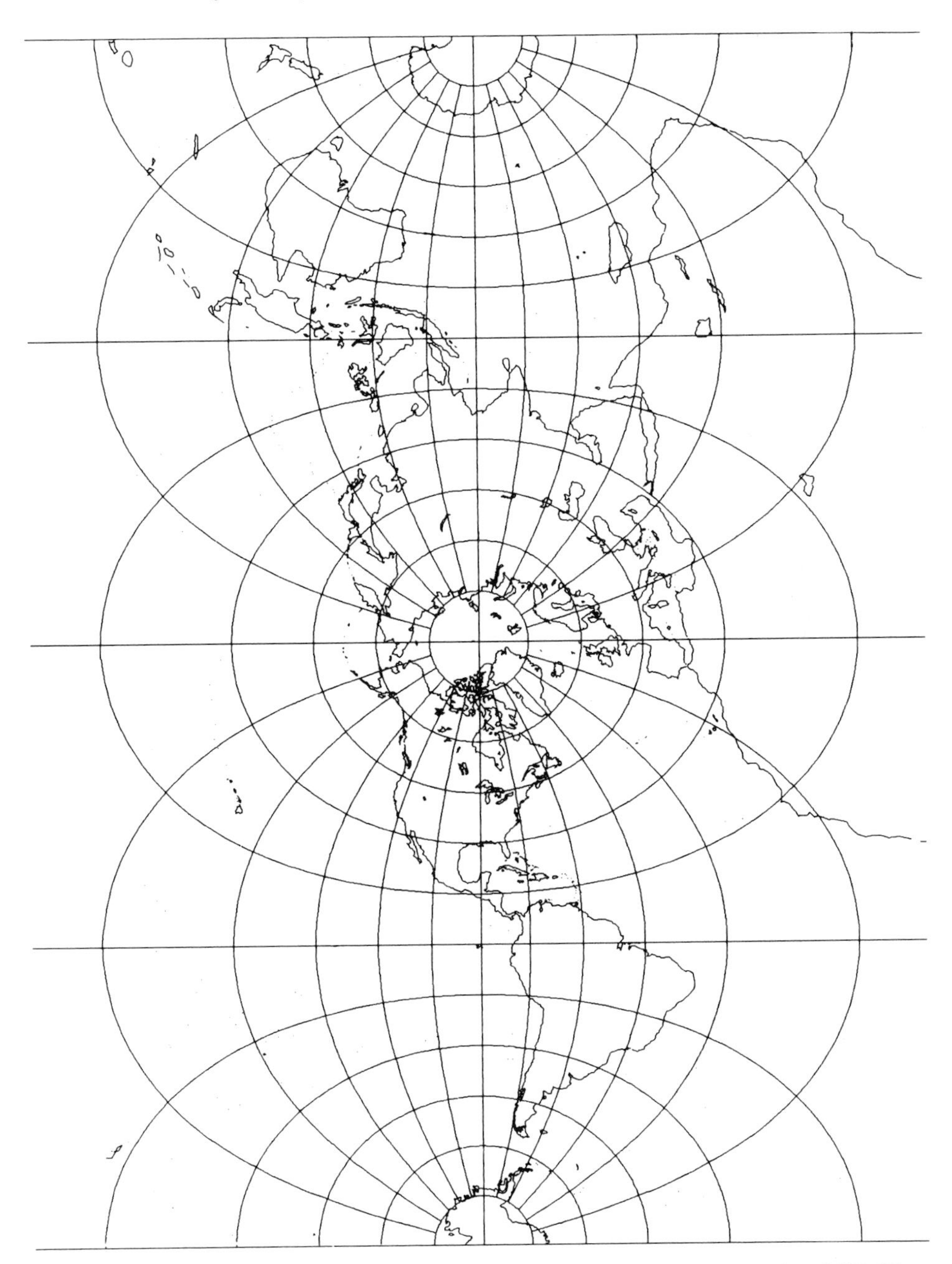

Figure 5.3 Transverse Mercator projection of most of the world, central meridian 90° E–W. 15° graticule.

5.1.3 Gauss–Krüger projection

In 1820–30 Gauss developed and published a double conformal projection preserving scale along the central meridian, and it was used in practice for calculating the Hannover triangulation. The theory of this projection was also published by Oskar

Schreiber in 1866. Detailed investigations of this projection were conducted by Louis Krüger and described by him in 1912 and 1919. He suggested a method of direct transformation of the ellipsoid onto the plane instead of the above double projection. Since then this projection has been called the Gauss–Krüger projection.

The Gauss–Krüger projection is determined by three conditions: it is conformal, it is symmetrical about the central meridian, and it preserves linear scale along the central meridian.

Any conformal projection can be expressed by an analytical function of a complex variable

$$y + ix = f(q + il) \tag{5.3}$$

where $l = \lambda - \lambda_0$, λ_0 is the longitude of the central meridian, q is the isometric latitude corresponding to geodetic latitute ϕ, and i is $\sqrt{-1}$.

Assuming that l is small, let us expand (5.3) into a Taylor series:

$$y + ix = f(q) + \frac{df(q)}{dq}(il) + \frac{d^2 f(q)}{dq^2}\frac{(il)^2}{2!} + \frac{d^3 f(q)}{dq^3}\frac{(il)^3}{3!}\cdots \tag{5.4}$$

Since $i^2 = -1$, $i^3 = -i$, $i^4 = 1$, etc., let us separate real and imaginary terms in (5.4); then

$$\begin{aligned} x &= \lambda\frac{df(q)}{dq} - \frac{l^3}{6}\frac{d^3 f(q)}{dq^3} + \frac{l^5}{120}\frac{d^5 f(q)}{dq^5} - \cdots \\ y &= f(q) - \frac{l^2}{2}\frac{d^2 f(q)}{dq^2} + \frac{l^4}{24}\frac{d^4 f(q)}{dq^4} - \cdots \end{aligned} \tag{5.5}$$

where $f(q)$ is the function $f(q + il)$ at the central meridian $l = 0$, and is called the projection characteristic.

From the third condition, $m_0 = 1$; consequently,

$$f(q) = s_m \tag{5.6}$$

Using (5.6) we can find the terms of equations (5.5) by successive differentiation:

$$\frac{df(q)}{dq} = \frac{ds_m}{dq} = \frac{ds_m}{d\phi}\frac{d\phi}{dq}$$

$$\frac{d^2 f(q)}{dq^2} = \frac{d^2 s_m}{dq^2} = \frac{d}{d\phi}\left(\frac{ds_m}{dq}\right)\frac{d\phi}{dq}$$

$$\frac{d^3 f(q)}{dq^3} = \frac{d^3 s_m}{dq^3} = \frac{d}{d\phi}\left(\frac{d^2 s_m}{dq^2}\right)\frac{d\phi}{dq}$$

but

$$\frac{ds_m}{d\phi} = M,\quad \frac{d\phi}{dq} = \frac{r}{M} = \frac{N\cos\phi}{M} = (1 + e'^2\cos^2\phi)\cos\phi$$

$$= (1 + \eta^2)\cos\phi,\quad dr/d\phi = -M\sin\phi$$

Therefore,

$$\frac{ds_m}{dq} = r, \quad \frac{d^2 s_m}{dq^2} = -r \sin \phi$$

$$\frac{d^3 s_m}{dq^3} = -N \cos^3 \phi(1 - \tan^2 \phi + \eta^2), \ldots$$

Rectangular coordinate formulas (5.5) then take the form

$$x = lN \cos \phi + \frac{l^3}{6} N \cos^3 \phi(1 - \tan^2 \phi + \eta^2) + \frac{l^5}{120} N \cos^5 \phi(5 - 18 \tan^2 \phi$$
$$+ \tan^4 \phi + 14\eta^2 - 58\eta^2 \tan^2 \phi) + \cdots \tag{5.7}$$
$$y = s_m + \frac{l^2}{2} N \sin \phi \cos \phi + \frac{l^4}{24} N \cos^3 \phi \sin \phi(5 - \tan^2 \phi + 9\eta^2) + \cdots$$

The difference of longitude l is expressed in radians in these formulas. When the longitude difference is equal to 3°30′, the accuracy of calculation is up to 0.001 m.

To calculate local linear scale factors and meridian convergence, we may use the formulas

$$m = n = \frac{1}{r} \sqrt{x_l^2 + y_l^2} \tag{5.8}$$

$$\tan \gamma = x_l / y_l \tag{5.9}$$

Introducing the values of the derivatives from equations (5.7) into (5.8) and (5.9) we obtain

$$m = 1 + \frac{l^2}{2} \cos^2 \phi(1 + \eta^2) + \frac{l^4}{24} \cos^4 \phi(5 - 4 \tan^2 \phi) + \cdots \tag{5.10}$$

or approximately $m = 1 + 0.000\,152\,3(l^\circ)^2 \cos^2 \phi$ for l in degrees,

$$\gamma = l \sin \phi + \frac{l^3}{3} \sin \phi \cos^2 \phi(1 + 3\eta^2) + \cdots \tag{5.11}$$

The Gauss–Krüger projection is not strictly conformal because of the truncation of terms; only one Cauchy–Riemann condition is satisfied, but if a sufficient number of terms in equations (5.7), (5.10), and (5.11) are provided, then it will be practically conformal.

The required number of terms increases with an increase in the width of the zone or an increase in the nominal scale of the map to be designed.

In accordance with a resolution of the USSR's Third Geodetic Assembly, the Gauss–Krüger projection was used in the Soviet Union beginning in 1928 for surveying projects and the design of maps at scales of 1 : 200 000 or larger. Beginning in 1939 it was also used for 1 : 500 000 maps.

On the Gauss–Krüger projection as used in the former Soviet Union and Russia, the ellipsoid is projected onto the plane in zones bounded by meridians: 6° zones are used for 1 : 10 000–1 : 500 000 maps, and 3° zones are used for 1 : 2000–1 : 5000 maps.

In the United States, the Gauss–Krüger transverse Mercator projection has been used extensively for two important grid systems, both of which differ from the system used in Russia and in other parts of Europe, in that the linear scale factor along the central meridian is made less than 1.0. These systems are the Universal Transverse Mercator projection and grid system and the State Plane Coordinate System, both discussed further in section 5.1.5.

5.1.4 Map sheet division and nomenclature for Soviet topographic and sheet maps

The official numbering of zones (6° wide in longitude) for 1 : 1 000 000 scale maps in the former Soviet Union begins east from the Greenwich meridian, and numbers are thus 30 less than the corresponding zones for the 1 : 1 000 000 scale International Map of the World series. The origin of rectangular coordinates for each zone is at the point of intersection of the central meridian of the zone with the equator. The division and nomenclature of the 1 : 1 000 000 map sheets is the basis for Soviet topographic map descriptions. The sizes of the sheets are given in Table 5.1.

For each quadrangle of the map at a scale of 1 : 1 000 000 there are $2 \times 2 = 4$ additional quadrangles at a scale of 1 : 500 000, using as sheet names the first four letters of the Russian alphabet, A, B, V, G. There are also $6 \times 6 = 36$ quadrangles at a scale of 1 : 200 000, with the sheets given the Roman numerals I to XXXVI, and $12 \times 12 = 144$ quadrangles at a scale of 1 : 100 000, with the numbers 1 to 144.

The 1 : 100 000 sheet is the basis for the sheet division and nomenclature of maps at larger scales. On one quadrangle of each map at a scale of 1 : 100 000 there are $2 \times 2 = 4$ map sheets at a scale of 1 : 50 000, with the sheets named A, B, C, D in Roman letters. Map sheets at a scale of 1 : 50 000 contain $2 \times 2 = 4$ map sheets at a scale of 1 : 25 000, with the sheets named a, b, v, g in Russian. A 1 : 25 000 map sheet contains four 1 : 10 000 map sheets, numbered 1, 2, 3, 4. Besides these a map sheet at a scale of 1 : 100 000 contains $16 \times 16 = 256$ map sheets at a scale of 1 : 5000, with the sheets numbered from (1) to (256) in parentheses. A 1 : 5000 map sheet contains nine 1 : 2000 map sheets, named with consecutive lower-case letters of the Russian alphabet from a to i, within the same parentheses.

As an example, we give the nomenclature of the northeastern map sheets of the 1 : 1 000 000 sheet covering Moscow at successive scales of 1 : 500 000 to 1 : 2000: N-37-B; N-37-VI; N-37-12; N-37-12-B; N-37-12-B-b; N-37-12-B-b-2; N-37-12-(16); N-37-12-(16-v), using the corresponding Russian letters in each case except for the Roman numeral and Roman letters N and all B's except the first B. The topographic map sheets of these series in the latitude zone 60°–76° are doubled in longitude range, and the longitude range to the north of latitude 76° is quadrupled.

Table 5.1 Sizes of map quadrangles in Russia

	Sheet margin dimensions			Sheet margin dimensions	
Nominal map scale	Along parallel	Along meridian	Nominal map scape	Along parallel	Along meridian
1 : 1 000 000	4°00′	6°00′	1 : 50 000	10′00″	15′00″
1 : 500 000	2°00′	3°00′	1 : 25 000	5′00″	7′30″
1 : 200 000	0°40′	1°00′	1 : 10 000	2′30″	3′45″
1 : 100 000	0°20′	0°30′	1 : 5000	1′15″	1′52.5″
			1 : 2000	0′25″	0′37.5″

5.1.5 Other transverse Mercator projections

For the design of topographic maps, other countries, especially the United States, currently use the Gauss–Krüger projection with 6° meridian zones in a form called the Universal Transverse Mercator (UTM) projection, which was adopted by the US Army Map Service in 1947 for their use in worldwide mapping.

The local linear scale factor m_0 for each zone along the central meridian is equal to 0.9996 rather than the usual Gauss–Krüger 1.0000, thus reducing the overall scale errors of the mapped regions. Hence, the projection formulas take the form of (5.7)–(5.10), with coefficient $k = 0.9996$ used as a multiplier, i.e.

$$x_{\mathrm{UTM}} = kx_{\mathrm{GK}}, \quad y_{\mathrm{UTM}} = ky_{\mathrm{GK}}, \quad m_{\mathrm{UTM}} = km_{\mathrm{GK}}, \quad \gamma_{\mathrm{UTM}} = \gamma_{\mathrm{GK}}$$

Isocols of zero distortion on the UTM projection are nearly straight and parallel to the central meridian at a distance of about 180 km to either side.

For other narrow zones based on the transverse Mercator projection, the Gauss–Krüger projection formulas may be used with the appropriate value for m_0 or k to obtain accurate rectangular coordinates. This includes the State Plane Coordinate System used since the 1930s for states in the United States with predominant north–south extent. Although different truncations of series and logarithmic calculations may have been used, the results are practically identical.

The Gauss–Boaga projection is the same as the Gauss–Krüger, except that the series is further truncated, and the central scale factor is 0.9996 (like the UTM). It was developed by Giovanni Boaga and has been adopted for Italy.

The Gauss–Schreiber projection was developed about 1880 and used for the Prussian land survey. It consists of a double conformal transformation of the ellipsoid onto the sphere and then of the sphere onto the plane using the spherical transverse Mercator projection. The central meridian varies in scale.

5.1.6 Gauss–Krüger projection for a wide zone

To obtain a wide-zone Gauss–Krüger projection we can use several methods. Below is a description of the method used by Krüger and considered in detail in works by V. V. Kavrayskiy, Solov'ev, and Morozov (see also Bugayevskiy 1982b).

Using this method the projection is developed by triple projection: (1) projecting the surface of the ellipsoid conformally onto the surface of the sphere, as described first by Mollweide in 1807; (2) transforming the conformal sphere onto a plane using the Gauss–Lambert conformal projection; and (3) carrying out a conformal transformation of the projection obtained to preserve the scale along the central meridian.

The first two stages of the transformation are considered in sections 1.7.3 and 5.1.2. As a result we get $x = R\eta$ and $y = R\xi$, where

$$\xi = \arctan(\tan \phi' \sec \lambda')$$

$$\eta = \frac{1}{2} \ln \frac{1 + \cos \phi' \sin \lambda'}{1 - \cos \phi' \sin \lambda'} = \frac{1}{2} \ln \frac{1+t}{1-t}$$

where ϕ', λ' are determined by the Mollweide method (section 1.7.3).

For carrying out the third transformation, i.e. transformation of coordinates from the Gauss–Lambert projection to the Gauss–Krüger projection, an analytical function is used:

$$y + ix = f(\xi + i\eta)$$

For points along the central meridian this function takes the form

$$y_0 = f(\xi_0) = f(\phi')$$

From the condition given for the Gauss–Krüger projection, scale is preserved along the central meridian, i.e. $y_0 = s_m$, where s_m is the arc length of the meridian from the equator to the given parallel. Using the known formulas for the relationship of the arc length of the meridian s_m to geodetic latitude ϕ on the ellipsoid and for latitudes ϕ and ϕ', after transformation we get

$$y_0 = s_m = R(\phi' + \alpha_2 \sin 2\phi' + \alpha_4 \sin 4\phi' + \cdots)$$

where

$$R = \frac{a}{1 + n'}\left(1 + \frac{n'^2}{4} + \frac{n'^4}{64} + \cdots\right)$$

$$\alpha_2 = \tfrac{1}{2}n' - \tfrac{2}{3}n'^2 + \tfrac{15}{16}n'^3 + \cdots$$

$$\alpha_4 = \tfrac{13}{48}n'^2 - \tfrac{3}{5}n'^3 + \cdots$$

where $n' = (a - b)/(a + b)$ and a, b are semiaxes of the reference ellipsoid (compare equation (1.145)). For the Krasovskiy ellipsoid, $R = 6\,367\,558.4969$ m, $\alpha_2 = 0.000\,837\,611\,8$, and $\alpha_4 = 0.000\,000\,760\,6$.

Using the analytical function for the general case we should write $y + ix$ instead of y_0 and the value of $\xi + i\eta$ instead of $\xi = \phi'$; then the formulas for the rectangular coordinates of the desired Gauss–Krüger strictly conformal projection can be presented in the form

$$x = R(\eta + \alpha_2 \cos 2\xi \sinh 2\eta + \alpha_4 \cos 4\xi \sinh 4\eta + \cdots)$$

$$y = R(\xi + \alpha_2 \sin 2\xi \cosh 2\eta + \alpha_4 \sin 4\xi \cosh 4\eta + \cdots)$$

where

$$\sinh 2\eta = \frac{2t}{z}, \quad \cosh 2\eta = \frac{1 + t^2}{z}, \quad t = \cos \phi' \sin \lambda'$$

$$z = 1 - t^2$$

From these formulas, at longitude $\lambda \leq 30°$ the errors of the calculated rectangular projection coordinates are less than 0.1 m.

The local scale factors for this projection are equal to

$$m = m_1 m_2 m_3$$

where m_1, m_2, m_3 are local linear scale factors for this conformal transformation.

In the general case,

$$m = \frac{R \cos \phi'}{N \cos \phi} \sqrt{\frac{x_\xi^2 + y_\xi^2}{z}}$$

Applying Morozov's notations we find that

$$m = \frac{H \cos \phi'}{\cos \phi} \sqrt{1 + e'^2 \cos^2 \phi} \sqrt{\frac{x_\xi^2 + y_\xi^2}{z}}$$

where $H = 0.994\,977\,825$ and $e'^2 = 0.006\,738\,525\,4$ for the Krasovskiy ellipsoid.

Partial derivatives can be calculated to sufficient accuracy from the expressions

$$x_\xi = -2 \sinh 2\eta \sin 2\xi(\alpha_2 + 8\alpha_4 \cosh 2\eta \cos 2\xi)$$

$$y_\xi = 1 + 2\alpha_2 \cos 2\xi \cosh 2\eta + 4\alpha_4(2 \cosh 2\eta - 1)(2 \cos^2 2\xi - 1)$$

The meridian convergence γ on this projection will be

$$\gamma = \gamma_1 + \gamma_2$$

where

$$\gamma_1 = \arctan(\sin \phi' \tan \lambda'), \quad \gamma_2 = -x_\xi/y_\xi$$

x_ξ, y_ξ are the above partial derivatives.

The advantage of this method for developing the Gauss–Krüger projection is that the formulas obtained, being comparatively simple, make it possible to create the projection in a practical manner for any longitude difference (except for the special point at $\phi = 0$, $\lambda = \pm 90°$ and its vicinity).

5.2 *Projections used for maps at scales of 1 : 1 000 000 and 1 : 2 500 000*

5.2.1 Modified polyconic projection (IMW) and its use for maps at a scale of 1 : 1 000 000

The formulas and main characteristics for this projection are given in section 4.3.2. An advantage of the modified polyconic projection used for individual quadrangles is the small value of distortion, but there are angular gaps of 25.1′ when mounting four sheets into a mosaicked block. This projection was used to represent regions of land and islands (978 sheets) in an International Map of the World (IMW) series at a scale of 1 : 1 000 000; sheets of this series were not made for ocean areas. The United States participated, but the Soviet Union and several other countries did not join the Convention designing the series. In the 1960s the projection was changed from the modified polyconic to the Lambert conformal conic, but the entire project was finally abandoned in the 1980s.

5.2.2 Projections for a world map at a scale of 1 : 2 500 000

A map at a scale of 1 : 2 500 000 is the unified Soviet *Karta Mira (World Map)* for the entire surface of the Earth (including oceans) with a unified scale, arrangement, design, contents, legend, and rules for transliteration. The map is composed of six zones on the Krasovskiy ellipsoid. Two polar zones (from $\pm 90°$ to $\pm 60°$ in latitude) are constructed on the polar azimuthal equidistant projection with the standard parallel $\pm 76°$, and for which the local scale factor along meridians equals 0.99. Four other zones (two for each hemisphere) are based on the equidistant conic projection:

the first zone from $\pm 24°$ to $\pm 64°$ with standard parallels $\pm 32°$ and $\pm 64°$, and the second zone from 0 to $\pm 24°$ with standard parallels $\pm 4°$ and $\pm 21°$.

Maximum distortion occurs along parallels with latitudes $\pm 60°$: $v_n = 3.7$ percent, $v_p = 2.6$ percent, $\omega = 2.6°$; along latitudes $\pm 48°$: $v_n = v_p = -3.9$ percent, $\omega = 2.2°$; along latitudes $\pm 24°$: $v_n = v_p = 4.0$ percent, $\omega = 2.3°$. Within the second zone $v_n = v_p = 1$ percent, $\omega \leq 0.7°$, where v_n and v_p are scale errors along parallels and in area, respectively (see sections 1.6.3 and 1.6.4).

The nomenclature for a sheet includes the hemisphere, the identification of the 1 : 1 000 000 map sheet, the ordinal number, and the name of the sheet; for example, KRASNOYARSK NM-0 45-48 39, where NM are Roman letters.

5.3 Conformal projections of the ellipsoid used in geodesy

The Gauss–Krüger projection, the Lambert conformal conic projection, and the stereographic or conformal azimuthal projection are very often used in surveying. In view of the fact that the geodetic ranges are usually comparatively small in area on a given quadrangle, the projection formulas, as a rule, take the form of series.

5.3.1 Gauss–Krüger projection

The rectangular coordinates for the projection can be calculated from the formulas (Morozov 1979)

$$\begin{aligned} x &= b\theta_1 - a_2\theta_2 - b_3\theta_3 + a_4\theta_4 + b_5\theta_5 - a_6\theta_6 \\ y &= s_0 + b_1\psi_1 - a_2\psi_2 - b_3\psi_3 + a_4\psi_4 + b_5\psi_5 - a_6\psi_6 \end{aligned} \tag{5.12}$$

where ψ_i, θ_i are the terms of harmonic polynomials, determined from the expressions

$$\begin{aligned} &\psi_1 = \Delta q, \quad \theta_1 = l \\ &\psi_n = \Delta q \psi_{n-1} - l\theta_{n-1}, \quad \theta_n = \Delta q \theta_{n-1} + l\psi_{n-1} \end{aligned} \tag{5.13}$$

Here,

$$\Delta q = q - q_0, \quad l = \lambda - \lambda_0$$

$$\begin{aligned} &\Delta q = t_1\Delta\phi + t_2\Delta\phi^2 + t_3\Delta\phi^3 + t_4\Delta\phi^4 + t_5\Delta\phi^5 + t_6\Delta\phi^6 + \cdots \\ &t_1 = \frac{1}{V_0^2 \cos\phi_0}, \quad t_2 = \frac{\tan\phi_0}{2V_0^4\cos\phi_0}(1 + 3\eta_0^2) \\ &t_3 = \frac{1}{6\cos\phi_0}(1 + 2\tan^2\phi_0 + \eta_0^2 + 6\eta_0^4\tan^2\phi_0 - 3\eta_0^4 + \cdots) \\ &t_4 = \frac{\tan\phi_0}{24\cos\phi_0}(5 + 6\tan^2\phi_0 - \eta_0^2 - 6\eta_0^4\tan^2\phi_0 + 21\eta_0^4 + \cdots) \\ &t_5 = \frac{1}{120\cos\phi_0}(5 + 28\tan^2\phi_0 - \eta_0^2 + 24\tan^4\phi_0 + \cdots) \\ &t_6 = \frac{\tan\phi_0}{720\cos\phi_0}(61 + 180\tan^2\phi_0 + 120\tan^4\phi_0 + \cdots) \end{aligned} \tag{5.14}$$

$$
\begin{aligned}
b_1 &= N_0 \cos \phi_0 \\
b_3 &= \tfrac{1}{6} N_0 \cos^3 \phi_0 (\eta_0^2 - \tan^2 \phi_0) \\
b_5 &= \tfrac{1}{120} N_0 \cos^5 \phi_0 (5 - 18 \tan^2 \phi_0 + \tan^4 \phi_0 + 14\eta_0^2 - 58\eta_0^2 \tan^2 \phi_0) \qquad (5.15) \\
a_2 &= \tfrac{1}{2} N_0 \sin \phi_0 \cos \phi_0 \\
a_4 &= \tfrac{1}{24} N_0 \sin \phi_0 \cos^3 \phi_0 (5 - \tan^2 \phi_0 + 9\eta_0^2 + 4\eta_0^4) \\
a_6 &= \tfrac{1}{720} N_0 \sin \phi_0 \cos^5 \phi_0 (61 - 58 \tan^2 \phi_0 + \tan^4 \phi_0 + 270\eta_0^2 - 330\eta_0^2 \tan^2 \phi_0)
\end{aligned}
$$

where

$$V_0^2 = 1 + \eta_0^2, \quad \eta_0^2 = e'^2 \cos^2 \phi_0, \quad N_0 = a/(1 - e^2 \sin^2 \phi_0)^{1/2}$$

$$\Delta\phi = \phi - \phi_0, \quad e'^2 = (a/b)^2 - 1$$

The arc length of the meridian, coefficients b_i, a_i, and differences $\Delta\phi$, Δq are calculated for the latitude ϕ_0 selected within the limits of the region being mapped.

If $\Delta\phi$ and $l \leq 4°$, then calculation with formulas (5.12)–(5.15) gives an error of not more than 0.002–0.003 m.

Formulas for the local scale factor have the form:

- as a function of a rectangular coordinate

$$m = 1 + \frac{x^2}{2R^2} + \frac{x^4}{24R^4} + \frac{x^6}{720R^6} + \cdots \qquad (5.16)$$

where

$$1/R^2 = (1 + \eta^2)/N^2$$

- as a function of geodetic coordinates

$$m = 1 + \frac{\lambda^2}{2} \cos^2 \phi (1 + \eta^2) + \frac{\lambda^4}{24} \cos^4 \phi (5 - 4 \tan^2 \phi) + \cdots$$

Meridian convergence can be determined from the formula

$$\gamma = \lambda \sin \phi + \frac{\lambda^3}{3} \sin \phi \cos^2 \phi (1 + 3\eta^2) + \cdots \qquad (5.17)$$

If planimetric rectangular coordinates are given, geodetic coordinates are usually determined first from the formulas (see section 10.2.1), and then equation (5.17) is used to calculate γ.

The correction in seconds of arc to the direction of a geodetic line for the curvature of its projection on the plane between points 1 and 2 is expressed by the formula

$$\delta''_{12} = -\rho'' \frac{\Delta y}{2R_1^2} \left(x_1 + \frac{\Delta x}{3} - \frac{x_1^2}{2R_1^2} \Delta x - \frac{x_1^3}{3R_1^2} \right) - \frac{e^2 \sin 2\phi_1}{2R_1^3} x_1^2 \Delta x$$

where $\Delta x = x_2 - x_1$, $\Delta y = y_2 - y_1$; R_1 is calculated from equation (5.16).

The calculation of the length of a chord with proper correction of the length of the geodetic line for the scale of its projection on the plane can be accomplished using the formula

$$d = s\left(1 + \frac{x_m^2}{2R_m^2} + \frac{\Delta x^2}{24R_m^2} + \frac{x_m^4}{24R_m^4} + \frac{x_m^6}{720R_m^6}\right)$$

where $x_m = (x_1 + x_2)/2$; R_m is determined from equation (5.16) at the midpoint.

The inverse relationship takes the form

$$s = d\left(1 - \frac{x_m^2}{2R_m^2} - \frac{\Delta x^2}{24R_m^2} + \frac{5x_m^4}{24R_m^4} - \frac{61x_m^6}{720R_m^6}\right)$$

where

$$d = \sqrt{(x_2 - x_1)^2 + (y_2 - y_1)^2}$$

5.3.2 Lambert conformal conic projection

Rectangular coordinates for the projection may be determined from the formulas

$$x = \rho \sin \delta, \quad y = \rho_0 - \rho \cos \delta$$

where

$$\begin{aligned} \rho &= ke^{-\alpha q}, \quad \delta = \alpha\lambda = \gamma \\ k &= \rho_0 = N_0 \cot \phi_0, \quad \alpha = \sin \phi_0 \end{aligned} \tag{5.18}$$

γ is the meridian convergence; ρ_0, α, k are constant parameters. These formulas are designed for a single standard parallel ϕ_0. This is the form used in some countries, frequently with a scale factor less than 1 along ϕ_0 providing two unstated standard parallels. See section 3.1.2 for other forms.

Instead of calculating ρ from formula (5.18) it is possible to determine it from

$$\rho = \rho_0 - \Delta\rho$$

where it is easy to find small values of $\Delta\rho$ as follows:

- when using isometric coordinates, relating equations (1.31) and (1.32),

$$\Delta\rho = \rho_0[\alpha\Delta q - \tfrac{1}{2}(\alpha\Delta q)^2 + \tfrac{1}{6}(\alpha\Delta q)^3 - \tfrac{1}{24}(\alpha\Delta q)^4 + \tfrac{1}{120}(\alpha\Delta q)^5 - \cdots] \tag{5.19}$$

- when using rectangular coordinates,

$$\Delta\rho = y - \frac{x^2}{2\rho_0} + \frac{x^2 y}{2\rho_0^2} + \cdots$$

Inversion of series (5.19) gives

$$\alpha\Delta q = \frac{\Delta\rho}{\rho_0} + \frac{1}{2}\left(\frac{\Delta\rho}{\rho_0}\right)^2 + \frac{1}{3}\left(\frac{\Delta\rho}{\rho_0}\right)^3 + \cdots$$

Local scale factors are calculated with the formula

$$\begin{aligned} m = 1 &+ \frac{V_0^2}{2N_0^2}\Delta\rho^2 + \frac{V_0^2 \tan \phi_0}{6N_0^3}(1 - 4\eta_0^2)\Delta\rho^3 \\ &+ \frac{V_0^2}{24N_0^2}(1 + 3\tan^2 \phi_0 - 3\eta_0^2 + \cdots)\Delta\rho^4 + \cdots \end{aligned}$$

or

$$m = 1 + \frac{V_0^2}{2N_0^2}y^2 + \frac{V_0^2 \tan \phi_0}{6N_0^3}(1 - 4\eta_0^2)y^3 - \frac{V_0^2 \tan \phi_0}{2N_0^3}x^2 y + \cdots$$

where $V_0^2 = 1 + \eta_0^2$.

The reduction formula takes the form

$$\delta_{12} = \frac{\rho}{6R_0^2}(x_2 - x_1)(2y_1 + y_2) = \delta_0(x - x_1)(2y_1 + y_2)10^{-10}$$

$$d - s = \frac{s}{6R_0^2}(y_1^2 + y_1 y_2 + y_2^2)$$

where δ and ρ are expressed in seconds of arc,

$$R^2 = M_0 N_0 = \frac{N_0^2}{V_0^2}, \quad \delta_0 = \frac{\rho}{6R_0^2} 10^{10}$$

5.3.3 Roussilhe stereographic projection

Rectangular coordinates for this projection, suggested in 1922 by Henri Roussilhe of France, can be determined from the formulas

$$x = c_1\theta_1 + c_2\theta_2 + c_3\theta_3 + c_4\theta_4 + c_5\theta_5 + \cdots$$

$$y = c_1\psi_1 + c_2\psi_2 + c_3\psi_3 + c_4\psi_4 + c_5\psi_5 + \cdots$$

where

$$c_1 = N_0 \cos\phi_0, \quad c_2 = -\tfrac{1}{2}N_0 \sin\phi_0 \cos\phi_0$$

$$c_3 = -\tfrac{1}{12}N_0 \cos^3\phi_0(1 + \eta_0^2 - 2\tan^2\phi_0)$$

$$c_4 = \tfrac{1}{24}N_0 \sin\phi_0 \cos^3\phi_0(2 - \tan^2\phi_0 + 6\eta_0^2)$$

$$c_5 = \tfrac{1}{240}N_0 \cos^5\phi_0(2 - 11\tan^2\phi_0 + 2\tan^4\phi_0)$$

ψ_i, θ_i are determined from formulas (5.13). To transform from the ellipsoid to the plane, the following formulas are used:

- local scale factors:

$$m = 1 + \frac{x^2 + y^2}{4R_0^2} - \frac{\tan\phi_0}{R_0^3}(2\eta_0^2 - \eta_0^4)x^2 y + \cdots$$

- meridian convergence:

$$\gamma = \rho\frac{\tan\phi_0}{N_0}x + \rho\frac{1 + 2\tan^2\phi_0}{2N_0^2}xy + \rho\frac{\tan\phi_0(3 + 4\tan^2\phi_0)}{12N_0^3}(3y^2 - x^2)x + \cdots$$

- correction for the curvature:

$$\delta_{12} = \frac{\rho}{4R_0^2}(x_2 y_1 - x_1 y_2)$$

- correction for the scale:

$$d - s = \frac{s}{12R_0^2}(x_1^2 + x_1 x_2 + x_2^2 + y_1^2 + y_1 y_2 + y_2^2)$$

where γ, ρ, δ are expressed in seconds of arc, and $R_0^2 = M_0 N_0$.

6

Map projection research

Map projection research is related to the further development of theory and practice, improving the mathematical basis of a map, obtaining new sets and variations which possess definite advantages over known projections, and satisfying new cartographic requirements facing science and the economy.

All the possible methods of obtaining projections are based on solving direct and inverse problems of mathematical cartography.

6.1 Direct and inverse problems of mathematical cartography involved in the theory of direct transformation of surfaces onto a plane

The general equations for map projection, as noted above, have the form

$$x = f_1(\phi, \lambda), \quad y = f_2(\phi, \lambda)$$

Functions f_1, f_2 are obtained in accordance with the initial requirements or conditions imposed. They are called the transformation functions, and their application makes it possible to determine the particular formulas for local scale factors and other projection characteristics based on the equations of the general theory (see section 1.3).

These equations are given in the form of formulas for

1. local linear and area scale factors:

$$m = \frac{1}{M}(x_\phi^2 + y_\phi^2)^{1/2}, \quad n = \frac{1}{r}(x_\lambda^2 + y_\lambda^2)^{1/2}$$

$$a^2 + b^2 = m^2 + n^2, \quad ab = mn \cos \varepsilon$$

$$p = \frac{1}{Mr}(x_\lambda y_\phi - x_\phi y_\lambda) = mn \cos \varepsilon = ab$$

2. meridian convergence γ:

$$\gamma = \arctan\left(\frac{x_\phi}{y_\phi}\right)$$

3. angles i of intersection on the projection between meridians and parallels, and their deviations ε from a right angle:

$$i = \arctan\left(\frac{x_\lambda y_\phi - x_\phi y_\lambda}{x_\phi x_\lambda + y_\phi y_\lambda}\right), \quad \varepsilon = i - 90^\circ$$

4. maximum angular deformation ω:

$$\sin\frac{\omega}{2}=\frac{a-b}{a+b} \quad \text{or} \quad \tan\frac{\omega}{2}=\frac{1}{2}\sqrt{\frac{m^2+n^2}{p}-2}$$

5. the relationship between azimuths or directions β on the projection and α on the ellipsoid:

$$\cot\beta=\frac{e}{h}\frac{r}{M}\cot\alpha+\frac{f}{h}=\frac{m}{n}\csc i\cot\alpha+\cot i$$

6. azimuths of cardinal directions:

$$\tan 2\alpha_0=\frac{2mn\cos i}{m^2-n^2}, \quad \tan 2\beta_0=\frac{n^2\sin 2i}{m^2+n^2\cos 2i}$$

and other characteristics.

Solutions of the direct problem of mathematical cartography provide methods for analyzing map projections. First, based on the given conditions, the transformation functions f_1, f_2 are determined. Then, based on these functions, projection characteristics are determined, and the corresponding calculations are performed.

The advantage of these methods for determining map projections is the comparative simplicity of the mathematical apparatus used for them. For developing new projections, however, the capability of these methods is limited, and the properties of the projections are revealed only after determining and analyzing the transformation functions.

Solutions of the inverse problem of mathematical cartography provide methods for determining map projections when all or a portion of the projection characteristics are given first. On the basis of these, the transformation functions may be found; on the other hand, only rectangular coordinates and other information may be provided which do not include projection characteristics.

Equations for the direct transformation of the surface onto a plane, used for solving the inverse problem of mathematical cartography, are determined as follows. Let us denote

$$\mu=mM, \quad \nu=nr \tag{6.1}$$

Then from formulas (1.55) and (1.56), we get

$$\begin{aligned} x_\phi&=\mu\sin\gamma, \quad x_\lambda=\nu\cos(\gamma+\varepsilon)\\ y_\phi&=\mu\cos\gamma, \quad y_\lambda=-\nu\sin(\gamma+\varepsilon)\end{aligned} \tag{6.2}$$

Substituting into the integrability condition of equations (6.2) derivatives from (6.2), we can obtain a set of first-order fundamental quasilinear equations containing partial derivatives:

$$\begin{aligned}\gamma_\phi&=-\varepsilon_\phi-\frac{\mu_\lambda}{\nu}\sec\varepsilon-\frac{\nu_\phi}{\nu}\tan\varepsilon\\ \gamma_\lambda&=\frac{\mu_\lambda}{\mu}\tan\varepsilon+\frac{\nu_\phi}{\mu}\sec\varepsilon\end{aligned} \tag{6.3}$$

Meshcheryakov called this the Euler–Urmayev set, which is indeterminant, since it consists of two equations with four unknowns.

Meshcheryakov considered all the possible variants for complete determinations of equations (6.3) and suggested on this basis a genetic classification of projections, differing from each other in the form of the differential equations which describe them. By introducing additional functions

$$m = z_1(\phi, \lambda), \quad n = z_2(\phi, \lambda), \quad \varepsilon = z_3(\phi, \lambda), \quad \gamma = z_4(\phi, \lambda)$$

he proved that all together 15 variables are involved in a complete determination.

The advantage of these methods for developing map projections is the possibility of using them to obtain a whole set of map projections, as well as accomplishing research on projections with these methods on the basis of the given desired properties.

However, to find these projections it is necessary to solve differential equations with partial derivatives of the first order involving elliptic, hyperbolic, parabolic, and combined functions, which is generally a rather complicated problem with cumbersome calculations.

6.2 *Equations for inverse transformation*

When calculating map projections we can proceed not only from equations (1.36) and (6.3) but also from the equations for parallels and meridians

$$\phi = f_1(x, y), \quad \lambda = f_2(x, y)$$

Using them, one can effect inverse transformation, in which the desired coordinates x, y of points on the projection are the arguments and the geodetic coordinates ϕ, λ are their functions.

N. A. Urmayev obtained formulas giving the relationship between partial derivatives $x_\phi, x_\lambda, y_\phi, y_\lambda$ for direct and $\phi_x, \phi_y, \lambda_x, \lambda_y$ for inverse transformation:

$$\begin{aligned} x_\phi &= -(1/J)\lambda_y, \quad y_\phi = (1/J)\lambda_x \\ x_\lambda &= (1/J)\phi_y, \quad y_\lambda = -(1/J)\phi_x \end{aligned} \tag{6.4}$$

where $J = 1/h = \phi_y \lambda_x - \phi_x \lambda_y$.

After substituting values of these partial derivatives in formulas for characteristics m, n, p, $\tan\gamma$, $\tan\varepsilon$, ... of the general theory of map projections, a set of differential equations of fundamental importance is obtained:

$$\begin{aligned} m^2 &= p^2 r^2(\lambda_x^2 + \lambda_y^2), \quad n^2 = p^2 M^2(\phi_x^2 + \phi_y^2) \\ \tan\gamma &= -\lambda_x/\lambda_y, \quad \tan(\gamma + \varepsilon) = \phi_x/\phi_y \\ p &= \frac{1}{Mr}\frac{1}{\phi_y \lambda_x - \phi_x \lambda_y}, \quad \tan\varepsilon = \frac{\phi_x \lambda_x + \phi_y \lambda_y}{\phi_y \lambda_x - \phi_x \lambda_y} \end{aligned} \tag{6.5}$$

Meshcheryakov called this the Tissot–Urmayev set. It can be used to obtain map projections based on solving both the direct and inverse problems of mathematical cartography.

If the equations for parallels and meridians are known or the conditions for obtaining their functions are given, the set of equations (6.5) makes it possible to solve the direct problem of mathematical cartography, i.e. to determine rectangular coordinates and the characteristics of the projection.

In case the projection characteristics or at least some of them are given, the same set (6.5) makes it possible to determine equations for parallels and meridians (or their numerical values) and the desired projections as a whole, based on solving the inverse problem of mathematical cartography. In that case set (6.5) is not linear; therefore, its solution is still more difficult than the solution of the Euler–Urmayev differential equations (6.3).

Taking this into account, Meshcheryakov obtained a set of equations analogous to that of Euler–Urmayev, but for inverse transformation:

$$(\ln v)_y \cos \gamma - (\ln v)_x \sin \gamma - \varepsilon_y \tan \varepsilon \cos \gamma + \varepsilon_x \tan \varepsilon \sin \gamma + \gamma_y \sin \gamma + \gamma_x \cos \gamma = 0$$

$$(\ln \mu)_y \sin(\gamma - \varepsilon) + (\ln \mu)_x \cos(\gamma - \varepsilon) + \varepsilon_y \frac{\cos \gamma}{\cos \varepsilon} - \varepsilon_x \frac{\sin \gamma}{\cos \varepsilon} - \gamma_y \cos(\gamma - \varepsilon) + \gamma_x \sin(\gamma - \varepsilon) = 0$$

Some specific examples of inverse formulas for common projections are given in section 10.2 and also in Snyder (1987b).

6.3 Map projection research by solving the direct problem of mathematical cartography

Map projections can be obtained by the classical analytical method (see Chapters 2, 3, 4, and 5), by perspective methods (see Chapters 1, 2, and 3), and also by other means, such as those below.

6.3.1 Combining equations of the initial projections to develop derivative projections

Projections of the same or various types can be combined. For example, Ginzburg and Malovichko suggested generalized formulas for azimuthal projections of the sphere (see section 3.2.6).

In a more generalized form these projections can be written

$$\rho = \frac{1}{k} R\left[L_1\left(\sin \frac{z}{k_1}\right)^{c_1} + L_2\left(\tan \frac{z}{k_2}\right)^{c_2} + L_3 \sin z + L_4 z + L_5 \tan z \right]$$

Given the corresponding values for the constant parameters, one can obtain various azimuthal projections with various levels of distortion.

When $c_1 = c_2 = 1$ (for all the following variants) and $k = 1$, $L_1 = k_1 = 2$, and $L_2 = L_3 = L_4 = L_5 = 0$, we obtain the Lambert azimuthal equal-area projection. When $k = 1$, $L_1 = L_3 = L_4 = L_5 = 0$ and $L_2 = k_2 = 2$, we obtain the stereographic projection. When $k = k_2 = 2$, $L_2 = L_4 = 1$, and $L_1 = L_3 = L_5 = 0$, we will get the Nell circular projection. When $k = L_4 = 1$ and $L_1 = L_2 = L_3 = L_5 = 0$, the azimuthal equidistant projection is produced. When $k = L_3 = 1$ and $L_1 = L_2 = L_4 = L_5 = 0$, we will get the orthographic projection. When $k = L_5 = 1$ and $L_1 = L_2 = L_3 = L_4 = 0$, we obtain the gnomonic projection. Projections such as the Airy and the Breusing geometric are not obtainable with this equation, however.

For cylindrical, conic and other types of projections the generalized formulas can take the form

$$X = k_1 x_1 + k_2 x_2, \quad Y = k_1 y_1 + k_2 y_2$$

where x_1, x_2, y_1, y_2 are formulas for the rectangular coordinates of one type of projection, but with different distortion, or of different projection types; k_1, k_2 (where $k_1 + k_2 = 1$) are constant parameters, the projection properties being dependent upon their relative weight. For example, Lisichanskiy obtained conformal and equal-area projections by combining pairs, respectively, of conic, cylindrical, and azimuthal projections. For equal-area projections of these combinations, the condition of equivalency was used, expressed in polar coordinates

$$x_a y_z - x_z y_a = R^2 \sin z$$

as well as Mayer's method, depending upon which one of the transformation functions (x or y) is given.

The resulting differential equation was solved by numerical methods for y-values, which were obtained from linear combinations of the initial projections. One of the Hammer projections and the Winkel tripel projection are also examples of different types of projection combinations. In the former projection, presented in 1900, abscissas are determined as the arithmetic mean values of the sinusoidal or Sanson–Flamsteed equal-area and the cylindrical equal-area projections:

$$x = R\lambda \cos^2\left(\frac{\phi}{2}\right)$$

Using the condition for equivalency and assuming that ordinates do not vary with longitude, we find that

$$y = 2R\left(\phi - \tan\frac{\phi}{2}\right)$$

The formulas for the Winkel tripel projection, presented by Oswald Winkel in 1921, are determined as the arithmetic mean values of coordinates for the equidistant cylindrical and Aitoff projections (see below):

$$x = \tfrac{1}{2}(Rk\lambda + x_A), \quad y = \tfrac{1}{2}(R\phi + y_A)$$

where k is a constant coefficient, and x_A, y_A are coordinates of the Aitoff projection. Setting the local scale factor along the equator at $n = 0.85$, V. V. Kavrayskiy obtained a value of $k = 0.7$.

The Winkel tripel projection is widely used for world maps in the United Kingdom and other countries. On this projection the percentage of area distortion is less than the degrees of angular distortion.

6.3.2 Aitoff method for derivative projections

In 1889, the Russian David Aitoff suggested a method for creating a projection for world maps on which all abscissas of the initial projection are doubled and meridians are denoted by doubling the corresponding longitudes.

He used the transverse or equatorial aspect of the azimuthal equidistant projection, the formulas for which, incorporating the transformations mentioned above,

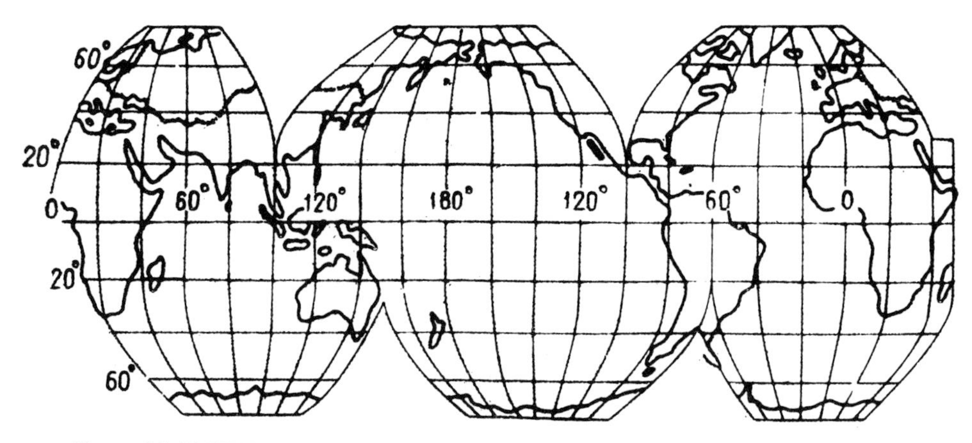

Figure 6.1 TsNIIGAiK compound interrupted equal-area projection. Exact design not available. 20° graticule.

take the form

$$x = 2Rz \sin a, \quad y = Rz \cos a$$

where z, a are polar spherical coordinates, determined by equations (1.13) with doubled longitudes.

Using this method, Hammer, beginning with the equatorial aspect of the Lambert azimuthal equal-area projection and doubling the abscissas and values of longitude, developed an equal-area projection for world maps, called the Hammer–Aitoff or just the Hammer projection:

$$x = \frac{B\sqrt{2}\,R \cos\phi \sin(\lambda/B)}{\eta}, \quad y = \frac{\sqrt{2}\,R \sin\phi}{\eta}$$

where

$$B = 2, \quad \eta = \sqrt{1 + \cos\phi \cos(\lambda/B)}$$

In 1926 Rosén chose a coefficient for B of 8/7, and later Eckert (by then Eckert-Greifendorff (1935)) used $B = 4$. M. D. Solov'ev developed formulas for the general case of the Hammer–Aitoff projection and considered a variant for which the east–west extension involved the coefficient $B = 1.6$.

There have also been variants developed for projections with extensions along not one but two directions. These are projections by Karl Siemon, Wagner, and E. Kremling (Ginzburg and Salmanova 1964, pp. 178–9, 243).

6.3.3 Connecting known projections

Several methods for connecting previously developed projections have been used.

V. V. Kavrayskiy proposed a combination of the Mercator conformal cylindrical projection, used for the zone between $\phi = \pm 70°$, and an attached equidistant cylindrical projection for higher latitudes.

To obtain a projection by N. A. Urmayev's method, it is assumed that two zones of the ellipsoid or sphere without common boundaries are plotted on different

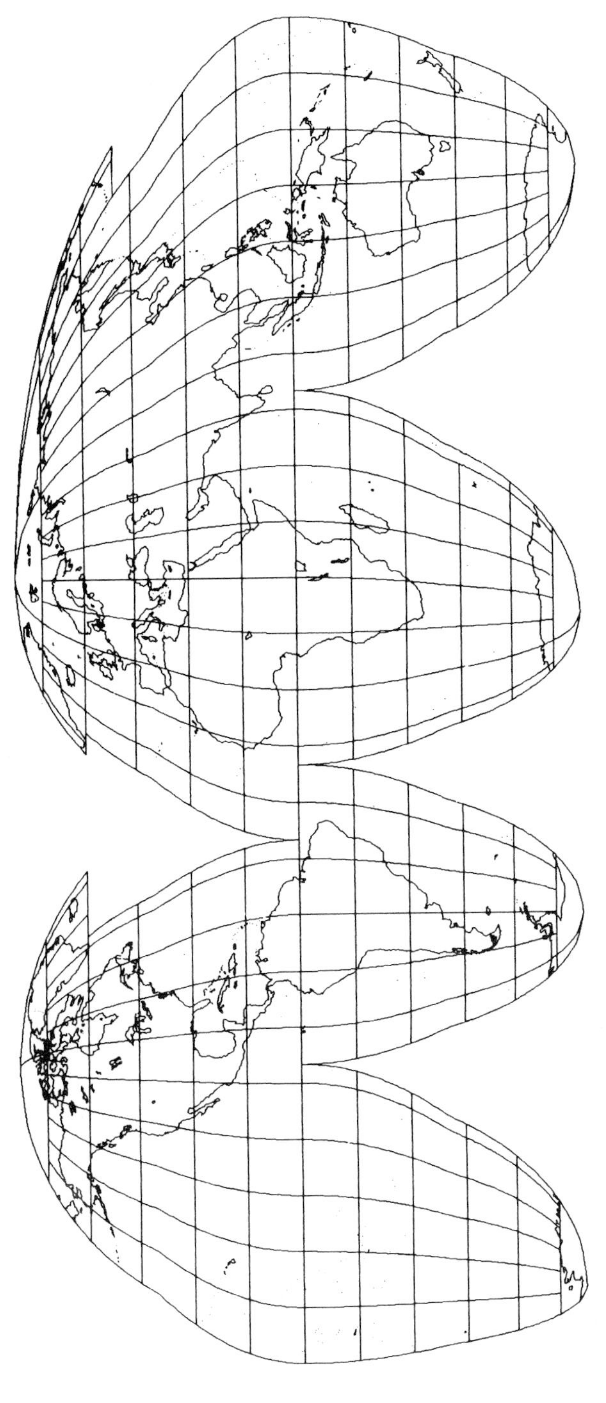

Figure 6.2 Goode homolosine equal-area projection, interrupted for continents. 15° graticule.

projections with smooth meridians and parallels. To determine the coordinates of points for an intermediate zone on the projection, an analytical relationship is established providing that three parallels of the same zone (in three zones) are tangential to common points.

The creation of a projection by Goode's method of interruption may be applied to any pseudocylindrical projection on which distortion near the central meridian is small and increases considerably with an increase of distance from it. To represent each continent (or else ocean), only the central part of the projection with its straight central meridian is used, and other portions are joined along the equator. Gaps occur through the oceans (or else continents) (Figure 6.1).

After first interrupting the Sanson–Flamsteed sinusoidal projection in this manner in 1916, Goode then used the equal-area Mollweide (or homolographic) pseudocylindrical projection to represent the world zone by zone. Later, for some world maps he suggested (Goode 1925) a projection called 'homolosine' (Figure 6.2), on which each zone or lobe consists of the sinusoidal projection (near the equator) and Mollweide projection (nearer the poles), joined along parallels with latitudes of approximately $\pm 40°44'$, at which the linear scale along parallels is the same. As a result, a noticeable meridian break appears at the joints of the two projections. The formulas are the same for the respective portions as those for the original projections (see section 2.2.2), except that the ordinates for the Mollweide are moved $0.052\,80R$ units closer to the equator to match the sinusoidal where they join.

Unlike the case with the Goode projections discussed above, the equator can be represented by a curved line in compound projections. These include the 'Regional', 'Tetrahedral', and 'Lotus' projections, developed by John Bartholomew in Scotland. For example, on the 'Lotus' projection used for world maps with gaps through landmasses, the graticule consists of an equidistant conic projection in the central part and a pseudoconic projection on each of the three petals or lobes.

Star projections possess analogous properties and are used as emblems in Russia and in the United States. Among them, for example, is the Berghaus star projection (Figure 6.3), with a compound graticule on which the central northern hemisphere is composed of the polar azimuthal equidistant projection. The southern hemisphere is shown as five identical radial parts of a pseudoconic projection on which parallels are circular arcs concentric with the parallels of the northern hemisphere and the meridians are straight and equally spaced along the equator.

6.3.4 Map projections using series and harmonic polynomials

Projections obtained using these methods belong to classes with adaptable isocols (see section 7.3), closely approaching Chebyshev's principles for conformal projection. These projections have a comparatively simple mathematical structure, and low values of distortion occur when they are used to map small and medium-sized regions.

6.3.5 Map projections developed by introducing additional constants into functions

A number of projection variations, especially pseudocylindrical and pseudoconic, were developed by F. A. Starostin and others of TsNIIGAiK. As an example, the formulas suggested by Ginzburg for calculating a conic projection with a given

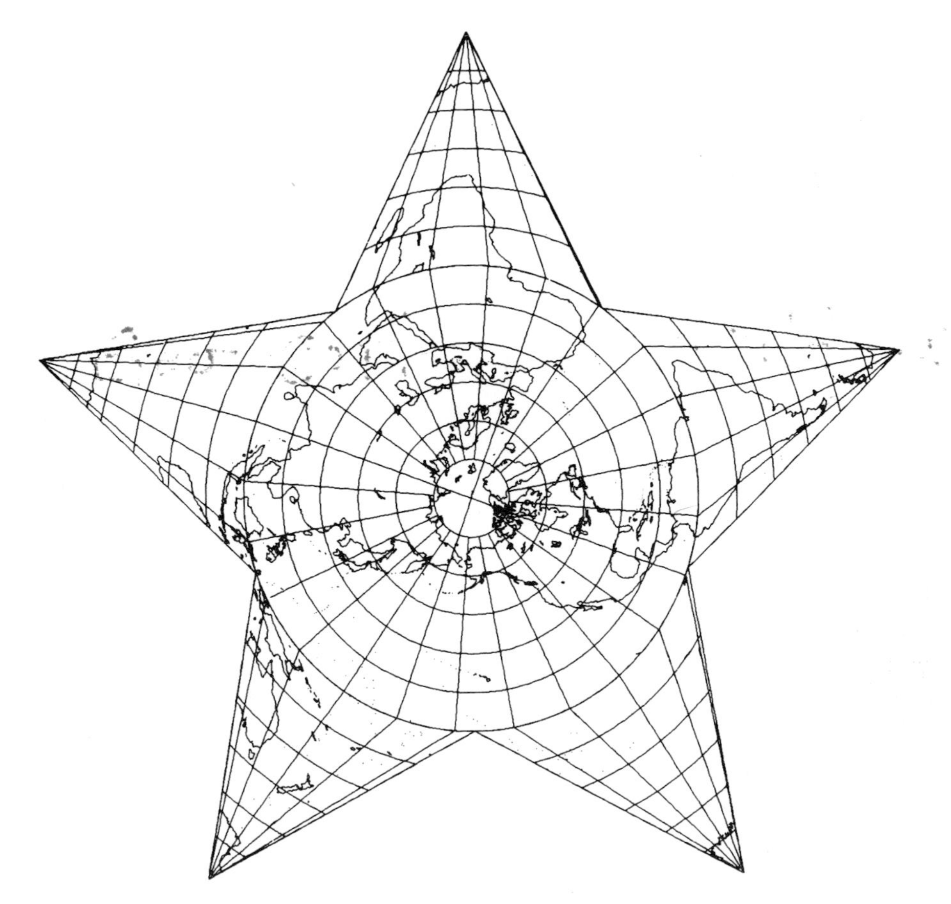

Figure 6.3 Berghaus star projection. 15° graticule.

distortion pattern are given:

$$\rho = C - \left(m_0 s + \frac{kM_0}{3} (\phi - \phi_0)^3 \right)$$

where C, k are constants and m_0 is the local linear scale factor along meridians at points with minimum scale along parallels.

6.3.6 Graphical and graphoanalytical methods of obtaining map projections

Graphical methods have seldom been used in recent years, especially since the advent of computers. We will briefly dwell on some projections created by these methods.

Al-Bīrūnī or Nicolosi projection

This is called a globular projection (Figure 6.4) because of its portrayal of a hemisphere bounded by a circle. An equatorially centered hemisphere is projected within

Figure 6.4 Al-Bīrūnī (Nicolosi) globular projection of eastern hemisphere. 15° graticule.

a circle of radius $\pi R/2$, with Earth radius R taken at the scale of the map, and on which two mutually perpendicular diameters are drawn. One of the diameters is taken to be the equator, the other is the central meridian, and the outer circle constitutes the outer meridians with longitude $\pm 90°$ from the central meridian.

The graticule is obtained by dividing both diameters and each quadrant of the circle into equal parts, and then drawing circular arcs through the three points, one along each of the three meridians for the parallels of latitude and through the points at the poles and the markings along the equator for the meridians.

The projection was suggested in the eleventh century by the Arab mathematician al-Bīrūnī. In 1660 G. B. Nicolosi of Italy again proposed it, and in the eighteenth century Aaron Arrowsmith in England applied this projection, leading to the names al-Bīrūnī, Nicolosi, Arrowsmith, English, and globular applied by various writers.

In view of the pattern of its graticule, the projection is one of the circular polyconic projections.

Apian projections

Two projections were suggested for equatorially centered hemispheres by German geographer Peter Apian in 1524. For the construction of each, a circle of arbitrary radius is drawn with two mutually perpendicular diameters representing the central meridian and the equator, as on the al-Bīrūnī projection above.

These diameters are divided into equal parts, and meridian lines are drawn through the two pole points and the points on the equator, as circular arcs on his first projection and as semiellipses on his second. The parallels of latitude are straight lines drawn through the points along the central meridian, orthogonal to it.

The pattern of the map graticule makes the projections nearly pseudocylindrical, except that meridians are not equally spaced on his first projection. The second projection was used independently by François Arago in the nineteenth century, and it is thus sometimes given his name.

Bacon projection

This was apparently suggested in the thirteenth century by Roger Bacon, although his design is not clear. In the sixteenth century maps based on the projection were prepared by Franciscus Monachus. The meridians are constructed in the same way as for the al-Bīrūnī and the first Apian projections. The parallels are parallel straight lines drawn orthogonal to the central meridian through points equally spaced along the outer ($\pm 90°$) meridians, and thus at a distance from the equator of

$$y_0 = \frac{\pi}{2} R \sin \phi$$

This is also a near-pseudocylindrical projection based on the pattern of its map graticule.

The al-Bīrūnī, Apian, and Bacon projections were intended for the design of maps of hemispheres. Mathematical descriptions, tables of scale factors, and distortion on these projections are given by Vitkovskiy (1907). Some formulas are also available in Snyder and Voxland (1989, p. 234).

Müffling projection

This projection was constructed using straightened arc segments of parallels and meridians (for 1 : 200 000 and larger-scale topographic maps) by the method of resection and intersection. The projection was used for quadrangle sheets in designing the topographic maps of the USSR (and earlier of Germany) before the use of the Gauss–Krüger projection. Within each sheet designed using the Müffling projection, distortion was insignificant, but in mosaicking blocks of sheets for mounting, linear and angular gaps of ε' appeared, where (in arc minutes)

$$\varepsilon' = \frac{1}{\rho'} \Delta\phi' \, \Delta\lambda' \cos \phi_{\text{cen}}$$

The trapezoidal pseudocylindrical projection is an analog of this projection (see section 5.1.1).

Graphoanalytical methods were also used for making cylindrical, conic, azimuthal, and other projections. At present this method is used mainly to create modified polyconic projections, based on approximating a sketch of the map graticule (see section 4.2.4).

The creation of these projections is based on the methods of inverse solution in mathematical cartography, which are considered in section 6.4.4.

6.3.7 Transforming an initial projection

As an illustration, let us provide the standard formulas for the projective transformation of one plane surface to another:

$$x = \frac{a_1 X + a_2 Y + a_3}{c_1 X + c_2 Y + c_3}, \quad y = \frac{b_1 X + b_2 Y + b_3}{c_1 X + c_2 Y + c_3}$$

where X, Y and x, y are rectangular coordinates of points of the initial and desired projections, respectively; a_i, b_i, c_i are constants determined from the conditions given. These formulas provide a linear transformation to 'stretch' one projection to fit another frame. Most modified projections involve a non-linear transformation.

6.3.8 Developing modified projections

Map projections can be obtained by modifying equations of known projections. As examples we will describe some of these projections.

Modified transverse Mercator projection

This projection was created within the US Geological Survey as a modification of the transverse Mercator projection in 1972, to use solely for maps of Alaska. It is so called because it consisted of placing UTM zones side by side and making adjustments. Actually it is closer to the ellipsoidal equidistant conic projection and has a constant scale factor of about 0.9992 along all meridians (Snyder 1987b, pp. 64–5). Approximate projection formulas have the form

$$x = \rho \sin \delta, \quad y = 1.561\,6640 - \rho \cos \delta$$

where

$$\delta = 0.862\,511\,1(\lambda^\circ + 150^\circ)$$

$$\rho = 4.132\,040\,2 - 0.044\,417\,3\phi^\circ + 0.006\,481\,6 \sin 2\phi$$

Miller cylindrical projection

This projection (Figure 6.5), with an arbitrary distortion pattern, was suggested by O. M. Miller (1942) for world maps. It is a modified Mercator projection, produced by using the Mercator formulas to calculate y for 0.8 of the latitude and then dividing the y-values by 0.8. It is not perspective but resembles the Gall perspective cylindrical projection (see section 2.1.7), but has no distortion along the equator, rather than along Gall's choice of $\pm 45^\circ$ latitude.

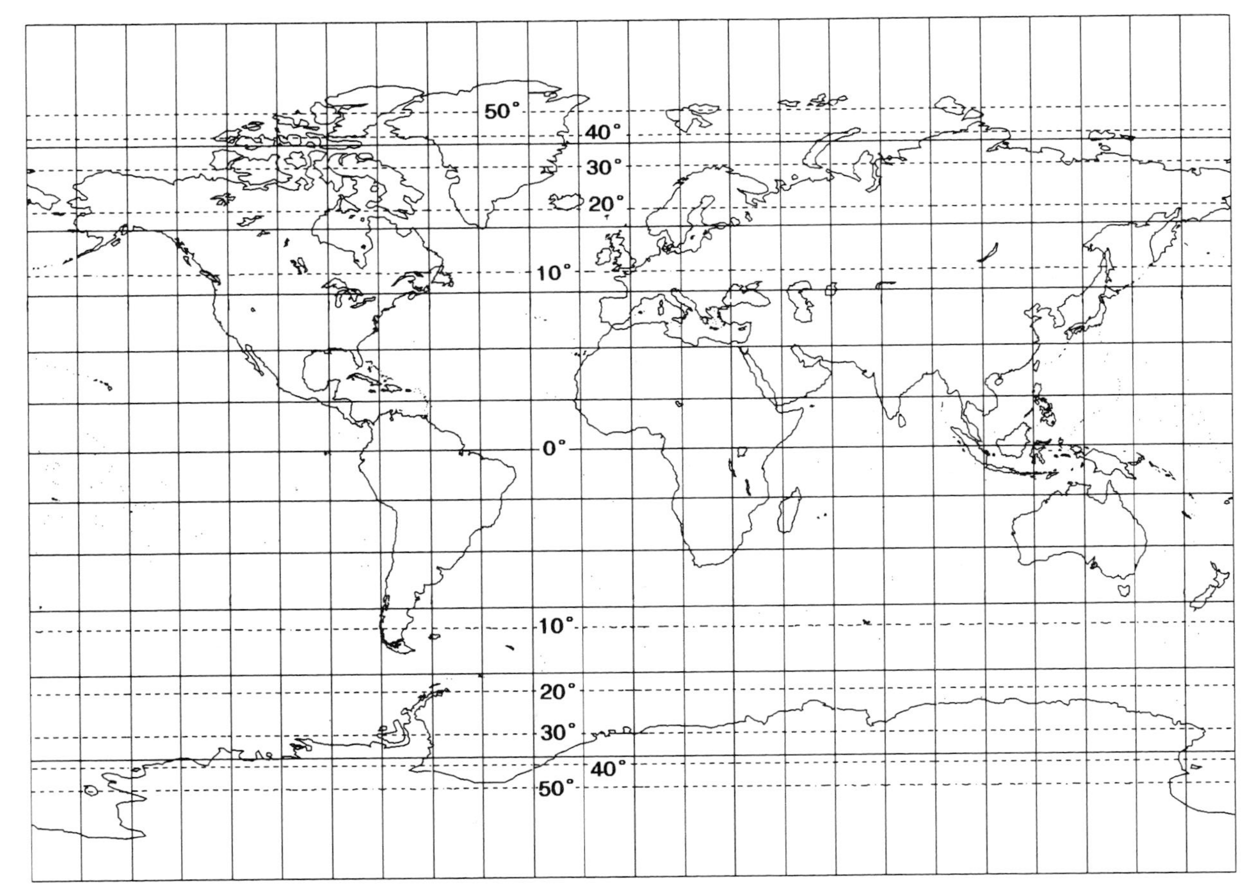

Figure 6.5 Miller cylindrical projection, with isocols for ω. 15° graticule.

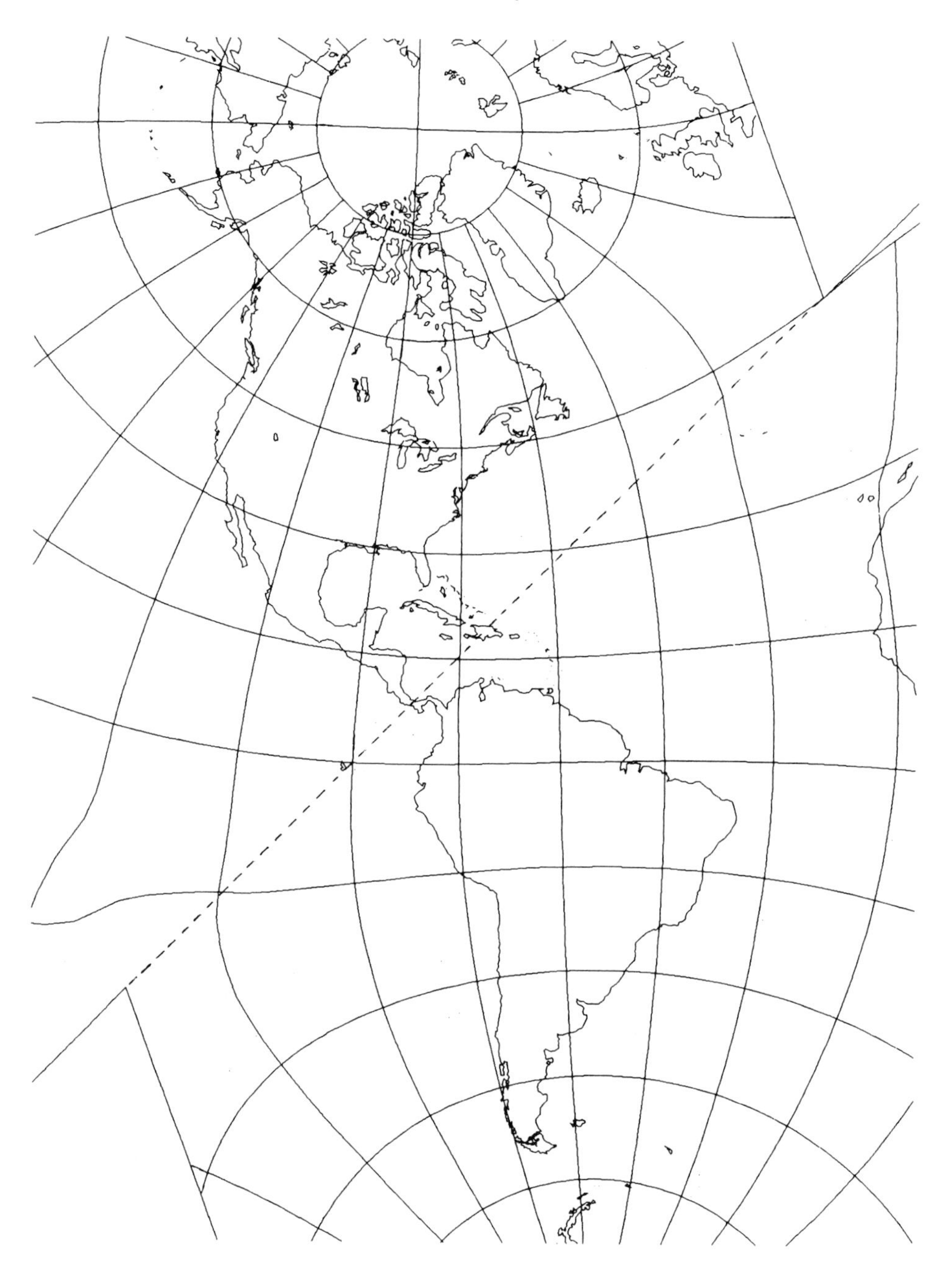

Figure 6.6 Bipolar oblique conformal conic projection of the Americas. 15° graticule.

Projection formulas have the form

$$x = R(\lambda - \lambda_0), \quad y = R[\ln \tan(\pi/4 + 0.4\phi)]/0.8$$

$$m = \sec 0.8\phi, \quad n = \sec \phi,$$

$$\sin(\omega/2) = (\cos 0.8\phi - \cos \phi)/(\cos 0.8\phi + \cos \phi)$$

Ordinates can also be calculated by the formulas

$$y = R[\operatorname{arcsinh}(\tan 0.8\phi)]/0.8$$

or

$$y = (R/1.6)\ln[(1 + \sin 0.8\phi)/(1 - \sin 0.8\phi)]$$

Double- and multiple-polar projections

Such projections either may provide a continuous transformation, or they may be compound. Examples of the first are the double- and triple-polar projections used to create anamorphous maps with modified geometry. A sufficiently detailed description of this method for creating a double-polar projection for these maps is given in Suvorov and Bugayevskiy (1987).

An example of other methods is a bipolar projection, its development supplementing the method described in section 6.3.5.

Bipolar oblique conic conformal projection of the sphere

This is a compound projection (Figure 6.6) comprising two conformal oblique conic projections. The poles of the two conic projections lie at the northerly end (for North America) and southerly end (for South America) of a great-circle arc 104° long linking the projections ($\phi_{01} = 20°$S, $\lambda_{01} = 110°$W; $\phi_{02} = 45°$N, $\lambda_{02} \approx 20°$W). The combination projection was suggested by O. M. Miller (1941) and W. A. Briesemeister of the American Geographical Society for designing a combined map of North and South America, because of the tendencies of the continents to curve in opposite directions. The linear scale on the two almucantars with zenithal distances of 31° and 73° for both the southern and northern parts is constant, but with a scale factor of 1.035. The parts are joined along the above great-circle arc, but there is a narrow band of non-conformality along this arc; the lengthy formulas are given in Snyder (1987b, pp. 116–23).

6.4 Map projection research by solving the inverse problem of mathematical cartography

Methods for solving this problem are based mainly on the use of the Euler–Urmayev and Tissot–Urmayev equations.

6.4.1 Deriving projections by solving the Euler–Urmayev equations

The approximate solution of the Euler–Urmayev differential equations (6.3), using the method of approximations, is described by Vakhrameyeva *et al.* (1986) for the general case. The solution of these equations is demonstrated with the development of a Chebyshev projection as an example of an application involving conformal projections (see section 7.2). Examples of solutions to these equations are also mentioned in section 7.5 while dealing with the problem of obtaining map projections with orthogonal map graticules.

Let us consider another method of obtaining conic projections with various distortion patterns and their generalization. For these projections the Euler–Urmayev equations can be written in the form

$$\gamma_\phi = -\mu_\lambda/\nu, \quad \gamma_\lambda = \nu_\phi/\mu \tag{6.6}$$

where, as in (6.1),

$$\mu = mM, \quad \nu = nr$$

To set up conditions for integrating equations (6.6), and taking into account the fact that on conic projections the local linear scale along meridians is a function only of the latitude, we find that

$$\nu_{\phi\phi}/\mu - \nu_\phi \mu_\phi/\mu^2 = 0 \tag{6.7}$$

The first integral of this equation is

$$(1/\mu)\nu_\phi = \alpha \tag{6.8}$$

For equal-area conic and generalized conic projections where the area scale factor along the central meridian is unity, we have, respectively,

$$m = r/\nu \quad \text{and} \quad m = r/\nu_0$$

Expressions (6.6) and (6.8) take the form

$$\gamma_\phi = 0, \quad \gamma_\lambda = \alpha$$

$$\frac{\nu\nu_\phi}{Mr} = -\alpha \quad \text{and} \quad \frac{\nu_0\,\nu_\phi}{Mr} = -\alpha$$

The integration of these equations gives the well-known equal-area conic projection and a modified conic projection close to equal area. Formulas for the latter have the form

$$\gamma = \delta = \alpha\lambda(1 + \tfrac{1}{3}b_2\lambda^2 + \tfrac{1}{5}b_4\lambda^4 + \cdots)$$

$$\nu^2 = r^2n^2$$

$$n^2 = 2\alpha(c - S)(1 + b_2\lambda^2 + b_4\lambda^4 + \cdots)/r^2$$

$$m^2 = r^2/[2\alpha(c - S)], \quad \rho^2 = 2(c - S)/\alpha$$

where α, c, b_i are constant parameters.

To obtain conformal projections, the differential equations (6.6) can be presented in isometric coordinates:

$$\gamma_q = -(\ln\mu)_\lambda = 0, \quad \gamma_\lambda = (\ln\mu)_q = \alpha$$

Integration of these equations gives

$$\mu = \alpha c/U^\alpha, \quad \gamma = \delta = \alpha\lambda$$

where U is found from (1.31)–(1.32); or in a more general form

$$\gamma = \delta = \alpha\lambda(1 + \tfrac{1}{3}b_2\lambda^2 + \tfrac{1}{5}b_4\lambda^4 + \cdots)$$

By representing equations (6.2) in an isometric form, we obtain equations in complete differentials, and after their integration we have the Lambert conformal conic projection and a modified conic projection that is close to conformal and on which the central meridian is a line of conformality:

$$\rho = c/U^\alpha, \quad \delta = \alpha\lambda(1 + \tfrac{1}{3}b_2\lambda^2 + \tfrac{1}{5}b_4\lambda^4 + \cdots)$$

$$m = \frac{\alpha c}{rU^\alpha}, \quad n = \frac{\alpha c}{rU^\alpha}(1 + b_2\lambda^2 + b_4\lambda^4 + \cdots)$$

When obtaining a generalized equidistant conic projection (equidistant along meridians), equations (6.6) take the form

$$\gamma_\phi = 0, \quad \gamma_\lambda = -v_\phi/M$$

The solution of equation (6.7), taking into account (6.1), can be given in the form

$$v = \alpha(c - s)f(\lambda)$$

where

$$f(\lambda) = 1 + b_2\lambda^2 + b_4\lambda^4 + \cdots$$

and further

$$\gamma = \delta = \alpha(\lambda + \tfrac{1}{3}b_2\lambda^3 + \tfrac{1}{5}b_4\lambda^5 + \cdots), \quad \rho = c - s$$

Formulas for local scale factors take the form

$$m = 1, \quad n = p = \frac{\alpha(1 + b_2\lambda^2 + \cdots)(c - s)}{r}$$

Rectangular coordinates for all these projections are calculated with the formulas common to conic projections (section 3.1.1).

Note that generalized cylindrical and azimuthal projections may be readily obtained with the same method.

6.4.2 Deriving projections by solving the Tissot–Urmayev equations

There has been no general solution of these equations so far. In 1953 N. A. Urmayev developed a theory of projections on which parallels are represented by equidistant curves, provided that the ratio of the local linear scale along parallels to the local area scale is a function only of the latitude, i.e. $n/p = f(\phi)$. To transform a sphere with a unit radius, one of the formulas of (6.5) takes the form

$$\phi_x^2 + \phi_y^2 = g^2(\phi) \tag{6.9}$$

To integrate this equation two methods were suggested. The first method used is the complete Lagrange integral (Urmayev 1953).

In the second method Urmayev, introducing the symbols $\phi_x = u_x g$, $\phi_y = u_y g$, and $t = \tan \tau$, presented equation (6.9) in the form

$$u_x^2 + u_y^2 = 1$$

and then

$$u_x = -t/\sqrt{1+t^2}, \quad u_y = \cos \tau = 1/\sqrt{1+t^2}$$

where τ is the angle formed by the normal to the parallel and the Y-axis.

Differentiating the first formula with respect to y and the second with respect to x we obtain, instead of the non-linear equation (6.9), a linear equation

$$t_y - tt_x = 0$$

Establishing a set of ordinary differential equations

$$dy = -dx/t = dt/0$$

we find the first integral $t = c_1$, and then the second and the general integrals

$$x + ty = c_2, \quad x + ty = f(t)$$

It follows from the formulas obtained that for these map projections, the parallels are equidistant curves and trajectories (meridians) orthogonal to them are a family of straight lines with one parameter t.

6.4.3 Deriving map projections with a given curvature of meridians, parallels, or a geodetic line

Equations for the curvature of meridians and parallels, respectively, can be written in the form

$$k_M = \gamma_\phi/\mu = -\frac{1}{\mu}\left((\mu_\lambda \sec \varepsilon + v_\phi \tan \varepsilon)\frac{1}{v} + \varepsilon_\phi\right)$$

$$k_P = (\gamma + \varepsilon)_\lambda/v = \frac{1}{v}\left((\mu_\lambda \tan \varepsilon + v_\phi \sec \varepsilon)\frac{1}{\mu} + \varepsilon_\lambda\right)$$

The average curvature of geodetic lines for map projections with any distortion pattern can be determined from the formula

$$k_{\text{ave}} = \frac{1}{2r}\left[\sin \phi\left(\frac{m}{n^2}\sin i - \frac{1+\cos i}{m \sin i}\right) + \frac{1+\cos i}{mn \sin i}\left(n_\phi \frac{r}{M} - m_\lambda\right) - \frac{i_\phi}{mnM}\right]$$

where i is again the angle of intersection on the projection between meridians and parallels.

Applying these formulas, one can obtain various methods for finding projections, depending on the given properties. Let us consider two methods of finding map projections from their given properties.

The method of obtaining a projection with equidistant parallels preserving scale along a given parallel with $\phi_0 = 50°$ was suggested by N. A. Urmayev.

The equation

$$k_P = 1/\rho = a_0 + a_1 s + a_2 s^2 + \cdots$$

in a particular case will have the form

$$k_P = \frac{1}{a} + \frac{6}{5}\frac{s^2}{a^3}$$

Taking into account the fact that $d\tau/ds = k_P$, after integrating we get

$$\tau = \frac{s}{a} + \frac{2}{5}\left(\frac{s}{a}\right)^3$$

where $a = 2R$ (with a scale of 1 : 10 000 000, $a = 127.4223$ cm, using a sphere equal in area to that of the Krasovskiy ellipsoid); s is the length of the arc of the parallel.

Assuming that for a given parallel ($\phi_0 = 50°$) $n = 1$, Urmayev compiled a table of values of s, k_P, and τ for points along this parallel at equal intervals of longitude.

Using these data and the known relationships

$$dx/ds = \cos\tau, \quad dy/ds = \sin\tau$$

the values of the integrals are determined by numerical methods:

$$x = \int \cos\tau\, ds, \quad y = \int \sin\tau\, ds$$

This results in rectangular coordinates of points along the given parallel. The coordinates of points along the other parallels are calculated with the formulas

$$x = x_0 + u\sin\tau, \quad y = y_0 + u\cos\tau$$

where $u = R$ arc $\Delta\lambda$; $\Delta\lambda = 10°$. Local linear scale factors along the parallels are determined from the formula

$$n = (\cos\phi_0/\cos\phi)[1 - k_P R(\phi - \phi_0)]$$

where k_P is the curvature of the given parallel.

The second method is that for obtaining conformal projections with a given curvature of meridians and parallels. On conformal projections

$$K_M = [\ln\mu]'_\lambda/\mu = \frac{\left(-\sum_{i=2}^{k} A_i\tau_i + \sum_{i=1}^{k} C_i T_i\right)}{\exp\left(\sum_{i=0}^{k} A_i\psi_i + \sum_{i=1}^{k} C_i\theta_i\right)}$$

$$K_P = -[\ln\mu]'_q/\mu = \frac{-\left(\sum_{i=1}^{k} A_i T_i + \sum_{i=2}^{k} C_i\tau_i\right)}{\exp\left(\sum_{i=0}^{k} A_i\psi_i + \sum_{i=1}^{k} C_i\theta_i\right)} \tag{6.10}$$

where K_M, K_P are the given values of curvature of the plotted meridians and parallels at a point on the projection.

The problem of deriving a projection is reduced to calculating the constant coefficients A_i, C_i in the solution of the Laplace equation (7.1). However, finding the values of the coefficients directly from formulas (6.10) using the given curvature of the meridians and parallels is difficult; therefore, values A_i, C_i can be iterated using the following approximate formulas:

$$1 + \ln K_M = -\sum_{i=0}^{k} A_i'(\psi_i + \tau_i) - \sum_{i=1}^{k} C_i'(\theta_i - T_i)$$

$$\Delta \ln K_M = -\sum_{i=0}^{k} \Delta A_i'(\psi_i + \tau_i/[\ln \mu]_\lambda') - \sum_{i=1}^{k} \Delta C_i'(\theta_i - T_i/[\Delta \ln \mu]_\lambda') \qquad (6.11)$$

$$1 + \ln K_P = -\sum_{i=0}^{k} A_i'(\psi_i + T_i) - \sum_{i=1}^{k} C_i'(\theta_i + \tau_i)$$

$$\Delta \ln K_P = -\sum_{i=0}^{k} \Delta A_i'(\psi_i + \tau_i/[\ln \mu]_q') - \sum_{i=1}^{k} \Delta C_i'(\theta_i + T_i/[\Delta \ln \mu]_q') \qquad (6.12)$$

The sequence of obtaining a conformal projection by iteration, e.g. from the given curvature of parallels K_P, is as follows:

1. Given the values of curvature K_P, compile and solve the set of equations (6.11); as a result, we obtain A_i', C_i'.
2. Applying these values, calculate the derivatives $[\ln \mu]_q'$, and from formulas (6.10) calculate the value of the curvature of the parallel K_P. Then

$$\Delta \ln K_P = \ln K_P - \ln K_P'$$

3. Determine the corrections to coefficients A_i', C_i' by solving the set of equations (6.12), and then calculate more precise values of the coefficients

$$A_i = A_i' + \Delta A_i', \quad C_i = C_i' + \Delta C_i'$$

4. Using the values obtained for coefficients A_i, C_i, the calculations are repeated until we obtain $\Delta \ln K_P \le \varepsilon$, where ε is a permissible limit determined by the desired accuracy of the computation.

After obtaining the final values of the coefficients, we continue to calculate the projection using the known formulas (see section 7.2).

6.4.4 Deriving projections from approximation sketches of map graticules

The problem of developing these projections is solved in two stages. The first one proceeds from the purpose of the map, the results of studies of the map graticule, and the distortion on projections used for the available maps. A model of the map graticule is constructed (usually on graph paper), satisfying the given specific requirements as much as possible. Using numerical methods, we calculate scale, distortion, and other parameters of the model and if necessary introduce the required corrections or construct a new model.

The second stage involves smoothing of the rectangular coordinates of the intersections of meridians and parallels on the sketch (normally by correlation). The approximation sketch of the map graticule is then made from these smoothed coordinates.

Various approximation relationships can be used for this purpose, e.g. the polynomials dealt with in the works of Boginskiy (1972), Bugaveyskiy and Portnov (1984), and others, as applied to cartographic transformations and their development. For example, when Boginskiy applied the method to finding projections with a random distortion pattern, he made substantial use of the algebraic power polynomials

$$x = \sum_{i=0}^{k} \sum_{j=0}^{k} a_{ij} \phi^i \lambda^j, \quad y = \sum_{i=0}^{k} \sum_{j=0}^{k} b_{ij} \phi^i \lambda^j$$

where ϕ, λ are geographical coordinates and a_{ij}, b_{ij} are constant coefficients.

Taking into account the manner in which the poles are represented and the symmetry of the map graticule of the new projection with respect to the central meridian and the equator, these polynomials are given the corresponding specific forms. In particular, when creating projections symmetrical about the central meridian, the y-values are calculated from the even power functions of the longitudes, and the x-values are calculated from the odd power functions of the longitudes. For projections symmetrical about the equator, the y-values are based on the odd power functions of the latitudes, and the x-values are based on the even power functions of the latitudes.

To create projections with the poles represented as points or straight lines for which the ordinate is a function only of the latitude, x in the first case is calculated from a polynomial with zero values at the poles, and in the second case as a function of two arguments (see section 1.2).

7

Best and ideal map projections; projections satisfying given conditions of representation

7.1 General conditions for the best and ideal projections

Determining the best and ideal projections is the most important independent problem in the theory and practice of creating and approximating map projections, as it is concerned not only with the direct improvement of the mathematical basis of maps, but also with finding the limits of possible improvement in the properties of map projections.

The best projections can be found either from a list of a particular type, or from the entire limitless set of projections. In the first case it is necessary to determine first from what particular class we are to select the best projection, and what criteria should serve as the basis for choosing the class.

All 'best' projections can be divided into minimax and variational (or least-squares) types of projections from the viewpoint of minimal distortion on maps. For a minimax type Chebyshev's criterion holds: the best projection for a given region is the one for which the maximum value of the logarithm of one of the various types of scale factors is a minimum.

In the case of a variational-type projection, the determination is reduced to solving variational problems on the extreme condition. When doing so the criteria suggested by Airy, Jordan, Klingach, V. V. Kavrayskiy, Konusova, and others allow estimation of projection distortion both at a single point and for the whole area represented.

However, in cartographic practice it often happens that the controlling factor for map projection selection is not the value and pattern of distortion, but some other factors or their combinations. Therefore, imposing the problem in a broad sense, we should note that the best projections can be of two types:

1. The best projections are those minimizing and optimizing distortion of a minimax or variational type.
2. The best projections are those satisfying in an optimum way an entire group of requirements for projections in accordance with the particular purpose of the map being designed (e.g. graticule simplicity, distortion values, etc.).

In the latter case the desired quality of the projection may be achieved by using generalized criteria (see section 10.3).

Of the various categories of best map projections, the problem of creating projections providing minimum distortion has been essentially completely solved for conformal projections, but is not adequately solved for other types of projections.

Resolution for conformal projections is based on finding Chebyshev projections using the inverse problem of mathematical cartography. Conformal projections that are close to ideal can also be created by solving the direct problem of mathematical cartography, e.g. through methods developed by L. A. Vakhrameyeva.

Solving the problem of creating the best equal-area projections is in its initial stages. There are no works at present which have even a general solution of this problem. Meshcheryakov considered the particular case of creating the best Euler projections, i.e. equal-area projections with an orthogonal map graticule, but the author considered only the variant in which the initial conditions were referenced to the straight central (zero) meridian. Dyer and Snyder (1989) applied iterative functions to develop general formulas for minimum error equal-area projections, with a design for Alaska as an example.

Lisichanskiy (1970a, 1970b) developed combined systems of conformal and equal-area azimuthal–cylindrical and azimuthal–conic projections possessing certain advantages. These projections are shown in the works of V. V. Kavrayskiy and Meshcheryakov not to be the best possible solutions.

Yuzefovich suggested an entire class of map projections in which $m = n^k$, where k is a constant for the given projection variant. Topchilov and Yuzefovich considered some particular solutions creating projections that are close to conformal.

An analysis of all the methods known for determining the best projections shows that at present two principal methods for solving this problem have been formulated. For one of them, used to develop the best conformal projections, the theorem for these projections was first formulated and then proved.

In the other method, the set of Euler–Urmayev equations and a corresponding set of complete characteristics were used, followed by a determination of some particular types of the best projections satisfying the limitations imposed. While noting the possibility of creating new types of best projections through this method, we should point out that it is impossible to solve this problem, stated in such a way, for the general case.

To solve the problem of developing the best map projections, it is necessary to formulate and prove the theorems for the best projections having various distortion patterns, analogous to the Chebyshev theorem for the best conformal projections, and to develop methods of calculation on this basis.

When discussing the derivation of ideal projections the following should be kept in mind. If one proceeds only from the condition of proving minimum distortion, then according to V. V. Kavrayskiy's definition the ideal projections are those, out of the entire group, on which a minimum of linear distortion is established within the limits of the region being represented. In this case, ideal projections may be of minimax and variational (least-squares) types. Rather interesting and useful is research by Novikova on the problem of devising such ideal variational-type projections, and by Hojovec, who in Prague in 1976 gave an approximate variational-type solution for representing a geographical quadrangle.

Considering this problem in a general sense, the ideal projections out of the entire set are those for which optimum satisfaction of all the requirements for map projections are fulfilled to make maps for specific purposes. In other words, if we bear in mind providing not only minimum distortion on maps but also optimum satisfaction of the whole group of requirements, then there are no ideal projections equally satisfying all classes of usage. They must be developed for every specific task.

The problem of developing these projections is extremely difficult. There are no concrete solutions so far.

7.2 Chebyshev projections

In 1856 Chebyshev presented the theorem for the best conformal projection. According to this theorem the best conformal projection for mapping a specific region is that on which the scale factor along the outline of the region is a constant value. This will provide the minimum scale variation within the outline. This theorem was proved by Grave in 1896. The Chebyshev type of projection was developed by V. V. Kavrayskiy, Dinchenko, Belonovskiy, M. M. Epshteyn, and others. A method of developing this type of projection in a practical manner was suggested by N. A. Urmayev in 1947.

The calculations for a Chebyshev projection include the solution of two problems:

1. finding values of local linear scale factors and of other projection characteristics at various points in the region to be mapped, with the given constant value of the local linear scale factor $m = n$ along the outline of this region;
2. determining rectangular coordinates x, y for points on the projection from the local scale factors available for points in the region to be mapped.

The former problem is reduced to the solution of Poisson's equation with zero boundary conditions or to the solution of Laplace's equation with given boundary conditions, i.e. solving the Dirichlet internal problem (μ is not a scale factor here)

$$\ln \mu_{qq} + \ln \mu_{\lambda\lambda} = 0 \tag{7.1}$$

with the given boundary conditions

$$\ln \mu |_B = \ln r_B \tag{7.2}$$

where q, λ are isometric coordinates, and B refers to the boundary;

$$r = N \cos \phi, \quad \ln m = \ln \mu - \ln r \tag{7.3}$$

To solve Laplace's equations (7.1), we use the function

$$\ln \mu = f(q + i\lambda)$$

which is to be continuous in the region being mapped, limited by the outline B.

We can obtain the solution to Laplace's equations in homogeneous harmonic polynomials for mapping regions with any outline:

$$\ln \mu = \sum_{i=0}^{k} a_i \psi_i + \sum_{i=1}^{k} b_i \theta_i \quad (i = 1, 2, 3, \ldots) \tag{7.4}$$

where ψ_i, θ_i are harmonic polynomials; a_i, b_i are constant parameters of the projection, determined from the conditions of the given boundary (7.2).

In the case where the region being mapped has an outline symmetrical about a straight central meridian, then equation (7.4) takes the form

$$\ln \mu = \sum_{i=1}^{k} a_i \psi_i \quad (i = 1, 2, 3, \ldots)$$

In this case, as previously mentioned, the boundary conditions are observed along the outline of the region being represented (7.2).

In the works of N. A. Urmayev it is shown that in order to solve Laplace's equation and, consequently, to find local scale factors at internal points of the region being represented, several methods may be used. These include Ritts' variational method, the graticule method, the method of finding a harmonic function which satisfies the boundary conditions in the best way, and the least-squares method. The last method is the most convenient and effective, especially for mapping regions with complex outlines.

Let us use the solution of Laplace's equation in the form that (7.4), incorporating (7.3), presents the formula for determining the natural logarithm of the local scale factor at the points of the projection under consideration:

$$\ln m = \ln \mu - \ln r = \sum_{i=0}^{k} a_i \psi_i + \sum_{i=1}^{k} b_i \theta_i - \ln r \tag{7.5}$$

Constant coefficients a_i, b_i can be found by minimizing the sum of the squares of the natural logarithms of the scale factors determined from formula (7.5) for various points along the outline, the number of which is more than the number of coefficients to be determined.

After determining the constant coefficients, we can calculate the value of local scale factors for internal points of the region to be mapped, i.e. solving the first part of the problem.

The other part of the problem is solved with the help of the differential equation

$$\mu^2 = x_q^2 + y_q^2$$

where $\mu = mr$; m is the local scale factor, and $r = N \cos \phi$, the radius of curvature of the parallel.

Various methods of determining rectangular coordinates for a Chebyshev projection are considered in the work of various authors, such as N. A. Urmayev, L. M. Bugayevskiy, Vilenkin, and others.

Let us use the method of linear approximation to derive Chebyshev projection formulas for the general case in which the region being mapped has an asymmetrical outline (Vakhrameyeva *et al.* 1986).

Equations (6.2) can be written in the form

$$\begin{aligned} x_q &= \mu \sin \gamma, \quad y_q = \mu \cos \gamma \\ x_\lambda &= \mu \cos \gamma, \quad y_\lambda = -\mu \sin \gamma \end{aligned} \tag{7.6}$$

where γ is the meridian convergence, which is unknown in the expression given; μ is a function that can easily be determined for every point, since the value of the local scale factor m for each point is known.

In order to determine the meridian convergence, we write Laplace's equation

$$\gamma_{qq} + \gamma_{\lambda\lambda} = 0$$

and the Cauchy–Riemann conditions equivalent to it in the form

$$\gamma_q = (\ln \mu)_\lambda, \quad \gamma_\lambda = -(\ln \mu)_q$$

After integrating these equations we find that

$$\gamma = -\sum_{i=1}^{k} a_i\theta_i + \sum_{i=1}^{k} b_i\psi_i$$

Now, we can obtain a numerical value of μ and γ for every internal point of the region.

As the projection under consideration is conformal,

$$y + ix = f(q + i\lambda)$$

In particular, we can write

$$\begin{aligned} x &= \sum_{i=1}^{k} m_i\theta_i + \sum_{i=1}^{k} n_i\psi_i \\ y &= \sum_{i=1}^{k} m_i\psi_i - \sum_{i=1}^{k} n_i\theta_i \end{aligned} \tag{7.7}$$

where m_i, n_i are constant coefficients.

In order to calculate them, let us differentiate equations (7.7):

$$\begin{aligned} x_q &= \sum_{i=1}^{k} n_i v_i + \sum_{i=1}^{k} m_i\tau_i \\ y_q &= \sum_{i=1}^{k} m_i v_i - \sum_{i=1}^{k} n_i\tau_i \end{aligned} \tag{7.8}$$

where $v_i = (\psi_i)_q = (\theta_i)_\lambda = i\psi_{i-1}$, and $\tau_i = (\theta_i)_q = -(\psi_i)_\lambda = i\theta_{i-1}$.

After introducing $\mu \cos\gamma = T''$, $-\mu \sin\gamma = P''$ into equations (7.6), from (7.6) and (7.8) we obtain

$$\begin{aligned} \sum_{i=1}^{k} im_i\psi_{i-1} - \sum_{i=1}^{k} in_i\theta_{i-1} &= T'' \\ \sum_{i=1}^{k} in_i\psi_{i-1} + \sum_{i=1}^{k} im_i\theta_{i-1} &= P'' \end{aligned} \tag{7.9}$$

where T'', P'', ψ_{i-1}, θ_{i-1} are known, and constant coefficients, m_i, n_i are unknown.

After computing a set of values in the form of (7.9), we can find these unknown coefficients using the method of least squares.

The projection created is a direct transformation of the surface of an ellipsoid onto the plane, and, according to the research of many scientists such as Grave, N. A. Urmayev, and V. V. Kavrayskiy, it provides minimum linear distortion within the limits of the region being represented and the best distribution of distortion, as well as the minimum average curvature of the plotted geodetic lines when compared with any other conformal projection.

The Chebyshev projection described by the following equation (Urmayev 1962) is of some interest:

$$\frac{1}{\mu} = 2\left[\cosh\frac{q}{2} + \cos\left(\frac{\pi}{4} - \frac{\lambda}{2}\right)\right]\left[\cosh\frac{q}{2} + \cos\left(\frac{\pi}{4} + \frac{\lambda}{2}\right)\right] \tag{7.10}$$

satisfying the boundary condition $\mu = 1/\cosh q$ when $\lambda = \pm\pi$, is designed for representing the entire surface of the Earth.

As applied to a particular case, this projection was investigated by Eisenlohr (1870, 1875), who suggested the following set of formulas for determining local scale factors:

$$m = \frac{\sin^2(u_1 + u_2)}{\sin^2(\phi/2)}, \quad \tan u_1 = \tan\frac{\omega}{2}\tan\frac{\pi - 2\lambda}{8}$$

$$\tan u_2 = \tan\frac{\omega}{2}\tan\frac{\pi + 2\lambda}{8}, \quad \sin\omega = \tan\frac{\phi}{2}$$

To determine rectangular coordinates of the projection from (7.10), the equation for μ_0 along the central meridian can be written as

$$\mu_0 = 1 \bigg/ \left[2\left(\cosh\frac{q}{2} + \cos\frac{\pi}{4}\right)^2\right]$$

Then for points along the central meridian and for any other point of the projection we have

$$y_0 = \int \mu_0 \, dq, \quad (y + ix) = f(q + i\lambda)$$

It should be noted, however, that the calculation of Chebyshev projections presents some problems which, with the use of a computer, can largely be overcome.

7.3 Conformal projections with adaptable isocols

7.3.1 Schols projection

An early form of a conformal projection with an oval isocol shaped to surround a given region on a map was suggested by Dutch geodesist C. M. Schols in 1882 (Kavrayskiy 1934, p. 65). Its formulas have the form

$$x = g_0 + g_1\lambda + g_2\lambda^2 + g_3\lambda^3, \quad y = f_0 + f_1\lambda + f_2\lambda^2 + f_3\lambda^3$$

where

$$\begin{aligned}
&f_3 = \frac{B}{3}\frac{N_0}{M_0} r_0 \cos^2\phi_0 = \frac{B}{3} f_3', \quad f_3' = \frac{N_0}{M_0} r_0 \cos^2\phi_0 \\
&g_3 = \tfrac{1}{6}[(1 - 2A) f_3' - r_0 \sin^2\phi_0] \\
&f_2 = \tfrac{1}{2} r_0 \sin\phi_0 + 3g_3\delta, \quad g_2 = -3f_3\delta, \\
&f_1 = -3f_3\delta^2, \quad g_1 = r_0 - r_0\sin\phi_0\delta - 3g_3\delta^2 \\
&f_0 = r_0\delta - \tfrac{1}{2} r_0 \sin\phi_0\delta^2 - g_3\delta^3, \quad g_0 = f_3\delta^3 \\
&\delta = \ln U - \ln U_0 \\
&A = \tfrac{1}{4}(1 - C\cos 2\alpha), \quad B = \tfrac{1}{4} C \sin 2\alpha \\
&C = \frac{a^2 - b^2}{a^2 + b^2} = \frac{1 - (b/a)^2}{1 + (b/a)^2}
\end{aligned} \tag{7.11}$$

where a, b are semiaxes of the second-order curve (an ellipse) approximating the outline of the region being mapped; α is the azimuth of axis a, measured from the y-axis, $\ln U$ is determined from equation (1.31), and ϕ_0, λ_0 are the latitude and longitude of the given center point.

Local scale factors can be determined from the approximate formula

$$m = 1 + Ay^2 - 2Bxy + (\tfrac{1}{2} - A)x^2 \tag{7.12}$$

7.3.2 Kavrayskiy adaptable conformal projection

Formulas for the rectangular coordinates of this projection by V. V. Kavrayskiy with oval isocols can be written in the form

$$X = \eta + \tfrac{1}{3}B(\xi^3 - 3\xi\eta^2) + \tfrac{1}{3}A(3\xi^2\eta - \eta^3)$$
$$Y = \xi + \tfrac{1}{3}A(\xi^3 - 3\xi\eta^2) - \tfrac{1}{3}B(3\xi^2\eta - \eta^3)$$

where $\eta = x - x_0$, $\xi = y - y_0$; x, y, x_0, y_0 are the Gauss–Krüger projection coordinates, respectively, at the current and central points; A, B are constant coefficients determined by the formulas

$$A = (1 - C\cos 2\alpha)/4R_0^2, \quad B = C\sin 2\alpha/4R_0^2$$
$$C = (1 - n'^2)/(1 + n'^2), \quad n' = b/a$$

a, b are semiaxes of the elliptical isocol; α is the rotation angle of its semimajor axis; and $R_0 = \sqrt{M_0 N_0}$ is the radius of the mean curvature of the Earth ellipsoid at the central point ϕ_0, λ_0 (Pavlov 1974).

7.3.3 Laborde projection

Jean Laborde, chief of the Service Géographique of Madagascar, developed this projection in 1926 for making maps and treating geodetic measurements on the long obliquely oriented island of Madagascar. This triple projection is obtained with the following steps.

First, the ellipsoid is conformally transformed onto the sphere:

$$\lambda' = \alpha(\lambda - \lambda_0), \quad \tan\left(\frac{\pi}{4} + \frac{\phi'}{2}\right) = CU^{\alpha}$$
$$\tan\phi'_0 = \sqrt{M_0/N_0}\tan\phi_0, \quad \alpha = \sin^2\phi_0 \csc^2\phi'_0$$
$$C = \tan\left(\frac{\pi}{4} + \frac{\phi'_0}{2}\right) U_0^{-\alpha}, \quad R_0 = 0.9995\sqrt{M_0 N_0}$$

where U is determined from equation (1.31). M_0 and N_0 are based on the International ellipsoid of 1924 ($a = 6\,378\,388$ m, $e^2 = 0.006\,722\,670\,02$ and equations (1.1)–(1.2)); $\phi_0 = 18.9°$S (21^{grads}), $\lambda_0 = 46°26'13.95''$E ($21^{\text{grads}}$E of the Paris Observatory).

The conformal sphere is then transformed onto a plane with the transverse Mercator projection using the Gauss–Schreiber approach:

$$x' = \ln\tan\left(\frac{\pi}{4} + \frac{\eta}{2}\right), \quad y' = \xi$$

where

$$\xi = \phi' - \phi'_0 + u', \quad \sin u' = \sin\phi' \tan\frac{\lambda'}{2}\tan\eta$$

$$\sin\eta = \cos\phi' \sin\lambda'$$

Finally rectangular coordinates on the projection are obtained using a third-order complex-algebra rotation to align the central line of the projection with the oblique direction of the island rather than with the central meridian of the transverse Mercator projection. Imaginary numbers are eliminated in this form of the equations:

$$x = R_0(x' + \tfrac{1}{3}By'^3 + Ax'y'^2 - Bx'^2y' - \tfrac{1}{3}Ax'^3)$$

$$y = R_0(y' + \tfrac{1}{3}Ay'^3 - Bx'y'^2 - Ax'^2y' + \tfrac{1}{3}Bx'^3)$$

where A, B are determined for a clockwise rotation of $\theta = 18.9°$ (21^{grads}):

$$A = (1 - \cos 2\theta)/4, \quad B = (\sin 2\theta)/4$$

'False' eastings and northings of 400 000 m for x and 800 000 m for y are finally added to the coordinates to make all of them positive for Madagascar. A. E. Young (1930) suggested a modification of the Laborde projection, but did not give final computation formulas.

7.3.4 Miller oblated stereographic conformal projection

Having oval isocols resembling those on the above projections, this projection was suggested by O. M. Miller (1953), crediting the work of Driencourt and Laborde (1932) for the general transformation. To calculate coordinates for this projection, first the regular formulas for the oblique spherical stereographic projection are used; they are of the form

$$x' = m' \cos\phi \sin(\lambda - \lambda_0)$$

$$y' = m'[\cos\phi_0 \sin\phi - \sin\phi_0 \cos\phi \cos(\lambda - \lambda_0)]$$

where

$$m' = 2/[1 + \sin\phi_0 \sin\phi + \cos\phi_0 \cos\phi \cos(\lambda - \lambda_0)]$$

Then the final coordinates x, y of a modified stereographic projection with oval isocols are determined with this rearranged set of third-order complex transformations (see section 7.3.5):

$$x = 0.9245x'\{1 - (0.2522/12)[3(y')^2 - (x')^2]\}$$

$$y = 0.9245y'\{1 + (0.2522/12)[3(x')^2 - (y')^2]\}$$

where ϕ_0, λ_0 are the latitude 18°N and longitude 20°E (later 18°E), respectively, of the center of the projection. The projection (Figure 7.1) initially enclosed a combined map of Europe and Africa with oval-shaped isocols, the major axis lying along the central meridian, using the above formulas. The projection and equations were augmented in 1955 to include other landmasses of the eastern hemisphere with oblique oval-shaped isocols (Sprinsky and Snyder 1986).

7.3.5 Projections using higher-order irregular isocols

The more general form of the just-preceding pair of formulas for complex conformal transformations, as given by Driencourt and Laborde (1932, p. 202), is

$$x + iy = R \sum_{j=1}^{k} (A_j + iB_j)(x' + iy')^j$$

where A_j, B_j are coefficients determined by the least-squares method from the conditions of minimal distortion within the region being mapped (Snyder 1987b, pp. 203–8). Laborde and Miller used $k = 3$ for their versions above, but others have used the power of modern computers to derive projections using higher-order trans-

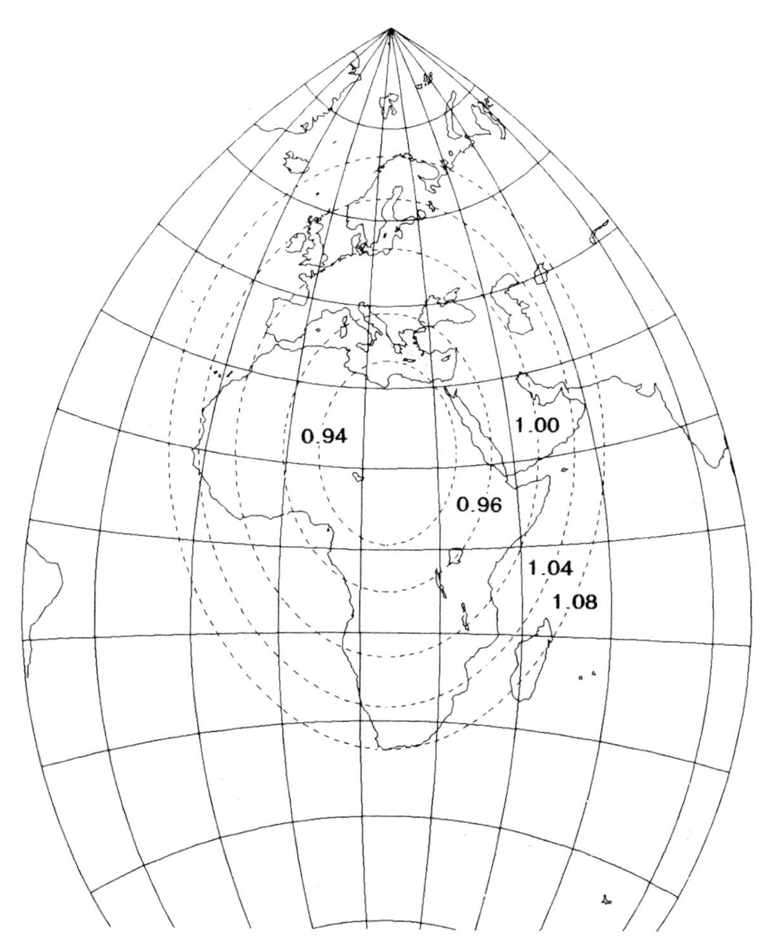

Figure 7.1 Miller oblated stereographic projection of Europe and Africa, centered at 18°N, 20°E, with isocols for m. 15° graticule.

formations from x', y' to x, y, resulting in isocols of irregular shapes suiting particular regions.

Reilly (1973) and Reilly and Bibby (1975–6) used $k = 6$ for a map of New Zealand, adopted for the New Zealand Map Grid, but they started with the Mercator as the base projection rather than the stereographic. Snyder used $k = 10$ for a map of the 50 US states, and $k = 6$ for a projection of the irregularly bounded state of Alaska (Figure 7.2); both of these projections use the oblique stereographic as the base, and the range of linear distortion was reduced to about a quarter of the range using standard projections. In addition Snyder (1984) used simplified forms of higher-order complex polynomials to develop conformal projections with isocols in the approximate shape of regular polygons and rectangles.

A much more mathematically complicated application of the complex equation given above was made by American physicist Mitchell Feigenbaum in 1992 to develop minimum error maps of each continent (except Antarctica) with an isocol along a smooth simplification of the outer coastlines including related islands. This projection was used in the 1992 *Hammond Atlas of the World*, and is called the Hammond optimal conformal projection.

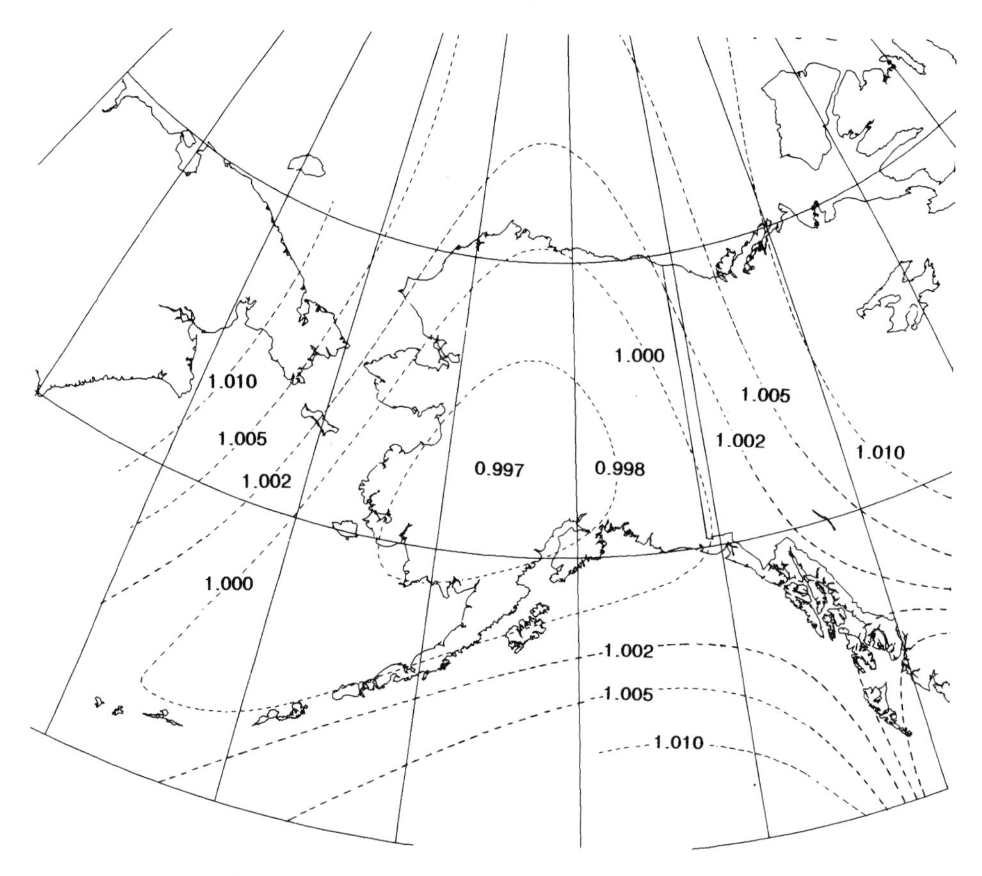

Figure 7.2 Modified–stereographic conformal projection of Alaska, with isocols for m. 10° graticule.

7.3.6 Projections using series and harmonic polynomials

The techniques for deriving these projections, conformal or nearly so depending on the type of distortion, were developed by L. A. Vakhrameyeva. On these projections isocols have oval, circular, or hyperbolic shapes, oriented differently with respect to the meridian of the central point (Vakhrameyeva *et al.* 1986).

Formulas for computing strictly conformal projections (with an example using a third-order harmonic polynomial) are

$$x = B_0 + A_1\lambda + B_2\lambda^2 + A_3\lambda^3, \quad y = A_0 + B_1\lambda + A_2\lambda^2 + B_3\lambda^3$$

where A_k, B_k are variable coefficients, obtained from the formulas

$$A_0 = a_0 + a_1(q - q_0) + a_2(q - q_0)^2 + a_3(q - q_0)^3$$

$$A_1 = \frac{dA_0}{dq}, \quad A_2 = -\frac{1}{2}\frac{d^2A_0}{dq^2}, \quad A_3 = -\frac{1}{6}\frac{d^3A_0}{dq^3}$$

$$B_0 = b_0 + b_1(q - q_0) + b_2(q - q_0)^2 + b_3(q - q_0)^3$$

$$B_1 = -\frac{dB_0}{dq}, \quad B_2 = -\frac{1}{2}\frac{d^2B_0}{dq^2}, \quad B_3 = \frac{1}{6}\frac{d^3B_0}{dq^3}$$

where q and q_0 are isometric latitudes of the current parallel and the parallel of the central point, respectively.

An analysis of the coefficients showed that it is convenient to give them the following values:

$$a_0 = 0, \quad a_1 = r_0, \quad a_2 = -\frac{r_0}{2}\sin\phi_0, \quad a_3 = r_0\left(\frac{A}{3}\cos^2\phi - \frac{\cos 2\phi_0}{6}\right)$$

To simplify the problem it is further desirable to determine coefficients b_k (with asymmetrical isocols) with the following formulas:

$$b_0 = b_1 = b_2 = 0, \quad b_3 = b = r_0(B/3)\cos^2\phi_0$$

In the formulas given, A and B (without subscripts) are defined numerical coefficients affecting the shape of isocols and their rotation with respect to the central meridian; their values can be obtained with the formulas

$$A = (1 - C\cos 2\alpha)/4, \quad B = (C\sin 2\alpha)/4$$

where $C = (a^2 - b^2)/(a^2 + b^2)$; α is the angle of isocol rotation with respect to the meridian of the central point; a and b are the semiaxes of the isocols.

The advantage of projections created with harmonic polynomials lies in the simplicity of derivation and the low distortion. The disadvantage of the projections recommended is that their isocols are only approximately rather than completely adaptable to the boundaries of regions being mapped. It is also impossible to represent near-polar regions on these projections.

Equations using series for conformal projections, with isocols symmetrical about the central meridian, can have the form

$$x = a_1 s_p + a_3 s_p^3 + \cdots, \quad y = a_0 + a_2 s_p^2 + \cdots$$

where s_p is the length of the arc of the parallel between the meridian of the central point and the current meridian, corresponding to the difference of longitude $(\lambda - \lambda_0)$; a_0 is a projection characteristic; a_1, a_2, a_3 are variable coefficients which can be computed from the following formulas:

$$a_0 = s_m + \frac{s_m^3}{kR_0^2}$$

$$a_1 = \frac{da_0}{ds_m} = 1 + \frac{3s_m^2}{kR_0^2}$$

$$a_2 = \frac{1}{2}\left(a_1\frac{\tan\phi}{N} - \frac{da_1}{ds_m}\right) = \frac{\tan\phi}{2N} - \frac{3s_m}{kR_0^2}$$

$$a_3 = -\frac{1}{3}\left(2a_2\frac{\tan\phi}{N} - \frac{da_2}{ds_m}\right) = -\frac{\tan^2\phi}{6R_0^2} + \frac{k-6}{6kR_0^2}$$

$$a_n = (-1)^n\frac{1}{n}\left((n-1)a_{n-1}\frac{\tan\phi}{N} - \frac{da_{n-1}}{ds_m}\right)$$

where s_m is the length of the arc of the meridian between the parallel of the central point and the current parallel; N is the radius of curvature of the first vertical section; R_0 is the mean radius of curvature $\sqrt{M_0 N_0}$ of the surface being mapped at the central point of the projection.

Formulas for rectangular coordinates have the form

$$x = \left(1 + \frac{3s_m^2}{kR_0^2}\right)s_p + \left(\frac{k-6}{6kR_0^2} - \frac{\tan^2\phi}{6R_0^2}\right)s_p^3 + \cdots$$

$$y = s_m + \frac{s_m^3}{kR_0^2} + \left(\frac{\tan\phi}{2N} - \frac{3s_m}{kR_0^2}\right)s_p^2 + \cdots$$

where k is a numerical coefficient; by changing its value it is possible to obtain seven groups of conformal projections (Table 7.1).

For projections with asymmetrical isocols,

$$x = b_0 + a_1 s_p + b_2 s_p^2 + a_3 s_p^3 + \cdots$$

$$y = a_0 + b_1 s_p + a_2 s_p^2 + b_3 s_p^3 + \cdots$$

Values of variable coefficients a_k are known, and those of b_k equal

$$b_0 = \frac{s_m^2}{kR_0^2}, \quad b_1 = \frac{3s_m^2}{kR_0^2}, \quad b_2 = \frac{3s_m}{kR_0^2}, \quad b_3 = \frac{1}{kR_0^2}$$

Formulas for the rectangular coordinates of projections with asymmetrical isocols have the form

$$x = \frac{s_m^3}{kR_0^2} + \left(1 + \frac{3s_m^2}{kR_0^2}\right)s_p - \frac{3s_m s_p^2}{kR_0^2} + \left(\frac{k-6}{6kR_0^2} - \frac{\tan^2\phi}{6R_0^2}\right)s_p^3 + \cdots$$

$$y = s_m + \frac{s_m^3}{kR_0^2} - \frac{3s_m^2 s_p}{kR_0^2} + \left(\frac{\tan\phi}{2R_0} - \frac{3s_m}{kR_0^2}\right)s_p^2 + \frac{s_p^3}{kR_0^2} + \cdots$$

Local scale factors for these projections are expressed by the formula

$$m = 1 + [6s_m^2 - 12s_m s_p + (k-6)s_p^2]/(2kR_0^2)$$

This projection can be used to design maps mainly for small regions at scales of 1:1 000 000 and smaller. With a decrease of map scale the boundary of the area being mapped correspondingly increases. Although these formulas are among the more complicated, computation of these projections is made practical by using a modern computer or similar facilities.

Table 7.1 Shapes of isocols for Vakhrameyeva projections

Values of k	Shape of isocols
0	Ovals extending along the central meridian; for small regions they are straight lines parallel to this meridian
$12 < k < \infty$	Elliptical curves extending along meridians
12	Circles
$6 < k < 12$	Elliptical curves extending along parallels
6	Circular arcs coinciding with parallels
$0 < k < 6$	Hyperbolic curves, the axis of symmetry coinciding with the meridian of the central point
$k < 0$	Hyperbolic curves, the axis of symmetry coinciding with the central parallel

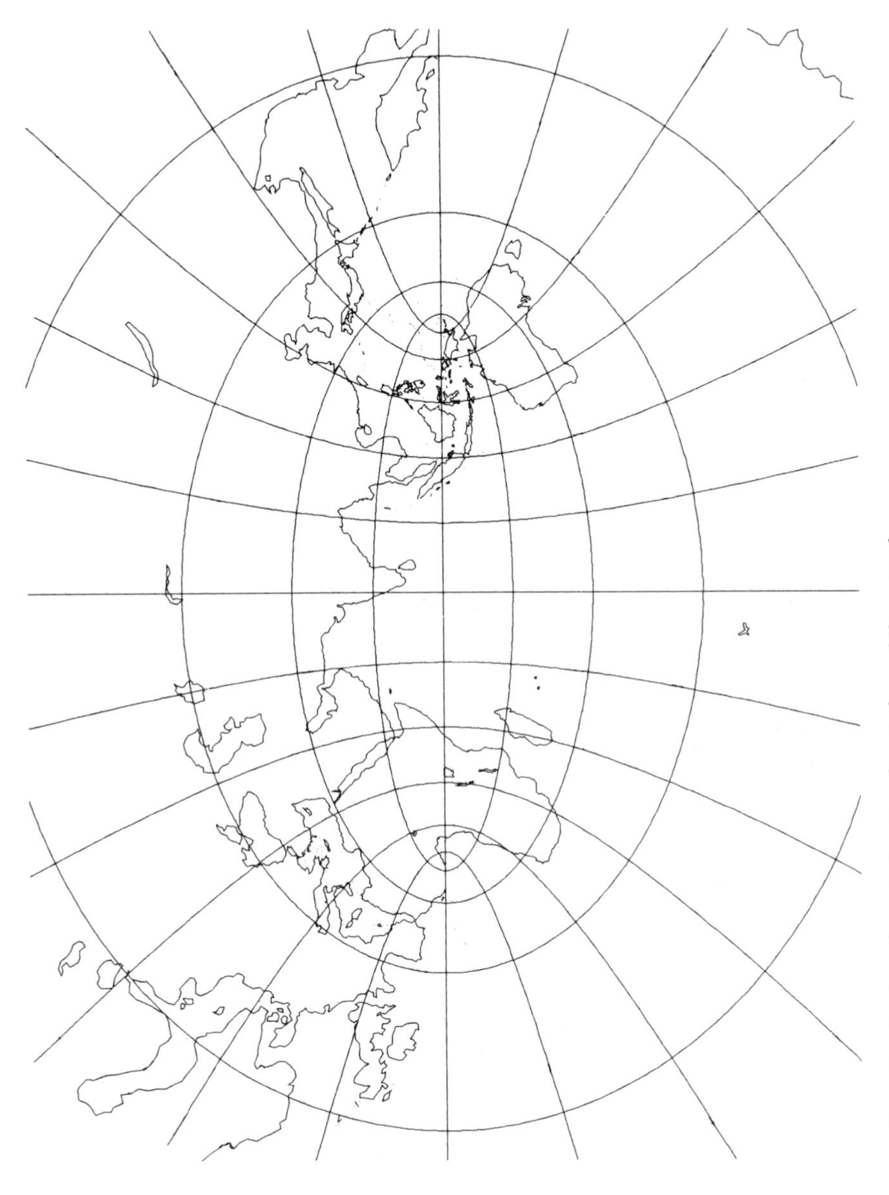

Figure 7.3 Littrow conformal projection of most of eastern hemisphere. 15° graticule.

The Lagrange projection also belongs to the group of adaptable projections. In particular cases where the outlines of regions to be mapped coincide with isocols of one of the various forms of this projection, the Lagrange projection falls into the category of the above conformal projections (see section 10.1).

It should be noted that the Schols, Kavrayskiy, Laborde, and Vakhrameyeva projections are especially appropriate for mapping small regions.

7.3.7 Littrow projection

Formulas for this projection (Figure 7.3) of a sphere with a unit radius can be given in the form

$$x = \sec\phi \sin\lambda, \quad y = \tan\phi \cos\lambda$$

or, using isometric latitude q in place of geodetic latitude ϕ,

$$x = \cosh q \sin\lambda, \quad y = \sinh q \cos\lambda$$

where

$$q = \ln\tan(\pi/4 + \phi/2)$$

$$\sinh q = \tfrac{1}{2}(e^{q} - e^{-q}), \quad \cosh q = \tfrac{1}{2}(e^{q} + e^{-q})$$

$$m = \frac{1}{\cos\phi\sqrt{2}}\sqrt{\cosh 2q + \cos 2\lambda}$$

$$= \sec\phi\sqrt{\sec^2\phi - \sin^2\lambda}$$

On this projection, which is conformal, parallels are ellipses and meridians are hyperbolas. An important navigational feature is that it is retroazimuthal; that is, the azimuth from any point to a point along the straight central meridian may be measured as the angle between a straight line connecting the two points and a vertical line (see also section 3.6).

7.4 *Conformal projections using elliptic functions*

Projections based on elliptic functions depend on the positions of the foci of spherical ellipses on the surface of the sphere (Adams 1925, pp. 90–103). C. S. Peirce in 1879, Émile Guyou in 1886, and O. S. Adams in 1925 presented related versions.

For the Peirce 'quincuncial' projection (Figure 7.4) all foci are situated on the equator at its intersections with meridians of longitude 90° apart, and intermediate coordinate formulas may be written in the following form:

$$\cos a = \cos\phi \cos(\pi/4 + \lambda)$$

$$\cos b = \cos\phi \cos(\pi/4 - \lambda) \tag{7.13}$$

$$\sin u = \sqrt{2}\cos[(a + b)/2], \quad \sin v = \sqrt{2}\sin[(a - b)/2]$$

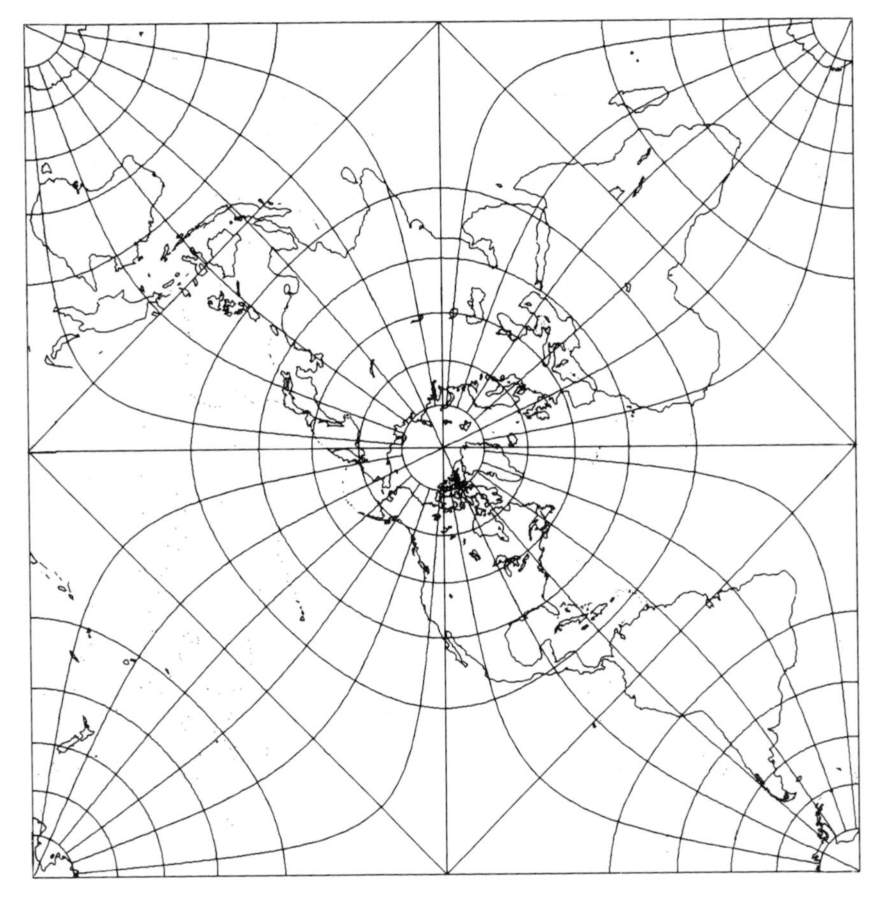

Figure 7.4 Peirce conformal quincuncial projection. 15° graticule.

For the Guyou projection, the foci are at latitude $\pm 45°$, and formulas for computing intermediate coordinates have the form

$$\cos a = \frac{\sqrt{2}}{2}(\sin\phi - \cos\phi\sin\lambda)$$

$$\cos b = \frac{\sqrt{2}}{2}(\sin\phi + \cos\phi\cos\lambda)$$

For the corresponding Adams projection one focus is situated at the pole and the other on the equator; formulas for coordinates have the form

$$a = 90° - \phi, \quad b = \arccos(\cos\phi\sin\lambda)$$

Isometric values u, v for all three coordinate systems are determined from equations (7.13).

Rectangular coordinates for the projections are determined from the formulas

$$x = \eta, \quad y = \xi$$

where

$$\eta = \int_0^v \frac{dv}{\sqrt{(1 - \frac{1}{2}\sin^2 v)}}, \quad \xi = \int_0^u \frac{du}{\sqrt{(1 - \frac{1}{2}\sin^2 u)}} \tag{7.14}$$

Equations (7.14) express elliptic functions of the first kind, values of which can be determined by known methods of numerical integration or by tables such as those compiled by Jahnke *et al.* (1968).

Like the other two, the Peirce 'quincuncial' projection gives a peculiar appearance to the surface of the entire Earth. The northern hemisphere is enclosed in a square, and the southern hemisphere is shown as a square which has been cut into four isosceles triangles along the two diagonals, each triangle appended to one side of the square representing the northern hemisphere. The five sections lead to the name quincuncial. There is no distortion at the geographical poles; maximum distortion occurs at the bends of the equator.

For the Guyou projection the world map consists of adjacent eastern and western hemispheres shaped as squares with straight outer meridians. There is no distortion at the points of intersection of the central meridian of each hemisphere with the equator. The maximum distortion occurs at points where the latitude $\pm 45°$ intersects the outer meridians. The Guyou projection is transverse to the Peirce projection and vice versa.

On the Adams projection described above, the Earth is represented by eastern and western hemispheres each in the shape of a square with the poles at the vertices. There is no distortion at the intersection of the two central meridians with the equator. Maximum distortion occurs at the poles and at the intersection of the equator with the outer meridian of the hemispheres. Adams (1925) also prepared conformal projections of the world using elliptic functions and bounded by a square, other regular polygons, and an ellipse.

All of the projections with square-shaped hemispheres may be tesselated, or mosaicked indefinitely. Models of map graticules for these projections with isocols are given in Vakhrameyeva *et al.* (1986) and without isocols in Snyder and Voxland (1989).

7.5 Quasiconformal transformation of flat regions; classes of equal-area projections closest to conformality

7.5.1 Quasiconformal transformation of flat regions

The chief requirements for these transformations are described by Meshcheryakov generally from the works of M. A. Lavrent'ev. Two planes are considered, the main equations are given, and it is shown that general quasiconformal transformations corresponding to those equations transform infinitesimal squares of one plane into parallelograms on the other plane (with an accuracy of up to infinitesimal first-order values).

It should be noted that this theory makes it possible to reduce the projection of arbitrary surfaces to quasiconformal transformations of flat regions. Specific developments of the theory of this transformation and of its practical application have so far not been applied to mathematical cartography.

7.5.2 Classes of equal-area projections closest to conformality

It was noted above that on Chebyshev projections, in contrast to other conformal projections, distortion of all types, including that of area, is a minimum. Hence, Chebyshev projections are the closest of all conformal projections to equal-area projections.

As we discussed previously, on equal-area projections

$$ab = mn \cos \varepsilon = 1 \tag{1.91}$$

where a, b are extreme linear scale factors at a given point.

Equation (1.91) alone is insufficient to derive a projection. When considering various methods of modifying this equation, remembering the condition of conformality $m = n$ and $\varepsilon = 0$, we can obtain various classes of equal-area projections close to conformality (Meshcheryakov 1968):

$$mn \cos \varepsilon = 1, \quad m - n = 0 \tag{7.15}$$

$$mn \cos \varepsilon = 1, \quad \varepsilon = 0 \tag{7.16}$$

In 1935 B. P. Ostashchenko-Kudryavtsev proposed in his report to the Soviet All-Union Astronomical–Geodetic Congress that semiconformal projections are those on which the equality of local scales $m = n$ is preserved. Later, while discussing equations (7.15)–(7.16), Meshcheryakov suggested that the projections defined by (7.15) be called semiconformal and the others, Euler projections; he also considered the problem of generalizing these projections.

Combining (7.15) and (7.16), we can write

$$P_1(m - n) + P_2 \varepsilon = 0$$

or

$$mn \cos \varepsilon = 1, \quad (m - n) + k\varepsilon = 0 \tag{7.17}$$

and

$$mn \cos \varepsilon = 1, \quad (n - m) + k\varepsilon = 0 \tag{7.18}$$

where P_1, P_2 are the relative weights of the conformality conditions, and $k = P_2/P_1$.

Equations (7.17)–(7.18) are called, respectively, the first and second classes of equal-area quasiconformal projections. Using these equations it is feasible to derive general formulas for local linear scale factors and other characteristics, among them the corresponding set of Euler–Urmayev equations. Of these projection classes only Euler projections are widely used (see section 7.7).

Oblated equal-area projection

Snyder (1988) presented a set of equal-area projections on which the isocols are shown as ovals, rectangles (Figure 7.5), or rhombi, thus enclosing a given region with low or minimum linear and angular distortion as well as no area distortion. They are thus equal-area counterparts of the Chebyshev conformal projections for regions of these non-circular shapes, except that the shapes of isocols are more limited. The transformation consists of two steps. First the rectangular coordinates are calculated for an oblique Lambert azimuthal equal-area projection of the sphere centered on the region, but relative to a Y-axis rotated by angle θ counterclockwise

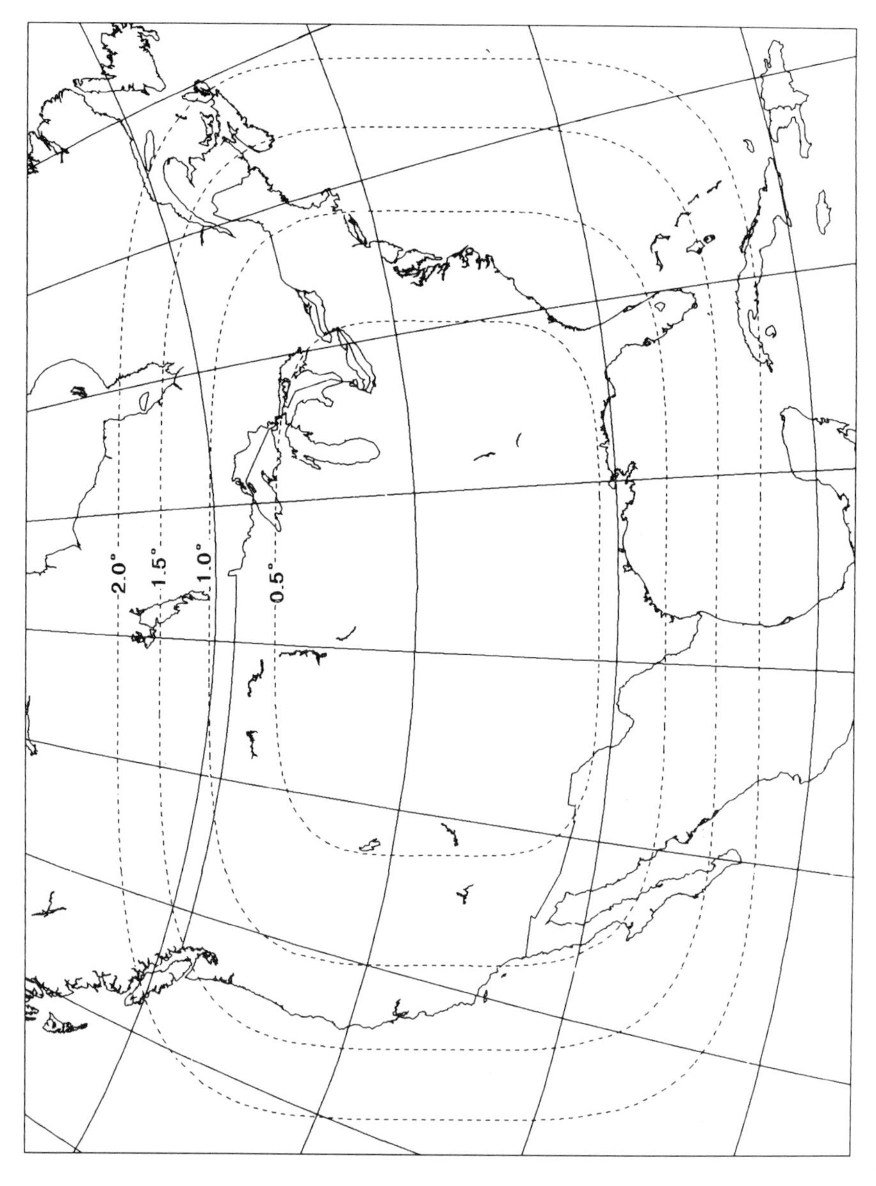

Figure 7.5 Oblated equal-area projection of the United States. Isocols for ω. 10° graticule.

from north so that the oval or rectangular isocols may be oblique:

$$z = \arccos[\sin \phi_0 \sin \phi + \cos \phi_0 \cos \phi \cos(\lambda - \lambda_0)]$$

$$a = \arctan_2\left(\frac{\cos \phi \sin(\lambda - \lambda_0)}{\cos \phi_0 \sin \phi - \sin \phi_0 \cos \phi \cos(\lambda - \lambda_0)}\right) + \theta$$

$$x' = 2 \sin(z/2) \sin a, \quad y' = 2 \sin(z/2) \cos a$$

where $\arctan_2$ is the arctan with quadrant adjustment, equivalent to ATAN2 of FORTRAN.

Second, the x'-values are compressed or expanded in a certain manner, and the y'-values are then expanded or compressed in inverse ratio to the change of spacing of the x'-values at the same point, so that the area of each element parallel to the Y-axis remains the same. A similar compression and expansion is applied to the new coordinates, with final formulas

$$M = \arcsin(x'/2), \quad N = \arcsin\left(\frac{y'}{2} \frac{\cos M}{\cos(2M/m)}\right)$$

$$x = mR \sin(2M/m)\left(\frac{\cos N}{\cos(2N/n)}\right)$$

$$y = nR \sin(2N/n)$$

where m, n are constant parameters, not scale factors, normally between 1.4 and 2.8, chosen for the desired shape of the isocols. (M and N are not symbols for curvature.) For rounded square isocols, $m = n = 1.82$; for circles $m = n = 2$ (the original Lambert azimuthal equal-area projection); for rectangles 1.5/1.0 in dimensions, extended along the Y-axis, $m = 1.54$, $n = 2.31$. In Figure 7.5, $m = 1.5$, $n = 2.4$, and $\theta = 90°$, with $\phi_0 = 39°$N, $\lambda_0 = 96°$W.

7.6 Projections with orthogonal map graticules

N. A. Urmayev considered equations for orthogonal projections of the sphere in a general form in 1947. As applied to projections for an ellipsoid of revolution, the Euler–Urmayev equations (6.3), when $\varepsilon = 0$, take the form

$$\gamma_\phi = -\mu_\lambda/\nu, \quad \gamma_\lambda = \nu_\phi/\mu \tag{7.19}$$

Hence, it is not difficult to obtain the second-order differential equation

$$\frac{1}{\nu} \mu_{\lambda\lambda} - \frac{1}{\nu^2} \mu_\lambda \nu_\lambda + \frac{1}{\mu} \nu_{\phi\phi} - \frac{1}{\mu^2} \nu_\phi \mu_\phi = 0$$

Declaring the linear scale to be a function only of latitude, Urmayev obtained equal-area and equidistant conic projections of the sphere from these equations; after imposing the condition that the linear scale along parallels is a function of both latitude and longitude, he obtained the generalized equidistant conic projection of the sphere with rounded isocols (see Chapter 3 and section 6.4.1).

Konusova (1973, 1975) and with co-author I. A. Bertik (1985) considered a number of the problems in creating orthogonal projections by solving Euler–Urmayev differential equations of various types. Among these are map projections with equidistant parallels, including generalized conic, azimuthal, and cylindrical projections.

For orthogonal projections with equidistant parallels, equations (7.19) were rewritten in the form

$$m = m(\phi), \quad Mm\gamma_\lambda + (rn)_\phi = 0$$

Incorporating the relationship between the radius of curvature ρ of parallels and the convergence γ of meridians

$$\rho = r_n/\gamma_\lambda$$

the expressions for the local linear scale factor were found to be

$$m = -\rho_\phi/M, \quad n = \gamma_\lambda(\rho/r)$$

The equations in exact differentials for the calculation of abscissa x and ordinate y of the projection were then determined.

Assuming that $\gamma = \gamma(\lambda)$, they obtained

$$x = (\rho - \rho_0)\sin \gamma + x_0(\lambda), \quad y = -(\rho - \rho_0)\cos \gamma + y_0(\lambda)$$

where ρ_0 is the radius of curvature of an initial parallel of latitude ϕ_0; $x_0(\lambda)$, $y_0(\lambda)$ are coordinates of points along this parallel; $\rho - \rho_0 = u(\phi)$ is a function whose form depends on the properties of the projection. On all these projections, the meridians are straight lines.

Examples of projections with an orthogonal graticule and curved meridians are the rectangular polyconic projections described in section 4.2.3 and Table 4.1.

Urmayev (1962) considered another method of defining orthogonal projections with an arbitrary distortion pattern. He used a given family of meridians or parallels. Giving that family a parameter F with the form

$$F(x, y) = 0 \tag{7.20}$$

after differentiating we can obtain

$$F_x(\delta x/\delta y) + F_y = 0 \tag{7.21}$$

where $\delta x/\delta y$ is the trigonometric tangent of the angle of the line tangent to family (7.20).

Denoting with dx/dy the tangent of the angle of the line tangent to the orthogonal curve, we have

$$(\delta x/\delta y)/(dx/dy) = -1$$

Now formula (7.21) takes the form

$$F_x - F_y(dx/dy) = 0 \tag{7.22}$$

It is this expression that is the differential equation for the curves orthogonal to family (7.20). To illustrate this, some known projections were derived by solving equation (7.22).

7.7 Euler projections

Euler projections, described by Leonhard Euler in 1777, are equal-area projections with an orthogonal map graticule. From this definition, on these projections

$$m = 1/n, \quad \varepsilon = 0$$

where m, n are the local linear scale factors along meridians and parallels, respectively, and ε is the deviation from a right angle of the angle of intersection of meridians and parallels on the projection.

Urmayev (1947) dealt with a number of propositions involving the theory of Euler projections of the sphere, and he derived an equal-area conic projection based on these propositions. Meshcheryakov (1968) proceeded with theoretical research on these projections and gave some examples of the theoretical propositions that he considered. He developed equal-area conic projections of the sphere, the Korkin–Grave projection (on which meridians are concentric circles and parallels are straight lines), and an Euler projection with the central meridian as a line of conformality along which there is no distortion of any kind.

In order to derive projections of the ellipsoid we introduce the following symbols:

$$S = \int_0^\phi Mr\, d\phi, \quad g = n^2r^2 \tag{7.23}$$

where $r = N \cos \phi$, M, and N are radii of curvature of the parallel, the meridian section, and the first vertical section, respectively.

Taking into account formulas for the local linear scale factor, by analogy with the above studies we obtain

$$x_s = -\frac{1}{\sqrt{g}} \sin \gamma, \quad x_\lambda = \sqrt{g} \cos \gamma$$

$$y_s = \frac{1}{\sqrt{g}} \cos \gamma, \quad y_\lambda = \sqrt{g} \sin \gamma$$

where γ is the meridian convergence.

On establishing conditions to integrate these equations, we get a known set of equations in first-order partial derivatives

$$2g^2\gamma_S + g_\lambda = 0, \quad g_S + 2\gamma_\lambda = 0 \tag{7.24}$$

and from this set we obtain a differential equation in second-order partial derivatives:

$$g^3 g_{SS} - g g_{\lambda\lambda} + 2g_\lambda^2 = 0 \tag{7.25}$$

Equations (7.24)–(7.25) provide the fundamental relationships in the theory of the Euler projections.

In the particular case where $g = f(S)$ from equation (7.25), using (7.23) we obtain the formulas for the known Albers equal-area conic projection of the ellipsoid:

$$\gamma = \alpha\lambda, \quad g = 2\alpha(c - S), \quad n^2 = \frac{2\alpha(c - S)}{r^2}, \quad \rho^2 = \frac{2}{\alpha}(c - S)$$

Meshcheryakov also showed that an Euler projection of the sphere can be developed by solving the differential equation in partial derivatives

$$\pm g g_S + g_\lambda = 0 \tag{7.26}$$

for which the common integrals are

$$F(g, S \mp g_\lambda) = 0 \quad \text{or} \quad f(g) = S \pm g_\lambda \tag{7.27}$$

Research proved that the possibilities of providing the optimum variant of Euler projections of the sphere or ellipsoid on the basis of equations (7.26)–(7.27) are very limited.

7.8 Two-point azimuthal projection

This projection is determined by the following conditions:

1. all arcs of great circles (orthodromes) on the projection are shown as straight lines;
2. angles at two given points, e.g. at two radio direction-finding stations, are not distorted (Figure 7.6).

The formulas for calculating the rectangular coordinates for the two-point azimuthal projection of the sphere with special focal points (ϕ_1, λ_1) and (ϕ_2, λ_2) at azimuths T_0 and U_0, respectively, take the form (Kavrayskiy 1934, pp. 181–2)

$$x = \frac{C \cot \phi \sin(\lambda - \lambda_0)}{1 + \cot \phi_0 \cot \phi \cos(\lambda - \lambda_0)}$$

$$y = -\frac{A \cot \phi \cos(\lambda - \lambda_0) + B \cot \phi \sin(\lambda - \lambda_0)}{1 + \cot \phi_0 \cot \phi \cos(\lambda - \lambda_0)}$$

where

$$A = R \cos^2 \zeta \cos \alpha \sec \beta \csc^2 \phi_0$$

$$B = R \sin^2 \zeta \sin \alpha \cos \beta \csc \phi_0$$

$$C = R \cos \zeta \sec \alpha \cos \beta \csc \phi_0$$

R is the Earth's radius determined so that there is no linear distortion at the two focal points.

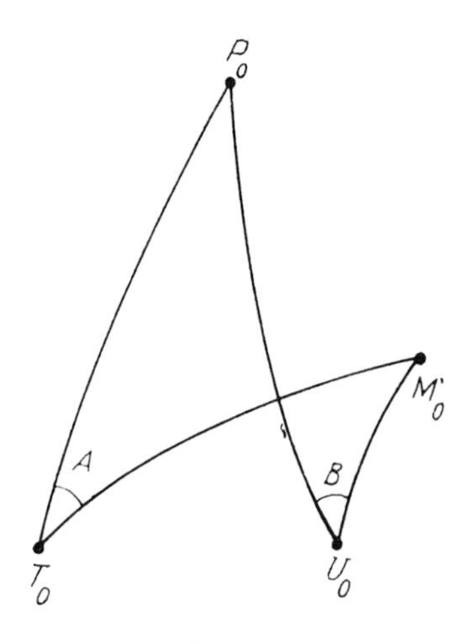

Figure 7.6 Construction of two-point azimuthal projection.

Values ζ, ϕ_0, λ_0, α, β are calculated from the following expressions:

$$\cos 2\zeta = \sin\phi_1 \sin\phi_2 + \cos\phi_1 \cos\phi_2 \cos(\lambda_2 - \lambda_1)$$

$$\cot U_0 = \tan\phi_1 \cos\phi_2 \csc(\lambda_2 - \lambda_1) - \sin\phi_2 \cot(\lambda_2 - \lambda_1)$$

$$\sin\phi_0 = \cos\zeta \sin\phi_2 + \sin\zeta \cos\phi_2 \cos U_0$$

$$\cot(\lambda_2 - \lambda_0) = \cot\zeta \cos\phi_2 \csc U_0 - \sin\phi_2 \cot U_0$$

$$\cot\alpha = \tan\phi_2 \cos\phi_0 \csc(\lambda_2 - \lambda_0) - \sin\phi_0 \cot(\lambda_2 - \lambda_0)$$

$$\tan\beta = \tan\alpha \sec\zeta$$

$$\lambda_0 = (\lambda_1 + \lambda_2)/2 - \arcsin(\sin\sigma \sin\delta \sin D/\cos\phi_0 \cos\zeta)$$

$$\sigma = (\phi_1 + \phi_2)/2, \quad \delta = (\phi_2 - \phi_1)/2, \quad D = (\lambda_2 - \lambda_1)/2$$

This projection is an affine transformation of a gnomonic projection; that is, it is a gnomonic projection uniformly compressed in a certain direction. It was first developed by Maurer (1914) in Germany and may be used for laying out radio bearings.

7.9 Two-point equidistant projection

Let PA, OA (Figure 7.7) be straight line segments equal to the great-circle or orthodromic distances from a given point $A(\phi, \lambda)$ to two fixed points $P(\phi_1, \lambda_1)$ and $O(\phi_0, \lambda_0)$. Let P be the pole of the polar coordinate system and O be the origin of a rectangular coordinate system XOY, the Y-axis lying along OP.

From these conditions,

$$\begin{aligned} \rho &= Rz = R\arccos[\sin\phi \sin\phi_1 + \cos\phi \cos\phi_1 \cos(\lambda - \lambda_1)] \\ q_0 &= R\arccos[\sin\phi_1 \sin\phi_0 + \cos\phi_1 \cos\phi_0 \cos(\lambda_1 - \lambda_0)] \\ \xi &= R\arccos[\sin\phi \sin\phi_0 + \cos\phi \cos\phi_0 \cos(\lambda - \lambda_0)] \end{aligned} \tag{7.28}$$

$$\delta = \arccos\left(\frac{\rho^2 + q_0^2 - \xi^2}{2\rho q_0}\right) \tag{7.29}$$

It is also necessary to know whether point A is to the right or left of the Y-axis. This may be determined by calculating the azimuths (clockwise from north) of

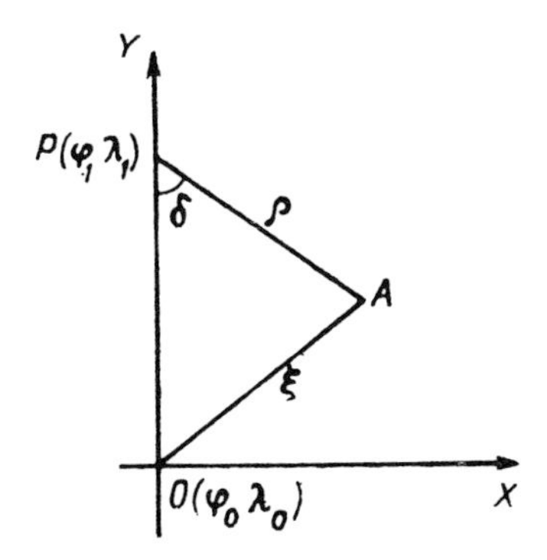

Figure 7.7 Construction of two-point equidistant projection.

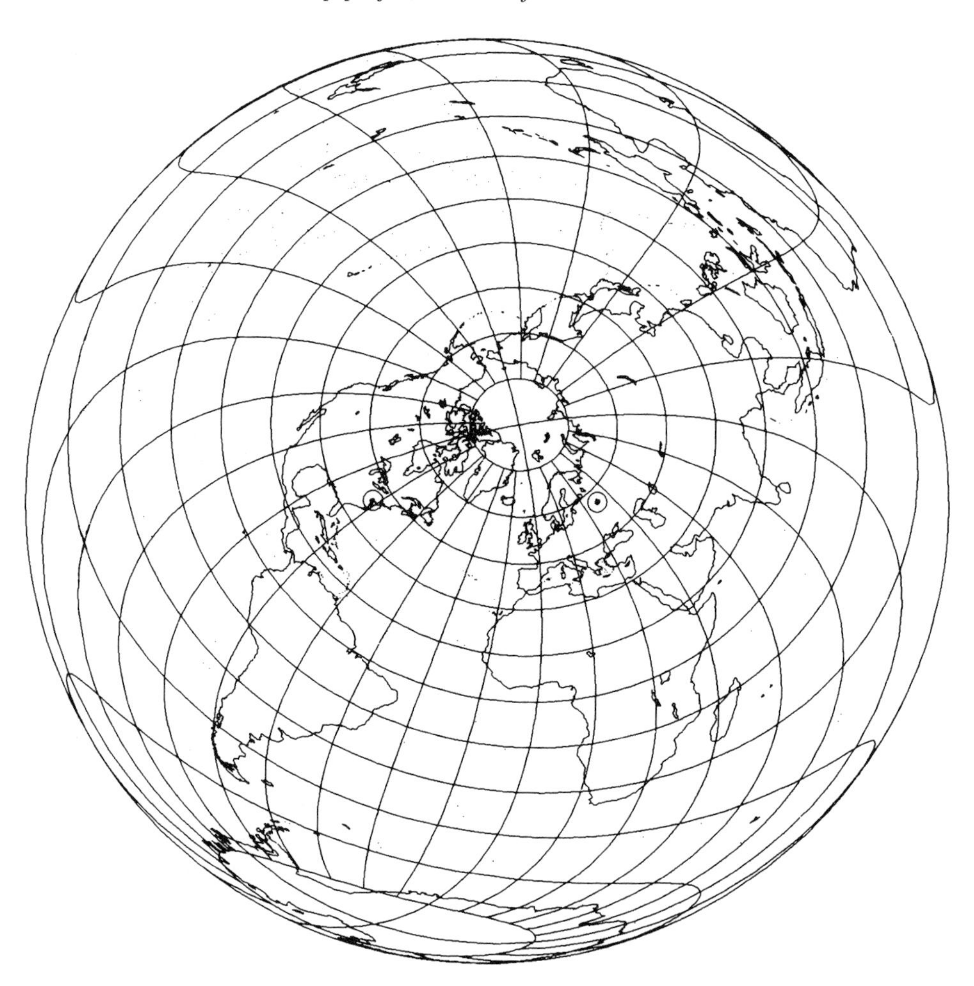

Figure 7.8 Two-point equidistant projection, focused on Washington, DC, and Moscow. 15° graticule.

points A (α) and O (β) as seen from P, using the following formulas:

$$\tan \alpha = \frac{\cos \phi \sin(\lambda - \lambda_1)}{\cos \phi_1 \sin \phi - \sin \phi_1 \cos \phi \cos(\lambda - \lambda_1)}$$

$$\tan \beta = \frac{\cos \phi_2 \sin(\lambda_2 - \lambda_1)}{\cos \phi_1 \sin \phi_2 - \sin \phi_1 \cos \phi_2 \cos(\lambda_2 - \lambda_1)}$$

If the denominator of these two equations is negative, 180° is added to or subtracted from the azimuth, so that it falls between $\pm 180°$. If α is less than β then δ is positive; otherwise δ is negative.

In the coordinate system adopted the rectangular coordinates of the projection are determined from the formulas

$$x = \rho \sin \delta, \quad y = q_0 - \rho \cos \delta$$

Local linear scale factors along the vertical and almucantar can be found from the following equations:

$$\mu_1 = \sec \varepsilon, \quad \mu_2 = (z/\sin z)\delta_\alpha$$

$$p = \mu_2, \quad \tan \varepsilon = z\delta_z$$

Here partial derivatives δ_α, δ_z, and others are determined by using equations (7.28)–(7.29) and the following equations:

$$\sin \phi = \sin z \cos \alpha \cos \phi_1 + \cos z \sin \phi_1$$

$$\sin(\lambda - \lambda_1) = \sin z \sin \alpha/\cos \phi$$

As the resulting formulas are very awkward, the local scale values can be determined by numerical methods from the rectangular coordinates of this projection. This projection (Figure 7.8) was also first presented by Maurer (1919a); C. F. Close followed independently in 1921, and developed it further.

7.10 Projections for anamorphous maps

The development of thematic cartography has necessitated the creation of anamorphous maps, with a resulting broadening of the functional capabilities of mapping and of the availability of a variety of detailed information concerning them.

Published works by L. I. Vasilevskiy, Vakhrameyeva, A. Berlyant, V. S. Tikunov, W. W. Bunge, Tobler, J.-C. Muller, Torsten Hägerstrand, Snyder, and others deal with various aspects of the creation of such maps. The study of anamorphous maps overlaps with that of cartograms, although each topic has unique phases.

Until recently, however, the general theory of projections for these maps has not been developed, and strict analytical relationships between geographical and rectangular coordinates for these projections had not been obtained.

In 1985–7 Yu. L. Bugayevskiy developed the principal concepts of this theory and of methods for making three classes of map projections for anamorphous maps: varivalent, variscaled, and with changed spatial geometry.

Equations for varivalent projections contain numerical values for various cartographic quantities expressed implicitly. General equations for these can be written in the following form:

$$x = f_1(\phi, \lambda, u), \quad y = f_2(\phi, \lambda, u), \quad u = f_3(\phi, \lambda)$$

where functions f_1, f_2 express relationships for the transformation of a surface to be mapped onto a plane, and f_3 characterizes the spatial distribution of some phenomenon of nature or society.

Variscaled projections are recommended for maps representing natural or societal objects and phenomena that are unevenly distributed in space. On these projections separate regions are contracted or expanded, and projections are designed with even sudden changes of linear and area scale. Mathematical definition is provided not for the whole projection, but for each separate meridian and parallel. On these projections separate portions of the region to be mapped can be shown with enlargements or reductions of two or more times the nominal scale.

General formulas for variscaled projections can be written in the following form (Bugayevskiy 1986):

$$X = \sum_{i=1}^{n} \sum_{j=1}^{m} p_{ij} x_{ij}, \quad Y = \sum_{i=1}^{n} \sum_{j=1}^{m} p_{ij} y_{ij}$$

where $p_{ij} = \delta_i \delta_j$ are delta functions

$$\delta_i = \begin{cases} 1 & i = k_1 \\ 0 & i \neq k_1 \end{cases} \qquad \delta_j = \begin{cases} 1 & j = k_2 \\ 0 & j \neq k_2 \end{cases}$$

k_1, k_2 are, respectively, the numbers of parallels and meridians being computed; x_{ij}, y_{ij} are rectangular coordinates of the points of intersection of parallel i and meridian j. They can be obtained from the simultaneous solution of the set

$$F_1(x_i, y_i, \lambda, A) = 0$$

$$F_2(x_i, y_i, \phi, B) = 0$$

where A and B are vectors of the constant parameters, the application of which makes it possible to provide a given expansion or contraction of the portrayal of a given region of the map.

Since the equations referred to above for meridians and parallels, as a rule, are unknown, they are derived using approximations.

Projections with changed spatial geometry differ greatly from the above projections for anamorphous maps. The difference is that the objects to be mapped are shown not only taking into account their geographical locations, but also with proper recognition of their functional relationship expressed in tons, monetary value, time, and other relevant units. This gives rise to a so-called transformation in Euclidean geometry to a given projection. The general equations for projections of this class can be written in the following form:

$$x = f_1(\phi, \lambda, T), \quad y = f_2(\phi, \lambda, T)$$

or

$$x = f_5(\phi_e, \lambda_e), \quad y = f_6(\phi_e, \lambda_e)$$

or

$$x = f_3(x', y', T), \quad y = f_4(x', y', T)$$

$$\phi_e = F_1(\phi, \lambda, T), \quad \lambda_e = F_2(\phi, \lambda, T)$$

where x', y' are rectangular coordinates on the original projection, T is a vector of the indexes to be mapped (time, cost, etc.); ϕ_e, λ_e are geographical coordinates transformed under the influence of vector T.

All the functions f_1–f_6 are continuous and unambiguous, and their Jacobians are not equal to zero. The above projections are for the improvement of thematic anamorphous maps.

An important feature of the research carried out is that it dispels the doubts of some cartographers about subjectivity and the mathematically indeterminate form of these maps. Now we can say with certainty that anamorphous maps, for which a theory and methods of deriving map projections have now been developed, possess all the characteristic features of cartographic products.

7.11 Isometric coordinates and conformal cylindrical projections of the triaxial ellipsoid

The squares of linear elements of the triaxial ellipsoid can be written in the form (see equation (1.127))

$$ds^2 = E\,du^2 + 2F\,du\,d\lambda + G\,d\lambda^2 \tag{7.30}$$

where, using the coordinate system discussed in 1.1.6 and adopted in (1.17)–(1.22),

$$\begin{aligned} E &= d^2 \sin^2 u + c^2 \cos^2 u, \quad G = (d^2 + d_\lambda^2)\cos^2 u \\ F &= -dd_\lambda \sin u \cos u, \quad H^2 = EG - F^2 \end{aligned} \tag{7.31}$$

Let us write an equation relating the squares of linear elements (7.30) and the plane and, following Jacobi, factor it into two components

$$dy \pm idx = (A \pm iB)\left[\sqrt{E}\,du + \left(\frac{F}{\sqrt{E}} \pm i\sqrt{G - \frac{F^2}{E}}\right) d\lambda\right] \tag{7.32}$$

where $i = \sqrt{-1}$.

Let us preassign values A and B so as to make the right-hand sides of these equations exact differentials. Then, letting $F = 0$ and

$$dy \pm idx = f_1(u, \lambda)\,du \pm if_2(u, \lambda)\sqrt{G}\,d\lambda \tag{7.33}$$

we carry out a transformation, the result of which is that this equation will correspond to expression (7.32) and ensure finding the desired projection.

Taking into account the above formulas (7.31) we rewrite (7.33) in the form

$$dy \pm idx = bf_2(u, \lambda)\cos u\left(\frac{f_1(u, \lambda)}{bf_2(u, \lambda)\cos u}\,du \pm i\,\frac{1}{b}\,(d^2 + d_\lambda^2)^{1/2}\,d\lambda\right) \tag{7.34}$$

Now let us consider all the terms of these equations. For a conformal projection we may write

$$f_2(u, \lambda)\cos u = 1/m \tag{7.35}$$

where m is the local linear scale factor on the projection. Combining the above formulas,

$$m = \sqrt{EG_0}/H = \sec\phi^0\sqrt{1 - p^2\sin^2\phi^0}\,\frac{\sqrt{1 + z^2}}{\sqrt{1 + z^2\cos^2\phi^0}} \tag{7.36}$$

where

$$G_0 = d^2 + d_\lambda^2 = d^2(1 + z^2)$$

Utilizing the squares of linear elements in isometric form in (1.29), from equations (7.35)–(7.36) we get

$$P = bf_2(u, \lambda)\cos u = b\cos\phi^0\,\frac{\sqrt{1 + z^2\cos^2\phi^0}}{\sqrt{(1 + z^2)(1 - p^2\sin^2\phi^0)}}$$

We impose the condition that on the projection to be derived, true scale is to be preserved along the equator; then from (7.33)

$$\eta = \frac{1}{b}\int_0^\lambda \sqrt{G_0}\, d\lambda = \frac{1}{b}\int_0^\lambda (d^2 + d_\lambda^2)^{1/2}\, d\lambda, \quad x = b\eta$$

Particular values of isometric longitudes η and abscissas x can be calculated with the formulas

$$\eta = B_0\lambda + B_5 \sin 2\lambda - B_6 \sin 4\lambda - B_7 \sin 6\lambda - B_8 \sin 8\lambda + \cdots$$

where

$$B_0 = 1 + \frac{1}{4}k^2 + \frac{13}{64}k^4 + \frac{326}{1600}k^6 + \frac{1877}{16\,384}k^8 + \cdots$$

$$B_5 = \frac{1}{4}k^2 + \frac{3}{16}k^4 + \frac{95}{512}k^6 + \frac{735}{4096}k^8 + \cdots$$

$$B_6 = \frac{1}{64}k^4 + \frac{5}{256}k^6 - \frac{121}{4096}k^8 + \cdots$$

$$B_7 = \frac{15}{512}k^6 - \frac{175}{8192}k^8 + \cdots$$

$$B_8 = \frac{121}{16\,384}k^8 + \cdots$$

where k is found from equations (1.18). In order to make $F = 0$ and move from equation (7.32) to equations (7.33) and (7.34), it is necessary to set some specific values for the expression

$$\Phi(u, \lambda) = \frac{f_1(u, \lambda)}{f_2(u, \lambda)\cos u} \tag{7.37}$$

Using for a first approximation

$$\Phi(u, \lambda)\, du = \frac{1}{\cos u}\sqrt{E}\, du = (d^2 \tan^2 u + c^2)^{1/2}\, du = d\, d\tau$$

taking into account (from equations (1.20))

$$\tan u = \frac{c}{d}\tan \phi^0, \quad du = \frac{c}{d}\cos^2 u \sec^2 \phi^0\, d\phi^0$$

and assuming that

$$M' = \frac{d(1 - p^2)}{(1 - p^2 \sin^2 \phi^0)^{3/2}}$$

$$N' = \frac{d}{(1 - p^2 \sin^2 \phi^0)^{1/2}}, \quad r' = N' \cos \phi^0 \tag{7.38}$$

we get

$$d\tau = (M'/r')\, d\phi^0 \tag{7.39}$$

Integrating equation (7.39) along the given meridian, for which d and p are constants,

$$\tau = \ln U', \quad U' = \tan\left(\frac{\pi}{4} + \frac{\phi^0}{2}\right) \Big/ \tan^p\left(\frac{\pi}{4} + \frac{\psi^0}{2}\right)$$

$$\psi^0 = \arcsin(p \sin \phi^0)$$

where

$$\tan \phi^0 = (1 + z^2)^{1/2} \tan \phi, \quad p^2 = 1 - (c/d)^2$$

From (7.34) in the given approximation

$$y = b\ d\ \tau = b\xi' \tag{7.40}$$

where ξ' is an approximate value of the isometric latitude. We use the expressions for scale in order to define more exactly function (7.37) and the values of isometric latitude ξ.

Taking into account (7.40) and (7.39), we find that

$$m' = \frac{1}{M'} y_{\phi'} = \frac{d}{M'} \tau_{\phi'} = (1 - p^2 \sin^2 \phi^0)^{1/2} \sec \phi^0 \tag{7.41}$$

A comparison of (7.41) and (7.36) shows that the two formulas do not differ greatly; hence, the values of ordinate y from (7.40) will be close to the values desired for Y on the conformal cylindrical projection being made. In view of the condition of projection symmetry, it is possible to write the approximating expression

$$Y = y + f(x, y)$$

as well as

$$m = (1/M)Y_\phi, \quad m' = (1/M')y_{\phi'}, \quad Y_\lambda = \sqrt{m^2G - G_0} \tag{7.42}$$

As an example we assume that

$$Y = y + a_1 y^k \tag{7.43}$$

Then from (7.42) we let

$$\delta = \left(\frac{m}{m'} - 1\right) = k a_1 y^{k-1}$$

Then equation (7.43) will take the form

$$Y = y + \frac{y\delta}{k} = y + \Delta y \tag{7.44}$$

We find parameter k (not the k used before equation (7.43)) from the condition that the values of local linear scale factors, calculated with numerical methods from (7.44) for each meridian, will be maximum close to their predicted theoretical values.

On finding the desired ordinate (7.44), it is not difficult to calculate isometric latitudes on the triaxial ellipsoid

$$\xi = Y/b$$

Now with isometric coordinates ζ, η it is possible to produce various map projections of the triaxial ellipsoid (see also L. M. Bugayevskiy 1987a). We should note that triaxial ellipsoids are accepted for Mars' satellites Phobos ($a = 13.5$ km, $b = 10.7$ km, $c = 9.6$ km) and Deimos ($a = 7.5$ km, $b = 6.0$ km, $c = 5.5$ km), and for Jupiter's satellite Amalthea ($a = 135$ km, $b = 85$ km, $c = 77.5$ km) (Tyuflin and Abalakin 1979).

7.12 Map projections for maps on globes

Globes can be made by pasting map gores onto balls or by making hemispheres. With the first approach, map projections of zones or gores bounded by meridians are constructed as wide as 30° in longitude ($\pm 15°$ from the straight central meridian) and with a latitude range of $\pm 70°$.

The projection for these zones can be a slightly modified ordinary polyconic, preserving distance along the central meridian and along all parallels, with some distortion along the outer meridian, but close to graphical accuracy. Projection formulas for a zone for ball globes may be given in the form (Ginzburg and Salmanova 1964, pp. 171–2)

$$x = \rho \sin \delta, \quad y = s + \rho(1 - \cos \delta)$$

$$\rho = kR \cot \phi, \quad \delta = \lambda(\sin \phi)/k$$

where k is a constant parameter, usually equal to 2, and s is the meridian distance of ϕ from the equator. If $\phi = 0$, $x = kR\lambda$. To allow for the deformation from the curvature of the ball while maps are being pasted onto it, projection coordinates are multiplied by a constant coefficient determined from experience. The azimuthal equidistant projection is used for polar caps (see section 3.2.4).

For the second approach, developed at TsNIIGAiK, thin film is used as a medium for cartographic representations. The film is capable of stretching uniformly in both longitudinal and transverse directions when transformed from a plane onto a hemisphere of the globe. This transformation is carried out by thermal treatment of the film and its formation into a hemisphere with special equipment.

The cartographic transformation printed on the film involves a modified azimuthal equidistant projection (Boginskiy 1985) with proper adjustment for the deformation or dimensional change that occurs in the course of the work:

$$x = \rho \sin a, \quad y = \rho \cos a,$$

$$\rho = kRz, \quad z = 90° - \phi, \quad a = -\lambda$$

where k is a constant parameter depending on the degree of deformation.

7.13 Mapping geodetic lines, loxodromes, small circles, and trajectory lines of artificial satellites of the Earth

Showing the position of special lines on maps includes the solution of the following problems:

1. determining the geodetic coordinates of intermediate points of these lines;
2. plotting these lines on maps using the rectangular coordinates of these points;

3. estimating the magnitude of deviation of these lines from straight lines or circles.

The principal problem is the first one, because calculating the projection from available formulas and solving subsequent problems is not difficult.

7.13.1 Determining geodetic coordinate points of position lines

Determining geodetic coordinates of intermediate points of geodetic line segments on an ellipsoid

This problem can be solved in two ways. The first is advisable for cases where the terminal points of the geodetic line segment are widely separated ($s > R/6$, or distances above 1000 km).

In that case we can solve the inverse geodetic problem on the ellipsoid, and then find the azimuth α_{12} from the first point (ϕ_1, λ_1) to the second point (ϕ_2, λ_2) and the distance s_{12} between them (Thomas 1970). After that, applying azimuth α_{12} at the first point and the lengths of segments $s_{ij} < s_{12}$, we solve the direct problem on the ellipsoid, e.g. by the Bessel method or using other approximation series (Thomas 1952, pp. 53–6; Thomas 1970; Morozov 1979). As a result we can find the geodetic coordinates of the first and subsequent intermediate points along the geodetic line.

The second method can be used for cases where the distance between the points is not very large ($s < R/6$). We determine the azimuth a_{12} of the normal section to a permissible accuracy with these formulas (Bugayevskiy 1986a) instead of azimuth α_{12} of the geodetic line:

$$\cot a_{12} = [\tan\phi_2 \cos\phi_1 \csc(\lambda_2 - \lambda_1) - \sin\phi_1 \cot(\lambda_2 - \lambda_1)]$$
$$+ e^2 \cos\phi_1 \csc(\lambda_2 - \lambda_1)\left(\frac{N_1}{N_2} \sec\phi_2 \sin\phi_1 - \tan\phi_2\right)$$

where

$$\frac{N_1}{N_2} = 1 + \frac{e^2}{2}(\sin^2\phi_1 - \sin^2\phi_2)$$
$$+ \frac{e^4}{4}\left(\frac{3}{2}\sin^4\phi_1 - \frac{1}{2}\sin^4\phi_2 - \sin^2\phi_1 \sin^2\phi_2\right) + \cdots$$

Applying Clairaut's formula, we can find a constant c from the latitude and azimuth at the first point:

$$c = r_1 \sin a_{12}$$

On selecting the latitudes ϕ_i of intermediate points and introducing the symbol

$$k = \cos\phi_1 \tan\phi_i\left[1 + e^2\left(\frac{N_1}{N_i}\csc\phi_i \sin\phi_1 - 1\right)\right]$$

we can find the longitudes of these points from the formula

$$\tan\frac{(\lambda_i - \lambda_1)}{2} = \frac{1}{(k + \sin\phi_1)}\left(\frac{\sqrt{r_i^2 - c^2}}{c} \pm \sqrt{\frac{r_i^2 - c^2}{c^2} + (\sin^2\phi_1 - k^2)}\right)$$

Calculation of intermediate points along orthodromes (great circles) on a sphere
Calculation may be accomplished using the formulas (Ginzburg and Salmanova 1964, p. 335)

$$\cot A = \cot \phi_1 \tan \phi_2 \csc(\lambda_2 - \lambda_1) - \cot(\lambda_2 - \lambda_1)$$

$$\tan \phi = \tan \phi_1 \csc A \sin[(A - \lambda_1) + \lambda]$$

where ϕ_1, λ_1 and ϕ_2, λ_2 are latitudes and longitudes of the initial and final points, respectively, of a segment of a great circle or orthodrome on a sphere.

Using the value of A (not an azimuth) from the first formula and choosing longitudes λ of intermediate points along the great circle, we can determine the latitudes of these intermediate points using the second formula.

Somewhat longer formulas may be used in the same manner, but α_{12} is the azimuth from point 1:

$$\tan \alpha_{12} = \frac{\cos \phi_2 \sin(\lambda_2 - \lambda_1)}{\cos \phi_1 \sin \phi_2 - \sin \phi_1 \cos \phi_2 \cos(\lambda_2 - \lambda_1)}$$

$$\tan \phi = \frac{\sin \phi_1 \sin \alpha_{12} \cos(\lambda_2 - \lambda_1) + \cos \alpha_{12} \sin(\lambda_2 - \lambda_1)}{\cos \phi_1 \sin \alpha_{12}}$$

Determination of intermediate points along a loxodrome on the sphere or ellipsoid
From the geodetic coordinates of each terminal point on the loxodrome or rhumb line we can calculate the rectangular coordinates of this point using the formulas for the Mercator projection:

$$x = r_0 \lambda, \quad y = r_0 \ln U$$

where

$$U = \frac{\tan(\pi/4 + \phi/2)}{\tan^e(\pi/4 + \psi/2)}, \quad \psi = \arcsin(e \sin \phi)$$

$$r_0 = N_0 \cos \phi_0, \quad e^2 = 1 - (b/a)^2$$

We calculate the direction angle (clockwise from north), using subscripts 1 and 2 as above,

$$\alpha_{r12} = \arctan\left(\frac{x_2 - x_1}{y_2 - y_1}\right)$$

Now, choosing latitudes ϕ_i for intermediate points along the loxodrome, we calculate $\ln U_i$ and then find the longitudes of these points from the formula

$$\lambda_i - \lambda_1 = \tan \alpha_{r12}(\ln U_i - \ln U_1)$$

Determining intermediate points along small circles
There may be three situations in solving this problem, depending on the initial data.

In the first situation the coordinates of the center point or pole of a small circle and its zenith distance from the pole are given, in the second case geographical coordinates of the pole and a point situated along the small circle are given, and in the third case the coordinates of three points situated along the small circle are given. The problem can be solved on the surface of either a sphere or ellipsoid, but we shall only provide the formulas for the sphere.

Let us consider these problems beginning with the first case. Given ϕ_0, λ_0, z and selecting azimuths $a = n\Delta a$ ($n = 0, 1, 2, \ldots$) with a defined step interval, we calculate

$$\begin{aligned} \sin \phi &= \sin z \cos a \cos \phi_0 + \cos z \sin \phi_0 \\ \sin(\lambda - \lambda_0) &= \sin z \sin a \sec \phi \end{aligned} \tag{7.45}$$

For the second case, locations ϕ, λ of points along the small circle are determined with the same formulas (7.45), but the zenith distance from the pole to the given point ϕ_1, λ_1 along the small circle is determined with the formula

$$\cos z = \sin \phi_1 \sin \phi_0 + \cos \phi_1 \cos \phi_0 \cos(\lambda_1 - \lambda_0) \tag{7.46}$$

In the third case, positions ϕ, λ along the small circle are found from equations (7.45), but first the pole location ϕ_0, λ_0 is found from the following formulas and the given points ϕ_n, λ_n ($n = 1, 2, 3$); then zenith distance z is found from (7.46):

$$\begin{aligned} \tan \lambda_0 = {} & [T(\cos \phi_1 \cos \lambda_1 - \cos \phi_3 \cos \lambda_3) - (\cos \phi_1 \cos \lambda_1 \\ & - \cos \phi_2 \cos \lambda_2)]/[T(\cos \phi_3 \sin \lambda_3 - \cos \phi_1 \sin \lambda_1) \\ & + (\cos \phi_1 \sin \lambda_1 - \cos \phi_2 \sin \lambda_2)] \end{aligned}$$

$$\begin{aligned} \tan \phi_0 &= \frac{\cos \phi_2 \cos(\lambda_2 - \lambda_0) - \cos \phi_1 \cos(\lambda_1 - \lambda_0)}{\sin \phi_1 - \sin \phi_2} \\ &= \frac{\cos \phi_3 \cos(\lambda_3 - \lambda_0) - \cos \phi_1 \cos(\lambda_1 - \lambda_0)}{\sin \phi_1 - \sin \phi_3} \end{aligned}$$

where

$$T = (\sin \phi_1 - \sin \phi_2)/(\sin \phi_1 - \sin \phi_3)$$

Showing groundtracks of artificial satellites of the Earth on maps

This problem is reduced to determining the geodetic coordinates of points along the trajectory or groundtrack from the given elements of the orbit, and then calculating the corresponding rectangular coordinates on a given map projection.

We can use the following Kepler two-body elements for undisturbed motion: i as orbital inclination; Ω as longitude of the ascending node, ω as the angle of pericenter from the ascending node; a as the semimajor axis of the elliptical orbit; e as its first eccentricity; and τ as the time of the satellite's passage through the pericenter (Figure 7.9).

These elements are functions of time for a disturbed motion, but they can be considered constant for a definite moment of time (or a short period of time). Assuming that for a given period of time they are known, let us determine the coordinates of points for the groundtrack under the satellite in the following sequence.

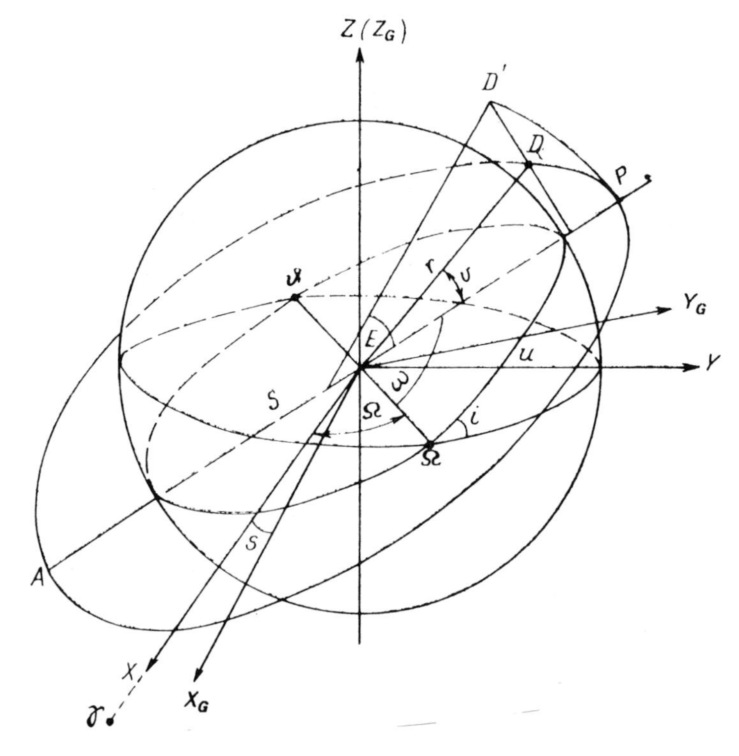

Figure 7.9 Elements of elliptical orbits.

Let us choose sequential values of the true anomaly v, or angle from the pericenter (Urmayev 1981), and calculate the values of eccentric anomalies

$$E = 2 \arctan[c_1 \tan(v/2)]$$

where

$$c_1 = \sqrt{(1 - e)/(1 + e)}$$

To calculate the radius, i.e. vector r, and angle u,

$$r = a(1 - e \cos E), \quad u = \omega + v$$

We find the spatial coordinates of a number of orbital points in an inertial system thus:

$$x = r(\cos \Omega \cos u - \sin \Omega \sin u \cos i)$$

$$y = r(\sin \Omega \cos u + \cos \Omega \sin u \cos i)$$

$$z = r \sin u \sin i$$

Let us denote the position vector of an object in the inertial coordinate system at epoch T_0 of an assigned inertial system, i.e.

$$\boldsymbol{r}_u = \begin{pmatrix} x \\ y \\ z \end{pmatrix}$$

and the position vector of the object in the Greenwich coordinate system

$$\boldsymbol{R} = \begin{pmatrix} X \\ Y \\ Z \end{pmatrix}$$

Let us transform the inertial coordinate system into a Greenwich system. In this case the transformation can be made with the formula (Urmayev 1981)

$$\boldsymbol{R} = \boldsymbol{S}\boldsymbol{r}_u$$

where

$$\boldsymbol{S} = \begin{pmatrix} \cos s & \sin s & 0 \\ -\sin s & \cos s & 0 \\ 0 & 0 & 1 \end{pmatrix}$$

s is true sidereal time in Greenwich, converted to degrees:

$$s = 15[s_0 + 1.002\,738(UT_1)]$$

s_0 is sidereal time in Greenwich at Greenwich mean midnight, and UT_1 is Greenwich mean time, both in hours; 1.002 738 is the ratio of an hour of sidereal time to mean solar time.

Let us now calculate geodetic coordinates ϕ, λ, H of orbital points along the trajectory (Morozov 1979):

$$\tan \phi = \frac{Z + e'^2 b \sin^3 \theta}{\rho - e^2 a \cos^3 \theta}, \quad \tan \lambda = \frac{Y}{X}$$

$$\tan \theta = \frac{Z}{\rho}\left(\frac{a}{b}\right), \quad H = \rho \sec \phi - N$$

$$\rho = \sqrt{X^2 + Y^2}, \quad N = a(1 - e^2 \sin^2 \phi)^{-1/2}$$

where a, b, e, e' refer to dimensions of the Earth ellipsoid. For the GRS 80 ellipsoid (other ellipsoids will give slight numerical differences),

$$a/b = 1.003\,364\,092, \quad e'^2 b = 42\,841.340 \text{ m}$$

$$e^2 a = 42\,697.701 \text{ m}, \quad a = 6\,378\,137 \text{ m}$$

In the 1970s, A. P. Colvocoresses of the US Geological Survey originated the concept and Snyder (1987b, pp. 214–29) developed the formulas for the space oblique Mercator (SOM) projection for the continuous plotting of imagery from Landsat, an artificial satellite devised for mapping the Earth from space. The projection is one of the standard formats used for these images. The groundtrack is shown at true scale on the projection, and the narrow band of sensing, approximately 185 km wide, is conformal within a few millionths of the scale factor. The formulas used assume an ellipsoidal Earth and a circular orbit for Landsat, for which five satellites were launched from 1972 to 1984. Landsats 1, 2, and 3 orbited at a nominal height of 919 km above the Earth's surface, while Landsats 4 and 5 nominally orbited at a height of 705 km. (A sixth launch failed in 1993, and the launching of a seventh has an uncertain future.)

The formulas for the SOM are lengthy, but make use of Fourier series, for which coefficients are calculated once to apply to a given orbit. A second series of projections, called satellite-tracking projections (Snyder 1987b, pp. 230–8), was developed showing groundtracks as straight lines on cylindrical or conic projections, but the projections have not been used in practice.

8

Numerical methods in mathematical cartography

Theoretical foundations and practical applications of numerical methods were treated by Urmayev (1953), as well as by Ginzburg and Salmanova (1964, pp. 263–97) and others.

When solving a problem in mathematical cartography, methods involving the theory of interpolation, numerical differentiation and integration, and approximation are used. Consequently, so-called finite differences are often used. Let the value be given for some function $f(t)$ corresponding to the values of argument t at equal step intervals (Table 8.1).

To calculate the first finite differences between entries in the second column of Table 8.1, values of f on a given line above are sequentially subtracted from values of f on the line below, and the differences are recorded in the third column (halfway between the lines of the second column), i.e.

$$f_{+2} - f_{+1} = f^{\mathrm{I}}_{+3/2}, \quad f_{+1} - f_0 = f^{\mathrm{I}}_{+1/2}, \quad f_0 - f_{-1} = f^{\mathrm{I}}_{-1/2}, \quad \cdots$$

The second and subsequent differences are obtained similarly:

$$f^{\mathrm{I}}_{+3/2} - f^{\mathrm{I}}_{+1/2} = f^{\mathrm{II}}_{+1}, \quad f^{\mathrm{I}}_{+1/2} - f^{\mathrm{I}}_{-1/2} = f^{\mathrm{II}}_0, \quad f^{\mathrm{I}}_{-1/2} - f^{\mathrm{I}}_{-3/2} = f^{\mathrm{II}}_{-1}$$

$$f^{\mathrm{II}}_{+1} - f^{\mathrm{II}}_0 = f^{\mathrm{III}}_{+1/2}, \quad f^{\mathrm{II}}_0 - f^{\mathrm{II}}_{-1} = f^{\mathrm{III}}_{-1/2}, \quad \cdots$$

Finite differences can be expressed as values of the function itself:

$$f^{\mathrm{I}}_{+1/2} = f_{+1} - f_0, \quad f^{\mathrm{II}}_{+1} = f_{+2} - 2f_{+1} + f_0$$

In a generalized form,

$$f^{(n)}_{+n/2} = f_{+n} - c^1_n f_{+(n-1)} + c^2_n f_{+(n-2)} - c^3_n f_{+(n-3)} + \cdots + (-1)^n f_0$$

where c^i_n are binomial coefficients.

Table 8.1 Differencing of a set of functions

		Differences			
Argument t	Function f	f^{I}	f^{II}	f^{III}	f^{IV}
t_{-2}	f_{-2}				
		$f^{\mathrm{I}}_{-3/2}$			
t_{-1}	f_{-1}		f^{II}_{-1}		
		$f^{\mathrm{I}}_{-1/2}$		$f^{\mathrm{III}}_{-1/2}$	
t_0	f_0		f^{II}_0		f^{IV}_0
		$f^{\mathrm{I}}_{+1/2}$		$f^{\mathrm{III}}_{+1/2}$	
t_{+1}	f_{+1}		f^{II}_{+1}		
		$f^{\mathrm{I}}_{+3/2}$			
t_{+2}	f_{+2}				

8.1 Interpolation and extrapolation

The theory of interpolation is used in mathematical cartography for determining the values of a function within a table (interpolation) or beyond the limits of the table (extrapolation) from their given values whether the intervals of the argument are equal or unequal. Even though computers have removed much of the need to interpolate when the function is analytical, there are other needs involving tables of empirical or iterated functions, interpolating with or without computers.

8.1.1 Interpolation formulas for equal argument intervals

Calculating the value of a function for any value of the argument can be performed with the following formulas:

1. Newton's formula (for differences downward along the diagonals):

$$f_n = f_0 + nf^{\mathrm{I}}_{1/2} + \frac{n(n-1)}{2!} f^{\mathrm{II}}_1 + \frac{n(n-1)(n-2)}{3!} f^{\mathrm{III}}_{3/2}$$
$$+ \frac{n(n-1)(n-2)(n-3)}{4!} f^{\mathrm{IV}}_2 + \cdots$$

2. Newton's formula (for differences upward along the diagonals):

$$f_n = f_0 + nf^{\mathrm{I}}_{-1/2} + \frac{n(n+1)}{2!} f^{\mathrm{II}}_{-1} + \frac{n(n+1)(n+2)}{3!} f^{\mathrm{III}}_{-3/2}$$
$$+ \frac{n(n+1)(n+2)(n+3)}{4!} f^{\mathrm{IV}}_{-2} + \cdots$$

3. Bessel's formula (for differences between the given and previous lines):

$$f_n = f_0 + nf^{\mathrm{I}}_{1/2} + \frac{n(n-1)}{2!} f^{\mathrm{II}}_{1/2} + \frac{n(n-\frac{1}{2})(n-1)}{3!} f^{\mathrm{III}}_{1/2}$$
$$+ \frac{(n+1)n(n-1)(n-2)}{4!} f^{\mathrm{IV}}_{1/2} + \cdots$$

$$f^{\mathrm{II}}_{1/2} = \tfrac{1}{2}(f^{\mathrm{II}}_0 + f^{\mathrm{II}}_1), \quad f^{\mathrm{IV}}_{1/2} = \tfrac{1}{2}(f^{\mathrm{IV}}_0 + f^{\mathrm{IV}}_1)$$

4. Stirling's formula (by differences along the given line):

$$f_t = f_0 + nf^{\mathrm{I}}_0 + \frac{n^2}{2!} f^{\mathrm{II}}_0 + \frac{n(n^2-1)}{3!} f^{\mathrm{III}}_0 + \frac{n^2(n^2-1)}{4!} f^{\mathrm{IV}}_0 + \cdots$$

$$f^{\mathrm{I}}_0 = \tfrac{1}{2}(f^{\mathrm{I}}_{-1/2} + f^{\mathrm{I}}_{1/2}), \quad f^{\mathrm{III}}_0 = \tfrac{1}{2}(f^{\mathrm{III}}_{-1/2} + f^{\mathrm{III}}_{1/2}) \cdots$$

where n is any integer or fraction representing the proportion of the argument interval. Formulas by Gauss, J. D. Everett, and Lagrange can also be used.

When solving many problems of mathematical cartography, it is sufficient to apply only a quadratic interpolation formula, for example, simplifying Bessel's equation above to the form

$$f_n = f_0 + nf^{\mathrm{I}}_{1/2} + \frac{n(n-1)}{2} f^{\mathrm{II}}_{1/2} = f_0 + nf^{\mathrm{I}}_{1/2} + \frac{n(n-1)}{4} (f^{\mathrm{II}}_0 + f^{\mathrm{II}}_1)$$

8.1.2 Determining values of functions with unequal argument intervals

For interpolation and extrapolation with either equal or unequal argument intervals, it is possible to apply the Stirling or Lagrange interpolation formulas, or various polynomials (see also sections 8.4 and 10.2).

The Lagrange interpolation formula can be written in the form

$$y = P(x) = \sum_{m=0}^{n} y_m P_m(x)$$

where

$$P_m(x) = \frac{(x - x_0)(x - x_1) \cdots (x - x_{m-1})(x - x_{m+1}) \cdots (x - x_n)}{(x_m - x_0)(x_m - x_1) \cdots (x_m - x_{m-1})(x_m - x_{m+1}) \cdots (x_m - x_n)}$$

y_m, x_m are values of the function and the argument at the given points.

8.2 *Numerical differentiation*

Numerical differentiation is used to calculate local scale factors and other characteristics of projections when the formulas are awkward or unavailable, but when values of rectangular coordinates for projections are given for graticule intersections. From these values and the formulas for local linear and area scale factors, etc. (see section 1.4), instead of the values of partial derivatives x_ϕ, x_λ, y_ϕ, y_λ that are usually calculated with analytical methods, the corresponding values of derivatives df/dt determined by numerical methods (separately for x and y at each point on the projection) are substituted. The formulas for derivatives using finite differences, carried out by differentiation of the above expressions (see section 8.1.1), can be written in the following forms.

Newton's formulas:

1. for differentiating diagonally downward,

$$\left(\frac{df}{dt}\right) = \frac{1}{\omega} (f^{\mathrm{I}}_{1/2} - \tfrac{1}{2} f^{\mathrm{II}}_1 + \tfrac{1}{3} f^{\mathrm{III}}_{3/2} - \tfrac{1}{4} f^{\mathrm{IV}}_2 + \tfrac{1}{5} f^{\mathrm{V}}_{5/2} - \cdots)$$

2. for differentiating diagonally upward,

$$\left(\frac{df}{dt}\right) = \frac{1}{\omega} (f^{\mathrm{I}}_{-1/2} + \tfrac{1}{2} f^{\mathrm{II}}_{-1} + \tfrac{1}{3} f^{\mathrm{III}}_{-3/2} + \tfrac{1}{4} f^{\mathrm{IV}}_{-2} + \tfrac{1}{5} f^{\mathrm{V}}_{-5/2} + \cdots)$$

Bessel's formulas (for differentiation by differences between the given and next lines):

$$\left(\frac{df}{dt}\right) = \frac{1}{\omega} (f^{\mathrm{I}}_{1/2} - \tfrac{1}{2} f^{\mathrm{II}}_{1/2} + \tfrac{1}{12} f^{\mathrm{III}}_{1/2} + \tfrac{1}{12} f^{\mathrm{IV}}_{1/2} - \tfrac{1}{120} f^{\mathrm{V}}_{1/2} + \cdots)$$

$$f^{\mathrm{II}}_{1/2} = \tfrac{1}{2}(f^{\mathrm{II}}_0 + f^{\mathrm{II}}_1), \quad f^{\mathrm{IV}}_{1/2} = \tfrac{1}{2}(f^{\mathrm{IV}}_0 + f^{\mathrm{IV}}_1), \quad \cdots$$

Stirling's formulas (for differentiating along the given line):

$$\left(\frac{df}{dt}\right) = \frac{1}{\omega}\left(f^{\mathrm{I}}_0 - \tfrac{1}{6}f^{\mathrm{III}}_0 + \tfrac{1}{30}f^{\mathrm{V}}_0 - \tfrac{1}{140}f^{\mathrm{VII}}_0 + \cdots\right)$$

$$f^{\mathrm{I}}_0 = \tfrac{1}{2}(f^{\mathrm{I}}_{-1/2} + f^{\mathrm{I}}_{1/2}), \quad f^{\mathrm{III}}_0 = \tfrac{1}{2}(f^{\mathrm{III}}_{-1/2} + f^{\mathrm{III}}_{1/2}), \quad \cdots$$

Derivatives for any point k can be determined directly from the values of a given function for a series of equally spaced points (along a line or a column) by using the formulas

$$\left(\frac{df}{dt}\right)_k = \frac{1}{12\omega}\left[(f_{k-2} - f_{k+2}) - 8(f_{k-1} - f_{k+1})\right]$$

$$\left(\frac{df}{dt}\right)_k = \frac{1}{12\omega}\left[3f_{k+1} + 10f_k - 18f_{k-1} + 6f_{k-2} - f_{k-3}\right]$$

$$\left(\frac{df}{dt}\right)_k = \frac{1}{12\omega}\left[f_{k+3} - 6f_{k+2} + 18f_{k+1} - 10f_k - 3f_{k-1}\right]$$

where ω is the interval of the argument in radians.

8.3 Numerical integration

Numerical integration is used to determine the rectangular coordinates of points on the projection from the given scale factors or distortion values. In the general case, this problem requires the integration of differential equations in the form of partial derivatives or ordinary differential equations and can be reduced to solving the Euler–Urmayev and Tissot–Urmayev equations (see section 6.4).

In a number of cases, e.g. when creating cylindrical and azimuthal projections, this problem can be reduced to the calculation of definite integrals. Here the Adams, Cowell, Gauss quadrature, and other methods of differencing can be used.

The Cowell formulas have the form

$$x_1 = x_0 + \Delta_{1/2}$$

$$x_{1+1} = x_1 + \Delta_{(1+1/2)}$$

$$\vdots$$

$$x_{n+1} = x_n + \Delta_{(n+1/2)}$$

$$\Delta_{k+1/2} = \omega(f_{(k+1/2)} + \sigma_{(k+1/2)})$$

where

$$f_{(k+1/2)} = \tfrac{1}{2}(f_k + f_{(k+1)})$$

and the initial values of the function (e.g. scale factors) at points k and $(k+1)$, i.e. at the given and subsequent points, are as follows:

$$\sigma_{(k+1/2)} = -\frac{1}{12}f^{\mathrm{II}}_{(k+1/2)} + \frac{11}{720}f^{\mathrm{IV}}_{(k+1/2)} - \frac{191}{60\,480}f^{\mathrm{VI}}_{(k+1/2)} + \cdots$$

$$f^{II}_{(k+1/2)} = \tfrac{1}{2}(f^{II}_{k} + f^{II}_{(k+1)}), \quad f^{IV}_{(k+1/2)} = \tfrac{1}{2}(f^{IV}_{k} + f^{IV}_{(k+1)}), \quad \ldots$$

Numerical integration, as well as numerical differentiation and interpolation, can be used not only for solving the problems mentioned above, but also for developing new map projections.

8.4 Approximation

Approximation functions, including polynomials, are used in mathematical cartography to design and transform map projections, or for mathematical descriptions of sketches of map graticules that were drawn to calculate values of functions at any point from their given values on regular or arbitrary graticules, etc. The main propositions concerning approximating polynomials and their applications are described in section 10.2.2, in Boginskiy (1972), and elsewhere.

9

Choice and identification of map projections

9.1 *Theoretical fundamentals of choosing a map projection*

When designing any map, the problem of choosing a map projection is of great importance. Analysis of the general problem involves finding an optimum solution of various problems relating to the map. The choice of map projection depends on many factors, which can be subdivided into three groups.

Factors characterizing the region being mapped are placed in the first group. These include the geographical position of the region, its dimensions, the shape of its outline (configuration), and the degree to which adjacent regions are also to be represented.

The second group includes the factors characterizing the map being designed and the methods and conditions of its use: the purpose and special features of the map, scale, contents, the problems to be solved with the map (cartometric, navigational, etc.), accuracy requirements, the nature of the display (on a table, on a wall, or in an atlas), cartographic information analysis (computer generated or not), conditions for working with the map (separately or in combination with other maps, mosaicked or not), requirements for representing the relative characteristics of the regions being mapped (geographical placement of regions with respect to each other), coverage requirements, outlines, the desire to show lines of communication and other links between regions, etc.

Factors characterizing the map projection to be used are placed in a third group. They include the type of distortion, requirements for minimizing distortion and the acceptable levels of distortion of distance, angles, and areas, the pattern of distortion distribution, and the curvature of the plotted graticule lines, geodetic lines, loxodromes, and/or other desired lines. In addition, the third group includes the ability of the projection to represent the entire region, requirements for orthogonality or limits of deviation from a right angle for the intersections of meridians and parallels, graticule equidistance, the manner of representing poles, the extent of symmetry of the graticule with respect to the central meridian and the equator, and relative lengths of the equator, central meridian, and the poles if the poles are represented by lines. Visual perception of the projection, how spherical it should appear, and the desire for repetition of portions of a world map also fall into the third group.

The choice of the map projection occurs in two steps. In the first step a list is made of projections (or their properties) from which the choice may be made. In the second step the desired projection is selected.

All the factors in the first group, as a rule, should be provided. Proper attention to them means, first of all, a choice of projections for which the central point and central lines (near which the scale changes slowly) are at the center of the region being mapped, and central lines are in the direction, if possible, of the greatest extent of these regions.

That is why the following are chosen for many maps:

1. normal cylindrical projections for regions situated near and symmetrically with respect to the equator, and extending east to west;
2. conic projections for regions situated near a latitude between the equator and either pole, and extending east to west;
3. normal (polar) azimuthal projections for polar regions;
4. transverse or oblique cylindrical projections for regions extending along meridians or verticals;
5. transverse or oblique azimuthal projections for nearly circular regions.

Thus, taking into account the factors in this first group makes it possible to narrow the list of possible projections (or other properties) from which a desirable projection can be chosen.

The second group of factors is the most important when solving the problem that has been imposed. It is from the conditions of this group that the relative importance of the factors in the third group is determined: which of the factors are the most important for the given situation and which factors can be ignored. When making this determination, some of the requirements (e.g. the desired pattern of projection distortion, its maximum values, the representation of the poles, symmetry or asymmetry of the graticule, intervals between meridians and parallels, feasibility of overlapping portions on the projection, and so on) should be taken into account in specific cases. This means that the projection should be chosen in this case only from a list of projections for which the imposed requirements are met completely, e.g. only among equal-area projections or only among projections with an orthogonal graticule.

Thus the factors that become essential in the particular case, in addition to the factors of the first group, make it possible to solve the first part of the problem, namely to prepare an appropriate list of projections (or their properties) from which a desirable projection may be chosen.

After separating all the factors to be taken into account, all other factors are arranged in an order of priority or hierarchy when selecting a particular projection. The choice can be performed automatically (see section 10.3) or by traditional methods, based on a comparative analysis of the various projections which may be used for designing a particular map.

When choosing projections by traditional methods, which is most widely done at present, a comparative analysis of map projections is made on the basis of the separate factors mentioned above.

As already noted, taking into account the factors of the first group makes it possible to arrange a list of projections from which a suitable projection can be selected. The effect of these factors on solving this problem is increased with an increase in the size of the region to be mapped.

To minimize and improve the distribution of distortion, especially when mapping large areas, in addition to taking into account the position of central points and lines and their relation to the geographical location of the region, one goal is to try to get isocols to coincide with a simplified outline of the region to be represented.

In the same way, one analyzes the effect of the purpose of the map, its content or type of specialization, its use, its cartographic information (with or without the help of a computer), the format of the publication, etc. Such an analysis is made for each particular case of map design.

For example, when designing maps for secondary-school students, attempts are made to draw map graticules symmetrical about the central meridian and to have meridians and parallels that are equally divided (or nearly so) with minimum curvature of the parallels. As school maps are not intended for measurement, the pattern, amount, and distribution of distortion are not subject to severe requirements. It is desirable that the map gives a visual perception of sphericity, and that the relative positions of the continents and oceans be conventional and common; on world maps the eastern extremities of Asia should normally be situated near the eastern edge of the map, and North and South America, near the western edge.

In choosing projections for maps where the principal cartographic information is represented by isolines, it is necessary to bear in mind the purpose and specialization of the maps, and the problems to be solved using them. In particular, if they are intended for making measurements of areas situated between isobars, isotherms, isogons, etc., it is recommended that equal-area or near-equal-area projections be used. If it is necessary to determine gradients of various phenomena such as magnetic declination and water salinity, or to interpolate values between isolines, it is important to use conformal projections, on which local scale does not depend on the direction.

In cases where large regions are to be shown and, consequently, length and area distortion will be rather large and cannot be neglected, it is advisable to choose projections on which it is easier to take into account this distortion, rather than using those with minimum linear distortion.

When designing small-scale maps intended for visual perception, it is more important to show the relative geographical positions of the regions, and to consider the shape of the map graticule, the sphericity effect, etc.

Additional information about some of the problems touched upon above, when choosing a map projection, is given in Ginzburg and Salmanova (1957, 1964, pp. 205–48), Bugayevskiy (1982a), Snyder (1987b, pp. 33–5), Snyder and Voxland (1989, pp. 5–7), and Maling (1992, pp. 245–65).

9.2 Distortion characteristics on projections of the former USSR, continents, oceans, hemispheres, and world maps

One should begin the choice of map projections with an analysis of projections analogous or similar to that desired for a given map design. In doing so one should utilize information available about various projection properties, such as in Table 9.1 (translated with minor revisions from Ginzburg and Salmanova, 1964, pp. 211–14, Table 23).

Table 9.1 Distortion of map projections used in the former USSR

Regions	Classes and groups of projections used in the former USSR	Limits of scale variations (%)		Maximum angular deformation (deg)
		Lengths	Areas	
1	2	3	4	5
	Maps of the former USSR			
The USSR as a whole:				
(a) without polar regions	Normal conic:			
	equidistant	5–6	5–6	2.5–3
	conformal	5–6	10–12	—
(b) including polar regions	Oblique equidistant cylindrical or close approximation	6–7	6–7	4–5
	Polyconic and others with small curvature of parallels	≥20	≥20	10–15
European part of the USSR	Normal equidistant conic	3.5	3.5	1–1.5
	Transverse conformal cylindrical (Gauss projection)	1.5	2.5	—
	Pseudoazimuthal (with oval isocols)	1.5	1.5	0.5–0.75
Western part of the USSR	Oblique azimuthal equidistant	2.5	2.5	1.25–1.5
	Maps of continents, parts of the world, or polar regions			
Europe, Australia	Oblique azimuthal equal area	4–5	—	2–2.5
South America	Oblique azimuthal equal area	6–7	—	3.5–4
North America, Africa	Oblique azimuthal equal area	9–10	—	5–6
Eurasian continent	Oblique azimuthal equal area	23–25	—	12–14
Polar regions (from the pole to the parallel of lat. 60°)	Polar azimuthal equidistant	5	5	2.5–3
	Polar azimuthal conformal	7	15	—
	Maps of largest oceans			
Atlantic Ocean	Pseudocylindrical equal area	45–55	—	22–25
	Transverse pseudoazimuthal (with oval isocols) with low area distortion	25–30	12–14	13–15
Atlantic and Arctic Oceans	Oblique pseudoazimuthal (with oval isocols) with low area distortion	37–42	16–18	18–20
Pacific Ocean	Pseudocylindrical equal area	85–105	—	35–40
	Urmayev pseudocylindrical (projection for ocean charts)	70–85	22–25	20–35
Pacific and Indian Oceans	Urmayev pseudocylindrical (projection for ocean charts)	70–85	22–25	30–35

Table 9.1 Continued

Regions	Classes and groups of projections used in the former USSR	Limits of scale variations (%)		Maximum angular deformation (deg)
		Lengths	Areas	
1	2	3	4	5
	Maps of hemispheres ($z_{max} = 90°$)			
Western and Eastern hemispheres or northern and southern hemispheres	Azimuthal:			
	equal area	100	—	39
	equidistant	57	57	26
	conformal	100	300	—
	with low area distortion	70	22	30
	Maps of large regions (60% of the Earth's surface) ($z_{max} = 100°–105°$)			
	Azimuthal:			
	equal area	≈150	—	45–50
	equidistant	75–80	75–80	28–33
	conformal	≈140	≈500	—
	World maps			
Excluding regions situated north and south of the parallels:	Normal equidistant cylindrical			
60°		100	100	39
75°		290	290	72
	Normal conformal cylindrical			
60°		100	300	—
75°		290	≈1400	—
	Pseudocylindrical equal area with pole lines	220–260	—	65–70
Extensions of the graticule along the equator	Kavrayskiy elliptical pseudocylindrical			
360°		150–170	90–100	50–55
≈400°–420°		180–200	90–100	55–60
For the whole surface of the Earth or a large region	Normal polyconic:			
	graticule symmetrical along equator	85–95	55–65	35–40
	asymmetrical along the equator	50–65	45–50	25–30
	With increase up to 400°–420° range of the graticule along the equator	85–115	60–70	35–45
	Pseudoazimuthal (with oval isocols); maps with central location of the continents	130–150	45–50	45–50

9.3 Projections for published maps

Various projections are used in Russia, the United States, and other countries. For economic and technological efficiency one attempts to standardize the use of projections, to use so-called typical projections when designing different maps of the same series or a group of the same types of regions. For example, many maps of continents are based on the Lambert azimuthal equal-area projection, and most navigational charts are designed using the Mercator projection.

9.4 Approximate hierarchy of requirements for map projections with varying perception and assessment of cartographic information

At present there are two principal systems or categories of maps: those for solving technical problems and those for general use. The former are intended to provide the greatest possible accuracy and availability of details for analysis, with sufficient general as well as local accuracy.

When designing maps for general usage, the variety of requirements imposed on maps should be borne in mind; hence, these maps can differ greatly in detail, completeness, and accuracy of representation, in the shape of the map graticule, in visual perception and clarity, etc.

Tables 9.1 and 9.2 provide data (primarily from works by Ginzburg and Salmanova (above) and Ledovskaya (1971), with revision by the current authors) on the principal projections used for published small-scale maps. They can be used as an aid in designing most new maps, especially those for general use.

In choosing map projections, all maps in the two systems, technical and general, described above (with a certain amount of convention) can be subdivided into five groups, differing in the method of perception and the estimation of cartographic information (Appendix 1).

Application of the data given in Appendix 1 makes it possible to obtain an idea of the magnitude of distortion which can be ignored, to develop an approximate hierarchy of requirements for map projections, and to get an idea of the character of projection distortion and scale.

9.5 Distortion requirements on various types of small-scale maps

When choosing projections, the factors of the second group mentioned above (section 9.1) are of primary importance. They characterize the map to be designed, its purpose, problems to be solved using it, and its particular contents.

Data concerning the distortion requirements of various small-scale maps, from Ginzburg and Salmanova (1964, pp. 219–21) with minor editing, can help take into account the factors in this group. The following types of maps will be considered.

Table 9.2 Map projections used in various parts of the world

Maps of large parts of or the whole former USSR

Projections of maps published in the former USSR:
Normal Krasovskiy and Kavrayskiy equidistant (rarely conformal) conics, oblique perspective cylindrical projections of Solov'ev and TsNIIGAiK. Oblique azimuthal. Transverse Mercator. TsNIIGAiK modified polyconic

Projections of maps published in the United States:
Normal equidistant and conformal conic projections. Transverse Mercator and transverse polyconic (by the National Geographic Society). Oblique Lambert azimuthal equal-area projection

Projections of maps published in other countries:
Normal equidistant, conformal, and equal-area conic projections, including those of Kavrayskiy and Krasovskiy (in the former socialist countries), and oblique azimuthal projections (generally equidistant and equal area)

Maps of large parts of and the whole United States

Projections of maps published in the former USSR:
Normal equidistant and occasionally conformal conic projections

Projections of maps published in the United States:
Normal equal-area and conformal conic projections. Occasionally polyconic, equidistant conic, and Chamberlin trimetric projections.

Projections of maps published in other countries:
Normal equal-area, conformal, and equidistant conic projections. Occasionally Chamberlin trimetric projection

Maps of large parts of the continents outside the former USSR

Projections of maps published in the former USSR:
Normal conic and cylindrical, oblique and transverse cylindrical, and oblique azimuthal projections – mostly conformal, rarely equidistant

Projections of maps published in the United States:
Normal equidistant, conformal, and equal-area conic projections; oblique Lambert azimuthal equal-area projection. Occasionally bipolar oblique conic conformal (for parts of South America), polyconic, Chamberlin trimetric, cylindrical equal-area (for central Africa), and transverse Mercator projections

Projections of maps published in other countries:
Normal conic, cylindrical (in low latitudes), and oblique azimuthal projections; very occasionally pseudocylindrical or pseudoconic. In the national cartography of a number of countries, other projections are used, such as Křovák's (double conformal) projection for maps of the former Czechoslovakia, oblique Lambert azimuthal equal-area projection for maps of India, oblique conformal conic projection for maps of Japan, and Laborde (triple-projection conformal) projection for maps of Madagascar

Maps of continents and parts of the world

Projections of maps published in the former USSR:
Lambert azimuthal equal-area projection: in oblique aspect for maps of Europe, Asia, North and South America, and Australia; in transverse (equatorial) aspect for maps of Africa; and the TsNIIGAiK azimuthal projection with low area distortion. Rarely, other azimuthals. For maps of Europe and Australia, the normal conic conformal and others as well. For Antarctica, chiefly the polar azimuthal equidistant projection. For maps of Africa, South America, and Australia as continents with surrounding waters in the *Morskoy Atlas* (*Naval Atlas*), the normal Mercator projection was used

Projections of maps published in the United States:
Oblique Lambert azimuthal equal-area, Bonne pseudoconic equal-area, conformal and equidistant conic, Chamberlin trimetric (by National Geographic Society), Hammond optimal conformal (by Hammond), Miller oblated stereographic (for Africa and Europe), polyconic projections

Table 9.2 Continued

Projections of maps published in other countries:
Oblique Lambert azimuthal equal-area projection; transverse (equatorial) aspect for maps of Africa. Rarely, other azimuthal projections, such as the azimuthal equidistant for several continents in the *Oxford Atlas*, and the Breusing geometric projection by Debes in 1895 for maps of North and South America. Maps of Africa, Australia, and South America are still published on the sinusoidal equal-area pseudocylindrical projection. Maps of Eurasia, Asia, and Europe on the Bonne pseudoconic projection. Also the normal Mercator, the Miller and Briesemeister bipolar oblique conic conformal for the Americas, as well as derived projections by Kremling, Wagner, and others

Maps of polar regions

Projections of maps published in the former USSR:
Polar azimuthal (chiefly equidistant and conformal) projections. In the *Morskoy Atlas* and in other specialized publications, also the polar Lambert azimuthal equal-area projection

Projections of maps published in the United States:
Polar azimuthal equidistant, equal-area, and conformal (by Hammond) projections

Projections of maps published in other countries:
Polar azimuthal equidistant, conformal, and equal-area projections, and in rare cases orthographic

Maps of seas and oceans (except the Arctic Ocean)

Projections of maps published in the former USSR:
Mercator projection. Oblique azimuthal projections. Normal pseudocylindrical equal-area and other projections (of Kavrayskiy, Urmayev, and others). For maps of the Pacific Ocean and combined maps of the Pacific and Indian Oceans, the Urmayev pseudocylindrical sinusoidal equal-area or near-equal-area projection is chiefly used. For the combined Atlantic and Arctic Oceans and the Atlantic Ocean alone, the TsNIIGAiK oblique pseudoazimuthal projection is mainly used

Projections of maps published in the United States:
Mercator, pseudocylindrical, and oblique Lambert azimuthal equal-area projections

Projections of maps published in other countries:
Mercator projection. The Atlantic Ocean is often represented on the Mollweide equal-area pseudocylindrical projection, sometimes on the Lambert azimuthal equal-area projection, in equatorial or oblique aspect. The Pacific Ocean is shown on equal-area pseudocylindrical projections, mainly the Mollweide and sinusoidal, rarely on the Winkel tripel, an Eckert pseudocylindrical, or the van der Grinten projection. A Wagner arbitrary projection is also used. In former socialist countries, the Urmayev pseudocylindrical projection is used. In the *Oxford Atlas*, and subsequently elsewhere, pseudoazimuthal projections with oval isocols are used

Maps of hemispheres and some larger parts of the Earth's surface

Projections of maps published in the former USSR:
Azimuthal equal-area and near-equal-area projections in various aspects; rarely, the azimuthal equidistant projection or TsNIIGAiK azimuthal projections showing sphericity

Projections of maps published in the United States:
Formerly globular and equatorial Lambert azimuthal equal-area projections. Hemispheres now rarely shown as published maps

Projections of maps published in other countries:
Lambert azimuthal equal-area projection (especially for maps of the eastern and western hemispheres). Equatorial azimuthal equidistant projection also used (especially in British publications) and in the past the similar-looking globular projection. For maps of the northern and southern hemispheres, azimuthal equidistant or equal-area projections are used more often

Table 9.2 Continued

Maps of the entire world

Projections of maps published in the former USSR:

For the usual 360° longitude range along the equator, the Mercator projection is often used; also TsNIIGAiK modified polyconics, primarily with graticules symmetrical about the equator; rarely, the Gall stereographic cylindrical projection, and equal-area and equidistant pseudocylindrical projections by Kavrayskiy, Urmayev, and others

With an increased range of longitude along the equator or at other latitudes, the Mercator projection or the TsNIIGAiK pseudocylindrical projection; rarely, TsNIIGAiK modified polyconic, the Gall stereographic, and Kavrayskiy pseudocylindrical projections are used. An Urmayev cylindrical projection was used in rare cases and, on some earlier maps, the van der Grinten circular projection and the Eckert VI pseudocylindrical projection, both uninterrupted and interrupted through oceans, designed using Goode's method

Projections of maps published in the United States:

Robinson, *The Times*, and Miller cylindrical arbitrary projections. Polar azimuthal equidistant projection. Briesemeister elliptical, interrupted Mollweide, Goode homolosine, and Eckert IV and VI equal-area projections. Mercator conformal projection. Van der Grinten arbitrary projection was used considerably in the past

Projections of maps published in other countries:

Mercator conformal projection, and equal-area Mollweide, Sanson–Flamsteed sinusoidal, and Hammer–Aitoff projections. The van der Grinten circular projection is more often used by Scandinavian countries and the People's Republic of China. The Gall stereographic cylindrical projection is used for maps by the United Kingdom and the former Czechoslovakia. In Austria, former East and West Germany, and Italy, the Eckert IV and VI equal-area and the Winkel tripel projections are used

In the former West Germany, the derivative projections of Siemon, Wagner, and Kremling have been used. The Winkel tripel projection is used in *The Times Atlas of the World*. In Oxford atlases, a pseudocylindrical modified Gall projection (see section 2.2.3) has been used. The oblique Mollweide and oblique Hammer–Aitoff equal-area projections have been used by Bartholomew (the 'Nordic' and 'Atlantis') in the United Kingdom. Combination graticules on which, in contrast to the usual variants of projections, the equator does not remain straight have been used by Bartholomew as 'Regional', 'Tetrahedral', and 'Lotus' projections

Geographical maps

On these maps one can compare regions with various physical–geographical characteristics, administrative units, and other features. One can also study the extent of shorelines, rivers, roads, lines of communication, and boundaries, as well as the shapes of relief features, the outline of shorelines, mountain ridges, and the direction of winds and currents. When satisfying various distortion requirements which are sometimes even contradictory, it is advisable to choose equidistant or nearly equidistant projections for geographical maps.

Physical maps

Representing relief and hydrography is of primary importance on these maps. These maps aid in the study and comparison of outlines and directions of river flow, valleys, and mountain ranges, as well as areas of basins, orographic (mountainous) features, etc. Hence, while equidistant or nearly equidistant projections are more suitable for the first set of factors, projections with low area distortion are most suitable for the second set, the final choice between these types of projections depending on the special nature of the various maps.

Climatic and meteorological maps

For climatic maps, from which patterns are studied with the help of isolines, it is important to preserve low area distortion; sometimes, however, it is not desirable to sacrifice the quality of appearance of outlines and shapes, as may be the case with equal-area projections. If it is necessary to interpolate many intermediate values of functions using isolines and to determine gradients, conformal projections are more suitable. Conformal projections are also advisable when the direction, velocity, and strength of winds are shown by arrows on some meteorological (weather) maps, or if the shapes of isobars and other isolines are of importance. In other cases equidistant or nearly equidistant projections are more suitable based on the set of requirements. Finally, if it is necessary to show great circles (orthodromes) as straight lines, the gnomonic projection should be used.

Geological maps

For many types of geological maps, on which bands with various geological structures are shown, the most important thing is to eliminate area distortion, the portrayal of outlines and shapes being of lesser importance, with the exception of some maps emphasizing geological faults. The same requirements are imposed on maps showing volcanic activity. Equal-area projections or those nearly so are used.

Tectonic, geomorphologic, and relief maps

For different types of tectonic and geomorphologic maps, as well as maps of the relief on the Earth's surface and on the ocean floors, projection requirements vary. When it is necessary to represent accurately the areas of regions with geological folds, various types of sedimentation, or basins, as well as the area between high plateaus, etc., it is recommended that projections be chosen with low area distortion. If it is important to show the directions of faults, mountain ranges, and ridges, the shapes and directions of rivers, and the shapes of valleys, plateaus, etc., an equidistant projection will be more suitable. In more difficult cases where it is necessary to avoid large distortion of areas, outlines, and shapes of objects, projections averaging distortion factors will be preferable.

Seismic maps

On those maps used for measuring distances from earthquake epicenters, it is important to avoid visible distortion of distance. On maps at relatively small scales and covering large areas, this requirement cannot be satisfied, so one should use projections on which it is possible to take into account the effect of scale distortion. To measure distances along orthodromes from a single seismic station, maps on the azimuthal equidistant projection are more suitable.

Maps of soils, the Earth's surface, and the ocean floors

When using these maps, it is sometimes necessary to determine areas of soil zones and of various types of earth; therefore, the requirements for distortion characteristics on projections are similar to those for geological maps.

Maps of flora and fauna
Representing areas is of primary importance for these as well, but to avoid relatively large distortion of outlines one should choose projections with slight area distortion rather than truly equal-area projections. The graticules of these maps should make it possible to compare areas of various latitude zones.

Administrative and political maps
Cartographic information is normally estimated visually from administrative and political maps of territories. If an administrative map is used to determine areas, then area distortion must not influence the results. The quality of area representation is of great importance for political maps as well, but the use of equal-area projections for political maps of the world or hemispheres can result in great distortion of the outlines and shapes of the territories of many countries. In a number of cases the most suitable projections are those closer to equidistant projections in distortion characteristics.

Historical maps
For most historical maps, it is of primary importance to show the dimensions of territories where historical events took place; the quality of outlines and shapes is of secondary importance, but projections with little distortion of area are still preferable to those that are equal area. On maps of geographical discoveries and travel, where routes of voyages around the world are shown, projections with little distortion of outlines are preferable.

Maps of population
On maps showing the distribution of peoples, population density is almost always compared with areas; hence, the use of equal-area projections is very important. This is especially true on maps where density is shown by dots. When the migration of peoples on maps of world population is shown, however, it is necessary to resort to projections with low distortion of direction.

Maps of service lines and communications
On projections that are equal area or with distortion characteristics close to equal area for general maps of continents, the Indian Ocean, and similar-sized regions, directions and the extent of roads and routes are distorted relatively little. To show service lines on maps of the largest oceans, hemispheres, and the world is more difficult, but in these cases the projections mentioned in Table 9.2 show the directions and lengths of routes generally better than equal-area or conformal projections. On sea charts, of course, service lines for ships are usually plotted on maps based on the Mercator projection, but air routes on this projection are poorly shown when they follow northern great circles. When studying the density of road networks it is better for projections to have low area distortion. Thus, for maps of service lines equidistant projections or projections with little area distortion should be chosen. In particular, for maps of the shortest distances from capitals and ports, azimuthal equidistant projections properly centered are needed.

Analogous requirements exist for projections used for maps showing communications such as telegraph, telephone, and radio.

Economic maps

In choosing projections for most economic maps, special attention should be given to how well area is represented, even at the expense of outlines and shapes. Projections close to equal area are often utilized on economic maps which have tinting or shading between outlines, e.g. maps of land cultivation, cattle breeding, or deposits of natural resources. The same distortion characteristics apply to maps of industrial production or other thematic purposes. To be consistent with other maps, many economic maps may be designed with equal-area projections. Only special types of such maps, e.g. world economic maps where traffic patterns for goods are shown, require projections that are closer to equidistant.

Navigational and aeronavigational charts

For navigational charts, conformal projections are required in most cases, centered with respect to the routes involved. For rhumb-line navigation, the normal Mercator projection is used almost exclusively. For great-circle navigation along a single route, the oblique Mercator projection is generally ideal. For great-circle navigation along many routes, the gnomonic projection centered near the preponderance of routes is standard, although its scale distortion well exceeds that of the Mercator projection.

9.6 Identification of a map projection from the shape of its graticule of meridians and parallels

It is highly desirable for maps to contain information about their projections, as it is sometimes impossible to identify the projection of a published map. This is especially true of large-scale maps, as the larger the scale, the smaller the size of the region shown on the map sheet, the less is the distortion of all types, and the more difficult it is to see differences between projections.

When identifying the class of the projection used, it is important to make a preliminary judgment, if possible, identifying it only by the appearance of the map graticule without measurements, i.e. whether it belongs to the normal conic, cylindrical, azimuthal, or other classes. Once the class of the projection is known, it is possible to make measurements and to determine the group of the projection according to its distortion characteristics. If it is difficult to identify the class of the projection from its appearance, it is necessary to make measurements.

An analysis of the map graticule should begin by identifying the pattern of meridians and parallels and the spacing between parallels along the central meridian and between meridians along the equator or central parallel. Then it is necessary to note whether the graticule is orthogonal, whether it is symmetrical, and any peculiarities of the representation of the poles.

When determining the pattern of meridians, first it is necessary to find out whether all of them or only the central one is a straight line, and whether the outer meridians are curves symmetrical about the central meridian. If the meridians are all straight lines, then it is necessary to know whether they are parallel or intersect at a common point. All these determinations are easily done graphically with the help of a straightedge.

When determining the pattern of parallels, the same problems arise. The symmetry of parallels is generally related to a straight equator; straight parallels can only be parallel.

In cases where meridians or parallels are curved, first it is necessary to find out whether they are circles (concentric or eccentric). If they are concentric circles, the spacing between each pair of adjacent parallels is constant, although it may or may not be equal to the spacing between the next pair.

Determining the projection group from the distortion characteristics in some cases can be combined with determining the class; however, in most cases research at this stage should be carried out by taking measurements from the maps. On a conformal projection, the graticule is orthogonal, and the intervals between parallels (or almucantars on oblique aspects) often increase from a central point or line to the edges of the map. It should be kept in mind, however, that these properties are also inherent to some arbitrary projections, e.g. cylindrical projections having nearly conformal distortion characteristics.

On the contrary, on equal-area projections distances between parallels (or almucantars) often decrease from the central point or line of the projection to the edges of the map.

One should note that on outer-perspective and orthographic projections, with arbitrary distortion characteristics, these distances decrease faster than on equal-area projections. Owing to the rapid decrease of distances between the map graticule lines, the representation acquires the effect of sphericity and perspective.

On projections equidistant along meridians (or verticals on oblique aspects), distances between parallels (or almucantars) do not change. The distances between parallels are also preserved on those projections where the scale along the central meridian is constant, e.g. on some pseudocylindrical projections.

If after studying the appearance of the graticule it is impossible to state what group the given projection belongs to, then it is necessary to take measurements from the map to determine the local scale along meridians and parallels. To do this, it is possible to measure the rectangular coordinates of map graticule intersections on an arbitrary system and to calculate local scale factors from formulas for numerical differentiation (see section 8.2).

After determining the class of the projection and its group according to distortion characteristics, the parameters of the given projection may be calculated and, hence, the form of its corresponding formulas. When preliminary analysis shows that the map being studied has been designed with an oblique or transverse coordinate system rather than a normal aspect, then the solution of these problems is more difficult.

In such cases one must study the curvature of meridians and parallels, measure various segments and rectangular coordinates of graticule intersections, calculate local scale factors, develop models with isocols, and conduct a comparative analysis of the data obtained with the data from known projections through special tables of projection characteristics, e.g. tables by Ginzburg and Salmanova (1964), Vakhrameyeva *et al.* (1986, p. 224), and Snyder (1987b). However, all of these studies do not necessarily give a unique answer for the projection class in question.

10

Problems and directions of automation in obtaining and applying map projections

Among the chief problems of automating mathematical cartography are the following:

1. Calculation of map projections on a computer.
2. Transformation of the map projections (cartographic representation) of base maps into desired projections.
3. Computerized selection of map projections, optimizing all the requirements when designing a particular map.
4. Automated preparation of new map projections in accordance with the requirements.
5. Computerized design of the nominal scale and arrangement of the map.
6. Identification of map projections in an automated environment.
7. Automatic identification and introduction of reduction in measurements from maps, making use of their mathematical basis.
8. Automated plotting of features using the mathematical basis.

The solutions to these problems are closely related to an adequate consideration of the foregoing theoretical and practical propositions. However, they also have specific aspects determined both by the methods involved in their solution and by the methods of applying a computer and input/output devices to the transformation. As the problem of complete automation in mathematical cartography has become extremely important, it is advisable not only to narrow the range of these problems but also to see that they are coordinated and comprehensively outlined.

10.1 Calculation of map projections on a computer

This problem can be solved in two ways. The first involves common methods, algorithms, and programs which make it possible to treat specific classes and types of projections as particular cases of some general solution.

In the other one, unique methods, algorithms, and programs are devised for every specific class (and in some cases for single variants of map projections). Software libraries are then arranged sequentially.

10.1.1 Calculating map projections on a computer using common algorithms and programs

When using this approach, it is necessary to define sets of map projections for which it is possible to develop common methods giving a theoretical basis for creating the corresponding algorithms and programs.

Without investigating the problem of defining such sets of projections in detail, as an example we shall consider the problem of developing common algorithms and programs for perspective and conformal projections to be plotted on a computer. After that, we consider calculations with these programs for any desired projections, based on the fact that it is quite possible to present them as particular cases of the general theory of the sets of projections being considered.

Calculation of perspective azimuthal projections using common methods
Rectangular coordinates x, y and local scale factors μ_1 along verticals and μ_2 along almucantars of perspective azimuthal projections of a sphere can be calculated with the formulas

$$x = \frac{(D \pm R)R \cos \phi \sin(\lambda - \lambda_0)}{D \pm R[\sin \phi \sin \phi_0 + \cos \phi \cos \phi_0 \cos(\lambda - \lambda_0)]}$$

$$y = \frac{(D \pm R)R[\sin \phi \cos \phi_0 - \cos \phi \sin \phi_0 \cos(\lambda - \lambda_0)]}{D \pm R[\sin \phi \sin \phi_0 + \cos \phi \cos \phi_0 \cos(\lambda - \lambda_0)]}$$

$$\mu_1 = \frac{(D \pm R)\{D[\sin \phi \sin \phi_0 + \cos \phi \cos \phi_0 \cos(\lambda - \lambda_0)] \pm R\}}{\{D \pm R[\sin \phi \sin \phi_0 + \cos \phi \cos \phi_0 \cos(\lambda - \lambda_0)]\}^2}$$

$$\mu_2 = \frac{D \pm R}{D \pm R[\sin \phi \sin \phi_0 + \cos \phi \cos \phi_0 \cos(\lambda - \lambda_0)]}$$

where D is the distance from the center of the sphere (radius R) to the point of view along the line through the projection center ϕ_0, λ_0. The formulas for local area scale factors and maximum angular deformation are, respectively,

$$p = \mu_1 \mu_2, \quad \sin(\omega/2) = (\mu_2 - \mu_1)/(\mu_2 + \mu_1)$$

These formulas make it possible to determine any particular variation of perspective azimuthal projections of the sphere. (See sections 3.3.1 and 3.3.2 for ellipsoidal formulas.) If in these formulas we substitute a '+' (plus) sign for the '±' sign, we will have a group of perspective azimuthal projections of the sphere with negative transformation. A '−' (minus) sign instead will result in perspective azimuthal projections with positive transformation (the projection of vertical aerospace photographs).

Calculation of Chebyshev and any other conformal projections using common methods
Every conformal projection possesses a distinct distortion pattern; in other words, isocols in every case have their own definite shapes. When developing a Chebyshev projection, one of its isocols should coincide with the outline of the region being mapped.

Hence, in order to devise various conformal projections, one can treat the corresponding isocols for these projections as contour lines on which there are constant values of local scale. In this way calculations for all conformal projections can be carried out using common methods, algorithms, and programs designed for calculating a Chebyshev projection (see section 7.2). The only difference will be in the preparation of the initial information characterizing the specific conformal projection desired.

The preparation for different conformal map projections includes the following.

For the conformal cylindrical projection (the Mercator projection), distortion is absent only along the standard parallels of latitude $\pm\phi_0$. After selecting a sufficient number of points along this parallel, their geodetic coordinates ϕ_0, λ are determined and recorded to provide the required initial information. This projection can be calculated retaining only coefficient a_0 (all other coefficients are assumed to be equal to zero) in equation (7.5).

On (Lambert) conformal conic projections, the local scale factor is equal to unity along one or two standard parallels ϕ_1 and ϕ_2, along which a sufficient number of points are selected and their geodetic coordinates ϕ_1, λ and ϕ_2, λ are recorded to provide the basic initial information. If only the first two coefficients a_0 and a_1 are retained in equation (7.5), the particular solution obtained will be that of a conformal conic projection.

On conformal azimuthal (stereographic) projections, isocols on the normal (polar) aspect coincide with parallels; for the oblique aspect, they coincide with almucantars which are concentric circles. On drawing a circle with a center $Q(\phi_0, \lambda_0)$ at the given pole on a map designed on this projection, points are selected along that circle; then their geographical coordinates ϕ, λ are determined and, in this case, they provide the initial information.

For the Lagrange projection, isocols are ovals. To obtain the initial information, an ellipse is drawn on a map designed on a conformal projection (preferably a stereographic projection). This ellipse should approximate as well as possible the outline of the region to be mapped. Then coordinates ϕ, λ of the points along the ellipse are obtained.

With methods analogous to these common methods, other conformal projections can be plotted as well.

10.1.2 Automatic calculation of map projections using methods, algorithms, and programs designed for individual projections

Formulas for rectangular coordinates, local scale factors, and other projection characteristics given in textbooks, manuals, and published papers, as well as those listed above, can be used to prepare the individual methods, algorithms, and programs which are then organized in software libraries.

10.2 Transformation of the map projection of a base map into a given projection

This question arises in many cases involving the design and investigation of maps.

Let

$$x = f_1(\phi, \lambda), \quad y = f_2(\phi, \lambda) \tag{10.1}$$

be the equations of the base map projection, and

$$X = F_1(\phi, \lambda), \quad Y = F_2(\phi, \lambda) \tag{10.2}$$

be the equations of the map to be designed.

From equations (10.1)

$$\phi = f_3(x, y), \quad \lambda = f_4(x, y) \tag{10.3}$$

Substituting (10.3) into (10.2),

$$\begin{aligned} X &= F_1[f_3(x, y), f_4(x, y)] = \Phi_1(x, y) \\ Y &= F_2[f_3(x, y), f_4(x, y)] = \Phi_2(x, y) \end{aligned} \tag{10.4}$$

It follows from these equations that there are two principal methods of transforming map projections. One, involving the determination of geographical from rectangular coordinates in advance, has some advantages over the other method, which is expressed by equations (10.4), as it is free of any limitations. With the second method, where there is a direct relationship between rectangular coordinate systems, various types of polynomials are often used to carry out transformations, but limitations may appear that result from differences in distortion of the projections involved, such as differences in representing geographical poles and in the nature of symmetry of the map graticules with respect to the central meridian and the equator.

10.2.1 Analytical transformation of map projections

This method uses analytical systems including a computer and peripheral devices for image output and input compatible with the computer, as well as corresponding software designed for rigorous solution of this problem.

If formulas (10.2) for the final projection are known (or may be obtained in advance), then the transformation can involve finding geodetic coordinates (10.3) of the initial points on the base projection from their rectangular coordinates and after that to calculate rectangular coordinates for the final projection, in accordance with the algorithms and programs developed.

Now we will cite rigorous formulas, without developing them, for calculating the geodetic coordinates of points from their rectangular coordinates ('inverse' transformations) for many of the most widely used projections of an ellipsoid.

Mercator conformal cylindrical projection

From the formulas for rectangular coordinates (see sections 1.7.3 and 2.1.2), we can obtain the longitude λ and isometric latitude q for each point:

$$\lambda = x/r_0, \quad q = \ln U = y/r_0$$

where x, y are rectangular coordinates of points on the projection; r_0 is the radius of curvature of the standard parallel at latitude ϕ_0.

Let us designate

$$\sin \phi' = \tanh q = (e_n^q - e_n^{-q})/(e_n^q + e_n^{-q}) = (U^2 - 1)/(U^2 + 1) \tag{10.5}$$

where U is calculated from (1.31)–(1.32); e_n is the base (2.718 ...) of natural logarithms; e (below) is the first eccentricity of the ellipsoid.

On expanding the formula for isometric latitude $\ln U$ into a Taylor series in terms of conformal latitude ϕ' and geodetic latitude ϕ,

$$\phi = \phi' + c_2 \sin 2\phi' + c_4 \sin 4\phi' + c_6 \sin 6\phi' + c_8 \sin 8\phi' + \cdots \tag{10.6}$$

where c_2, c_4, c_6, c_8 are constant coefficients:

$$c_2 = \left(\frac{e^2}{2} + \frac{5}{24}e^4 + \frac{e^6}{12} + \frac{13}{360}e^8 + \cdots\right)$$

$$c_4 = \left(\frac{7}{48}e^4 + \frac{29}{240}e^6 + \frac{811}{11\,520}e^8 + \cdots\right) \tag{10.7}$$

$$c_6 = \left(\frac{7}{120}e^6 + \frac{81}{1120}e^8 + \cdots\right), \quad c_8 = \left(\frac{4279}{161\,280}e^8 + \cdots\right)$$

Using this formula we can obtain

$$\sin\phi = \sin\phi'(b_0 + b_2\cos 2\phi' + b_4\cos 4\phi' + \cdots) \tag{10.8}$$

where

$$b_0 = \left(1 + \frac{e^2}{2} + \frac{21}{96}e^4 + \frac{115}{480}e^6 + \cdots\right)$$

$$b_2 = \left(\frac{e^2}{2} + \frac{10}{24}e^4 + \frac{7}{64}e^6 + \cdots\right) \tag{10.9}$$

$$b_4 = \left(\frac{7}{192}e^4 + \frac{237}{960}e^6 + \cdots\right)$$

Lambert conformal conic projection

From the projection formulas for rectangular coordinates (see section 5.3.2),

$$\lambda = \frac{1}{\alpha}\arctan\left(\frac{x}{\rho_0 - y}\right) \tag{10.10}$$

$$q = \ln U = \frac{1}{\alpha}\ln[k/\sqrt{(\rho_0 - y)^2 + x^2}] \tag{10.11}$$

or

$$U = [k/\sqrt{(\rho_0 - y)^2 + x^2}]^{1/\alpha} \tag{10.12}$$

where α, k are projection parameters; ρ_0 is the radius on the map from the North Pole to the parallel of origin of axes at the central meridian.

After calculating $\sin\phi'$ from (10.11) or (10.12) followed by (10.5), we determine the latitudes of the designated points from equations (10.6)–(10.7) or (10.8)–(10.9). It is not difficult to find the longitudes of these points from (10.10).

Lagrange conformal projection

Proceeding from the formulas for rectangular coordinates,

$$\lambda = \frac{1}{\alpha}\arcsin\{2kx/[(y^2 - k^2)^2 + x^2(x^2 + 2y^2 + 2k^2)]^{1/2}\} \tag{10.13}$$

$$q = \ln U = \frac{1}{\alpha}\left(\frac{1}{2}\ln\frac{(y+k)^2 + x^2}{(y-k)^2 + x^2} - \ln\beta\right) \tag{10.14}$$

or

$$U = \left(\frac{1}{\beta}\sqrt{\frac{(y+k)^2 + x^2}{(y-k)^2 + x^2}}\right)^{1/\alpha} \tag{10.15}$$

where α, β, k are the projection parameters (see section 4.2.1).

After finding $\sin\phi'$ from equation (10.14) or (10.15), followed by (10.5), we can calculate latitudes ϕ of the points from equations (10.6)–(10.7) or (10.8)–(10.9), and longitudes λ from (10.13).

Stereographic projection of the sphere

Here we have (see section 3.2.2)

$$z = 2\arctan\left(\frac{1}{2R\cos^2(z_k/2)}(x^2 + y^2)^{1/2}\right), \quad a = \arctan\frac{x}{y}$$

$$\phi = \arcsin(\sin z\cos a\cos\phi_0 + \cos z\sin\phi_0)$$

$$\lambda = \lambda_0 + \arcsin(\sin z\sin a\sec\phi)$$

where z, a are polar spherical coordinates with the pole at a given point ϕ_0, λ_0; R is the radius of the sphere, often assumed to be

$$R = \sqrt{M_0 N_0} = \frac{a(1-e^2)^{1/2}}{1 - e^2\sin^2\phi_0}$$

(here a is the semimajor axis); z_k is the zenith distance of the almucantar along which the local scale factor is unity.

Conformal azimuthal projection of the ellipsoid

For this projection, parameter $\alpha = 1$ for the above conic projection formulas. Geodetic coordinates of points can be determined to emphasize the polar regions from formulas (10.10)–(10.12), and for any other region from formulas (10.13)–(10.15).

Gauss–Krüger conformal projection

When applied to Krasovskiy's ellipsoid used in Russia and the former USSR, the formulas for determining geodetic coordinates can be rewritten in the following form (see also section 5.3.1):

$$\phi = \phi_y + \{[(a_{28}z'^2 - a_{26})z'^2 + a_{24}]z'^2 - 1\}z'^2 a_{22}$$

$$l = \lambda - \lambda_0 = \{[(b_{17}z'^2 + b_{15})z'^2 + b_{13}]z'^2 + 1\}z'$$

where

$$\phi_y = [(2382 \cos^2 \beta + 293\,609)\cos^2 \beta + 50\,221\,747]\sin \beta \cos \beta \times 10^{-10} + \beta$$

$$N_y = [(0.605 \sin^2 \phi_y + 107.155)\sin^2 \phi_y + 21\,346.142]\sin^2 \phi_y + 6\,378\,245$$

$$a_{22} = (0.003\,369\,263 \cos^2 \phi_y + 0.5)\sin \phi_y \cos \phi_y$$

$$a_{24} = [(0.005\,615\,4 - 0.000\,015\,1 \cos^2 \phi_y)\cos^2 \phi_y + 0.161\,612\,8]\cos^2 \phi_y + 0.25$$

$$a_{26} = [(0.003\,89 \cos^2 \phi_y + 0.043\,10)\cos^2 \phi_y - 0.001\,68]\cos^2 \phi_y + 0.125$$

$$a_{28} = [(0.013 \cos^2\phi_y + 0.008)\cos^2 \phi_y - 0.031]\cos^2 \phi_y + 0.078$$

$$b_{13} = (\tfrac{1}{6} - 0.001\,123\,09 \cos^2 \phi_y)\cos^2 \phi_y - \tfrac{1}{3}$$

$$b_{15} = [(0.008\,783 - 0.000\,112 \cos^2 \phi_y)\cos^2 \phi_y - \tfrac{1}{6}]\cos^2 \phi_y + 0.2$$

$$b_{17} = (0.1667 - 0.0361 \cos^2 \phi_y)\cos^2 \phi_y - 0.1429$$

$$\beta = \frac{y}{6\,367\,558.497}, \quad z' = \frac{x}{N_y \cos \phi_y}$$

The accuracy of determining geodetic coordinates is 0.0001″ when the longitude difference $l = 9°$.

Equidistant cylindrical projection of the ellipsoid

On this projection (see section 2.1.4) $\lambda = x/r_k$ is the longitude of a point on the projection; $s = y$ is the arc length of the meridian from the equator to a given parallel, determined with formulas (1.145). It should be noted that these ellipsoidal formulas apply only to the normal aspect (for the transverse aspect, or Cassini–Soldner projection, forward and inverse equations are given in Snyder (1987b, p. 95)).

Conversion of the series (1.145), when applied to Krasovskiy's ellipsoid, gives

$$\phi = \tau + \{50\,221\,746 + [293\,622 + (2350 + 22 \cos^2\tau)\cos^2 \tau]\cos^2 \tau\} \times 10^{-10} \sin \tau \cos \tau \tag{10.16}$$

where

$$\tau = \frac{s}{R}, \quad R = \frac{a}{1 + n'}\left(1 + \frac{n'^2}{4} + \frac{n'^4}{64} + \cdots\right) = 6\,367\,558.4969 \text{ m} \tag{10.17}$$

Equidistant conic projection of the ellipsoid

From the projection formulas (see sections 3.1.1 and 3.1.4) we can obtain longitudes λ for the projection points and arc lengths along the meridians from the equator to a given parallel:

$$\lambda = \frac{1}{\alpha} \arctan\left(\frac{x}{\rho_s - y}\right), \quad s = k - \sqrt{x^2 + (\rho_s - y)^2} \tag{10.18}$$

where α, k are the projection parameters; ρ_s is the radius of the polar distance of the southern parallel.

Using the values of s calculated from equations (10.18), we can determine the values of τ from (10.17) and after that the latitudes of the points desired from equation (10.16), again only for the normal aspect.

Azimuthal equidistant projection of the ellipsoid
Equations (10.16)–(10.18) apply to the normal (polar) aspect of this projection using the constant $\alpha = 1$.

Trapezoidal pseudocylindrical projection
From the projection formulas for rectangular coordinates (see section 5.1.1) we find that

$$\lambda = \frac{x}{a_1(a_2 - ky)}, \quad s = \frac{y}{k}$$

where a_1, a_2, k are the projection parameters.

Now, using the values of s, we can determine the values of τ from (10.17) and latitudes ϕ of the points from (10.16).

Bonne equal-area pseudoconic projection
For the projection (see section 3.4.2)

$$s = s_0 + N_0 \cot \phi_0 - \sqrt{(\rho_0 - y)^2 + x^2} \tag{10.19}$$

$$\lambda = \arctan\left(\frac{x}{\rho_0 - y}\right) \frac{\sqrt{(\rho_0 - y)^2 + x^2}}{N \cos \phi} \tag{10.20}$$

where ρ_0, ϕ_0 are given.

After obtaining the values of s from equation (10.19) we can find the values of τ from (10.17) and the latitudes of the desired points from (10.16). After that it is not difficult to find the longitudes from (10.20).

Ordinary polyconic projection
This calculation procedure is based on Snyder (1987b, p. 130), and formulas are given in the form of a Newton–Raphson iteration, converging to any desired degree of accuracy after several iterations. If $|\lambda| > 90°$, this iteration does not converge, but the projection should not be used in that range in any case. The formulas may be calculated in the following order, given a, e, ϕ_0, x, y. First s_0 is calculated from equation (1.145), with ϕ_0 for ϕ and s_0 for s.

If $y = -s_0$, the iteration is not applicable, but $\phi = 0$ and $\lambda = x/a$. If $y \neq -s_0$, the calculation is as follows:

$$A = (s_0 + y)/a, \quad B = x^2/a^2 + A^2$$

Using an initial value of $\phi_n = A$, the following calculations are made:

$$C = (1 - e^2 \sin^2 \phi_n)^{1/2} \tan \phi_n$$

Then s_n and s'_n are found from equation (1.145) and the following equation, respectively, using ϕ_n for ϕ and s_n for s in (1.145). Let $s_a = s_n/a$. Then

$$s'_n = 1 - \frac{e^2}{4} - \frac{3}{64}e^4 - \frac{5}{256}e^6 - \cdots - 2\left(\frac{3}{8}e^2 + \frac{3}{32}e^4 + \frac{45}{1024}e^6 + \cdots\right)\cos 2\phi_n$$

$$+ 4\left(\frac{15}{256}e^4 + \frac{45}{1024}e^6 + \cdots\right)\cos 4\phi_n - 6\left(\frac{35}{3072}e^6 + \cdots\right)\cos 6\phi_n + \cdots$$

$$\phi_{n+1} = \phi_n - [A(Cs_a + 1) - s_a - \tfrac{1}{2}(s_a^2 + B)C]/[e^2 \sin 2\phi_n(s_a^2 + B - 2As_a)/4C$$
$$+ (A - s_a)(Cs'_n - 2/\sin 2\phi_n) - s'_n]$$

Each value of ϕ_{n+1} is substituted in place of ϕ_n, and C, s_n, s'_n, and ϕ_{n+1} are recalculated. This process is repeated until ϕ_{n+1} varies from ϕ_n by less than a predetermined convergence value. Then ϕ equals the final ϕ_{n+1}.

$$\lambda = [\arcsin(xC/a)]/\sin\phi$$

using C calculated for the last ϕ_n.

Modified polyconic projection for the International Map of the World (IMW)
The formulas for this projection are

$$\frac{\phi - \phi_s}{4} = \left[y - \frac{\lambda^2}{2}\left((r_n \sin\phi_n - r_s \sin\phi_s)\frac{\phi - \phi_s}{4} + r_s \sin\phi_s\right)\right] \Big/$$
$$(s_n - s_s)(1 - 0.000\,609\,2 \cos^2 \phi_{\text{mid}})$$

$$\lambda = x \Big/ \left[r_s + (r_n - r_s)\frac{\phi - \phi_s}{4} - \frac{\lambda^2}{6}\right.$$
$$\left.\times\left(r_s \sin^2\phi_s + (r_n \sin^2\phi_n - r_s \sin^2\phi_s)\frac{\phi - \phi_s}{4}\right)\right] \tag{10.21}$$

The calculation involves iteration. After introducing ϕ_s, x, and y, we determine (from ϕ_s and $\phi_n = \phi_s + 4°$) all the terms in these formulas (see sections 4.3.1 and 4.3.2). Using $\lambda = (\phi - \phi_s)/4 = 0$ for the first iteration (on the right-hand side) we can find the values of $A^{(2)} = [(\phi - \phi_s)/4]^{(2)}$ and $\lambda^{(2)}$ for the second iteration. After substituting these values into the right-hand side of (10.21), we determine $[(\phi - \phi_s)/4]^{(3)} = A^{(3)}$ and $\lambda^{(3)}$ for the third iteration. The same calculations are repeated to get the proper accuracy; the values $\phi = \phi_s + 4A^{(n)}$ and $\lambda = \lambda^{(n)}$ are those desired, and they are obtained with the last iteration.

10.2.2 Transforming map projections using polynomials

The following is a generalized polynomial combining portions of harmonic and power-series algebraic polynomials:

$$T = \sum_{i=0}^{k_1} a_i\psi_i - \sum_{i=1}^{k_1} a'_i\theta_i + \sum_{i=0}^{k_2}\sum_{j=0}^{k_2} a''_{ij}\xi^i\eta^j + v$$
$$Q = \sum_{i=1}^{k_1} a_i\theta_i + \sum_{i=1}^{k_1} a'_i\psi_i + \sum_{i=0}^{k_2}\sum_{j=0}^{k_2} a'''_{ij}\xi^i\eta^j + v \tag{10.22}$$

where v are terms distinguishing polynomials from functionally dependent terms;

$$\psi_1 = \xi, \quad \theta_1 = \eta, \quad \psi_i = \xi\psi_{i-1} - \eta\theta_{i-1}, \quad \theta_i = \xi\theta_{i-1} + \eta\psi_{i-1},$$

$$T = \frac{1}{\mu_1} Y \quad \left(\text{or} \quad T = \frac{1}{\mu_1} \phi, \quad \text{or} \quad T = \frac{1}{\mu_1} q\right)$$

$$Q = \frac{1}{\mu_2} X \quad \left(\text{or} \quad Q = \frac{1}{\mu_2} \lambda\right)$$

X, Y, ϕ, λ, q are rectangular, geodetic, or isometric coordinates of the projection to be constructed;

$$\xi = \frac{1}{\mu_3} y \quad \left(\text{or} \quad \xi = \frac{1}{\mu_3} \phi', \quad \text{or} \quad \xi = \frac{1}{\mu_3} q'\right)$$

$$\eta = \frac{1}{\mu_4} x \quad \left(\text{or} \quad \eta = \frac{1}{\mu_4} \lambda'\right)$$

x, y, ϕ', λ', q' are rectangular, geodetic, or isometric coordinates of the original projection; μ_1, μ_2, μ_3, μ_4 are scale factors which are chosen so that the maximum values of T, Q, ξ, η are not greater than unity. In choosing them it is often assumed that $\mu_1 = \mu_2 = \cdots = \mu_0$; a_i, a_i', a_{ij}'', a_{ij}''' are constant coefficients determined by solving sets of equations similar to those in (10.22) by the least-squares method, the sets being formed for a sufficient number of ground-control points for which coordinates are determined in both coordinate systems.

When the map projection transformation (geodetic coordinates into rectangular coordinates or vice versa) is applied with the help of analytical systems such as computers, it is quite possible to apply a general transformation by using approximations, e.g. polynomials in (10.22).

To apply this transformation it is necessary to take into account differences in projection distortion characteristics, the shape of the plotted geographical poles (whether as points, straight lines, or curved segments), and the symmetry of the map graticule about the central meridian and equator (see section 1.2).

For example, when transforming one conformal projection into another, or various projections into conformal ones, it is advisable to use harmonic polynomials. To apply this transformation, all coefficients in equations (10.22) should be equated as follows: $a_{ij}'' = a_{ij}''' = 0$, $a_i' = a_i$.

When transforming projections with other types of distortion using polynomials, all kinds of coefficients are involved, but these polynomials do not allow strict adherence to the distortion characteristics of the projection being produced.

Transformation with analytical systems using approximations will be less accurate than that using the corresponding rigorous formulas; the accuracy will depend on many factors, such as the number of ground-control points, the geometry of their locations, the specific form of the polynomials used, the number of terms preserved, and so on. In many cases, the accuracy will be sufficient for practical use, and the mathematical functions and algorithms obtained will be simpler.

When analytical systems are not available, projection transformations can be accomplished with various other techniques: electronic and optomechanical differential transformers, optical cameras, mechanical devices (e.g. pantographs), and others.

In this situation the general transformation, expressed by equations (10.22), is replaced with a corresponding localized one which is given successive values. On small plots, entire regions may be converted depending on the ability to achieve the transformational accuracy desired.

For example, when using phototransformation it is possible to apply the following types of transformation:

1. similarity transformations

$$X = ax, \quad Y = ay$$

2. affine transformations

$$X = a_0 + a_1 x + a_2 y, \quad Y = b_0 + b_1 x + b_2 y$$

3. or homographic transformations

$$X = \frac{a_0 + a_1 x + a_2 y}{c_0 + c_1 x + c_2 y}, \quad Y = \frac{b_0 + b_1 x + b_2 y}{c_0 + c_1 x + c_2 y}$$

where a_n, b_n, c_n are constants determined to fit chosen points on each projection, by algebraic or (for a larger number of points) least-squares calculations.

10.3 Computerized selection of map projections

The computerized selection of map projections is chiefly carried out in accordance with the general fundamentals described in Chapter 9.

As mentioned above, the principal emphasis when determining the class of projections from which one must choose a particular projection will be on a group of factors characterizing the object of the mapping. There may also be other factors that are very important in the design of this particular map.

After singling out all the essential factors to be taken into account, the other factors of the given group or of other groups characterizing the projection to be produced are arranged according to their importance. Then the relative importance of each of them in selecting a particular projection is assigned.

For the selection process a generalized criterion for assessing the advantages of map projections at every point is established; this criterion can consist of a set of variables. In setting this up, let us initially list local criteria ε_i:

$$\varepsilon_1 = a/b - 1, \quad \varepsilon_2 = ab - 1, \quad \varepsilon_3 = \{[(a-1)^2 + (b-1)^2]/2\}^{1/2}$$

$$\varepsilon_4 = k_{\text{mid}}/k_{\text{midmax}} - 1, \quad \varepsilon_5 = \Delta d/\Delta d_{\max} - 1, \quad \varepsilon_6 = \Delta k_M/\Delta k_{M\,\max} - 1$$

$$\varepsilon_7 = \Delta k_P/\Delta k_{P\,\max} - 1, \quad \varepsilon_8 = C_T/C_{T\,\max} - 1, \quad \varepsilon_9 = \Delta\varepsilon/\Delta\varepsilon_{\max} - 1$$

and others.

In this listing, a, b are maximum and minimum local linear scale factors; k_{mid} is the mean curvature on the map of a geodetic line (along meridians and parallels); Δd is a value characterizing the deviation of a loxodrome from a straight line; $\Delta k_M = R_{Mj} - R_M$ is the difference between the curvature of a meridian at the jth point on the projection and its true value; Δk_P is the same for a parallel; C_T is a value characterizing the stereographicity of the projection, i.e. the degree to which shapes are correctly represented for the region being studied; $\Delta\varepsilon = \varepsilon_j - \varepsilon'$ is the difference between the declination of angle ε at the jth point on the projection and its true value.

When computing these values we can use formulas from the general theory of map projections. Each of these particular criteria characterizes the projection at every point.

If we consider a criterion for selecting the projection from the point of view of best satisfying the criteria mentioned, the best possible choice for a particular map would be one for which a function of the form

$$E_i^2 = \frac{1}{S} \int \varepsilon_i^2 \, ds$$

has the least value within the limits of the region being mapped. Here ε_i^2 consists of the local criteria mentioned above.

In order to determine the function E_i^2, it is sufficient to divide the region to be mapped into small plots and to compute the values of the local criteria at the mid-points of each one, and then to find their mean arithmetic values:

$$E_i^2 = \frac{1}{n} \sum_{k=1}^{n} \varepsilon_{ik}^2$$

where n is the total number of plots; k is the number of the plot at which values of ε_i^2 are computed.

Now a generalized criterion can be written in the form

$$E_{\text{gen}}^2 = \sum_{i=1}^{m} P_i E_i^2 \bigg/ \sum_{i=1}^{m} P_i$$

where P_i is a weighting of the various local criteria indicating their relative significance.

This generalized criterion takes into account a majority of the requirements possible for map projections. These requirements are given in a normal form and in relative values, which makes it possible to compare and simultaneously take into account the various requirements for projections.

In each particular case the significance of factors will be different, and the generalized criterion generated for the given map to be designed will include, as a rule, fewer requirements. For example, for maps from which cartographic information is generally determined and assessed visually, weight factors P_1, P_2, P_4, P_5, P_9 may be considered to be zero, and the generalized criterion, changing the numbering of local criteria and factors, can take the form

$$E^2 = \sum_{i=1}^{4} P_i E_i^2 \bigg/ \sum_{i=1}^{4} P_i$$

The problem of objectively determining the significance of factors and their proper arrangement when choosing projections for the design of particular maps needs to be investigated further.

The values of E_{gen}^2 are computed for the whole set of map projections from the generalized criterion, and by comparing these values a projection is chosen where E_{gen}^2 is a minimum. As a rule, this approach is sufficient for the creation of a specific map.

The choice of a map projection, as well as the choice of other parameters of the mathematical basis, begins with an understanding of the mathematical basis of map projections of analogous maps, developed earlier.

For the final choice of the projection it may be advisable to prepare a model of the graticule for this projection with the help of peripherals or on the computer. The model is of great importance when, after the above analysis, we have several variations of the projection for which the values of the function E_{gen}^2 are very close to each other.

10.4 Research on map projections in an automated environment to meet given requirements

Preparation of the initial data is an important problem in the automated research of map projections. It includes a clear-cut determination of factors in the three groups affecting the selection of or research concerning map projections, and then a formalized description of these factors.

The problem of research on projections can be solved in two ways. The first is analogous to the computerized selection of map projections.

The second one is more satisfactory, as it proceeds from the possibility of obtaining a projection directly on the basis of a given set of principal requirements for the map projection used to design a particular map.

As in the first case, the primary initial data and properties that the desired projection should satisfy are determined first. This is done by accounting for the factors of the first group and for significant factors of the second group.

In some cases these data are quite sufficient for determining the projection of the map being designed. For example, suppose the problem is to find (i.e. research) the best possible conformal projection, with minimum distortion. In this case the condition of conformality and the factors of the first group, determining the geographical position and configuration of the region being mapped, provide a unique solution: the desired projection will be a Chebyshev projection, made under the condition that there is constant local linear scale (see section 7.2) along the outline of the region to be mapped.

In the general case, the set of properties or group of main factors will also be the initial factors for obtaining the desired projection. On singling out the main factors, defining their significance and taking into account the geographical position of the region, its configuration and dimensions, and conditions specifying the desired type of transformation functions (e.g. whether there is symmetry of the map graticule), the desired map projection is determined based on the method involved in the research, including the best possible and ideal projections, partially considered in Chapters 6 and 7.

10.5 Identification of map projections in an automated environment

The identification or determination of a projection used for the design of a particular map includes the solution of four problems:

1. establishing the aspect and coordinate system used for computing the projection;
2. determination of the projection class;
3. determination of the distortion characteristics;
4. determination of the projection parameters.

The solution of these problems is possible only if the region shown on the map is large.

To solve the first problem, attention should first be given to the portrayal of the geographical pole(s) on the map. On the normal aspect of azimuthal projections, the geographical pole will be shown in the center of the region mapped; for other types, it will be shown at the northern or southern limits as a point or as a line; for some projections, such as conic, pseudoconic, or oblique azimuthal, it may not be shown within the limits of the map.

If initially it is seen that the map was designed with a normal aspect, then to determine its class it is sufficient to use the definition of the projection given in available classifications. It is possible to determine the distortion characteristics of the projection on the basis of local linear scale factors at the intersections of the map graticule; this can be done using numerical differentiation (see section 8.2). On determining the class and distortion characteristics of the projection being considered, the constants for its parameters can be found directly from the projection formulas or after some manipulation.

More difficult is the problem of identifying the map projection in cases where an oblique or equatorial aspect was used for its computation, or when some derivative or other complex projection was used. In these cases it is expedient to solve this problem with automated computations either in an environment interactive between user and computer with the help of a display and control board, or in a completely automated environment. Detailed studies of computer-assisted projection identification are given in Snyder (1985, pp. 6–13, 25–55).

10.6 Automated identification and reduction of measurements from maps using their mathematical basis

Quantitative characteristics may be determined from maps chiefly on the basis of cartometry. It is possible to determine the coordinates of points on maps, distances between points, azimuths or directions, areas of flat, concave, or convex surfaces, volumes of specified masses using contours, angles of inclination of slopes, etc. Cartometry can be carried out on large- and small-scale maps, manually or automatically. A comprehensive work on cartometry is by Maling (1989).

10.7 Automated plotting of elements of the mathematical basis

The elements of the mathematical basis include the map graticule, borders, and lines defining division into sheets, as well as insets and legends, all of which may be plotted onto maps. In addition, some elements of geodetic surveys, i.e. triangulation stations and related coordinate grids, are shown on some maps.

There are two principal methods for solving the problem of automated plotting. Using the first one, the elements of the mathematical and geodetic bases are automatically plotted on maps in conjunction with all the transformed cartographic information comprising the contents of the newly proposed map. With the second method, the elements of the mathematical basis are plotted before the other elements. Several software packages for plotting a few or scores of map projections and features are commercially available in the United States and elsewhere.

11

Summary of miscellaneous projections, with references

The following is some brief information about projections not mentioned or briefly described in the previous chapters.

- B. R. Adams (1984): Two arbitrary projections. The first is a transverse stereographic cylindrical projection obtained by the perspective projection of an ellipsoid onto the surface of a tangent elliptical cylinder. The second is a stereographic conic determined by perspective projection of an ellipsoid onto the surface of a tangent cone.
- H. C. Albers (1805): Equal-area conic projection of the sphere (see section 3.1.3). O. S. Adams derived the formulas for the ellipsoid, and suggested that the standard parallels be placed about 1/6 and 5/6 of the spacing between the northern and southern limiting parallels (Adams 1927). The formulas for the sphere may be written as follows:

$$x = \rho \sin \vartheta, \quad y = \rho_0 - \rho \cos \vartheta$$

where

$$\rho = R(C - 2\alpha \sin \phi)^{1/2}/\alpha$$

$$\vartheta = \alpha(\lambda - \lambda_0)$$

$$\rho_0 = R(C - 2\alpha \sin \phi_0)^{1/2}/\alpha$$

$$C = \cos^2 \phi_1 + 2\alpha \sin \phi_1$$

$$\alpha = (\sin \phi_1 + \sin \phi_2)/2$$

and ϕ_0, λ_0 are latitude and longitude, respectively, for the origin of the rectangular coordinates; ϕ_1, ϕ_2 are the standard parallels and α is the cone constant.
- N. V. Aleksandrova, V. V. Lashin, and V. G. Federyakov: Applying conformal and equidistant cylindrical projections for overlapping satellite scanner imagery (*Referativnyy Zhurnal* (*R. Zh.*) (*Abstract Journal*) *Geografiya* 1988, 7M39).*
- P. S. Anderson (1974): An 'oblique poly-cylindrical, orthographic azimuthal equidistant' projection for a world map.
- Marin Andreyev: Variants of orthogonal circular projections (Andreyev 1971, 1983). Use of hyperbolic functions for new projections (Andreyev 1972). Method of making variations of a perspective cylindrical projection (Andreyev 1974). Variants of azimuthal projections, made by generalizing perspective azimuthal projections with positive and negative representations (Andreyev 1975). Circular

* Hereafter referred to as *R. Zh.* This journal is also used for some of the descriptions in this chapter even if an original paper is referenced.

projections with slight area distortion, determined by either parameter m_0 or n_0 ($m_0 n_0 = 1$) as the scale factor at the central point (Andreyev 1984).

- Arnd: Arbitrary conic projection (*Tezarius* 1987).
- Arnd–Werner: Conformal projection (*Tezarius* 1987).
- Atlantis projection is a transverse Mollweide projection. See under Bartholomew.
- F. W. O. August: Epicycloidal projection representing the entire Earth's surface conformally with no singular points (where conformality would fail). The 180th meridians from center form a two-cusped epicycloid. The equator and the central meridian are straight; the central meridian is half the length of the equator (Schmid 1974; Snyder and Voxland 1989, pp. 186, 187, 235).
- G. Kh. Avetyan: Some problems of the general theory of equivalent projections (Avetyan 1970; *R. Zh.* 1971, 4M30). Quasiconformal transformation of the ellipsoid (Avetyan 1972).
- Azimuthaloid projections are obtained from transverse azimuthal projections through the Aitoff transformation, e.g. the Aitoff, Hammer–Aitoff, and Eckert–Greifendorff projections.
- P. L. Baetsle (1970): Choice of an optimal projection for representing a spherical segment.
- J. G. P. Baker (1986): 'Dinomic' world map projection, using Mercator projection between 45°N and S latitude with interrupted modified Mercator projection poleward. Gaps occur in the Atlantic and Indian Oceans. Poles are represented, respectively, by two points in the northern hemisphere and by three points in the southern hemisphere.
- M. Balthasart (1935): Cylindrical equal-area projection with two standard parallels at $\pm 55°$ latitude.
- János Baranyi (1968): Seven variants of pseudocylindrical projections, except for variable spacing of meridians, made from sketches of map projections.
- János Baranyi and Ferenc Karsay (1971, 1972): Pseudocylindrical projection for world maps.
- John Bartholomew: World maps based on combinations of the equidistant conic projection with two standard parallels (for the northern continents) and the Bonne projection interrupted in three or four places for southern landmasses. These are called the 'Kite' and 'Regional' projections and were introduced in the 1940s. His 'Lotus' projection of 1958 emphasizes the oceans, with the equidistant conic projection in the southern hemisphere and the interrupted Bonne projection for northern oceans. On his 'Tetrahedral' projection to emphasize land, the north polar azimuthal equidistant projection is used for the northern hemisphere only to the Tropic of Cancer, and then three interrupted Werner projections for southern lobes. To emphasize oceans, the opposite orientation was used. His Atlantis projection is an equal-area projection, used for a world map beginning in 1948. Centered on the Atlantic Ocean, it is a transverse Mollweide pseudocylindrical projection with a central meridian of 30°W longitude, and the North Pole displaced 45° from the center. His Nordic projection is a symmetrical oblique Hammer–Aitoff projection, centered at 45°N latitude with the Greenwich meridian along the vertical centerline. It appeared first in 1950. (See Snyder 1993, pp. 222–3, 238–9, 256–7.)
- Ye. Yu. Bayeva (1987): Polyconic projections with pole lines for world maps.
- P. D. Belonovskiy: Research into adaptable conformal projections (Ginzburg *et al.* 1955).

- Franciszek Biernacki (1975): Projection belongs to class of isoperimetric projections, including all equal-area and many arbitrary projections; it is characterized by $a > 1 > b$.
- S. W. Boggs (1929): Eumorphic projection, a pseudocylindrical equal-area projection formed by averaging the ordinates of the Mollweide and sinusoidal projections, deriving the abscissas to provide equal area, and making a slight adjustment in the ratio of the axes.
- V. M. Boginskiy (1967, 1969): Development of projections for small-scale maps.
- V. M. Boginskiy and G. A. Ginzburg (1971): A combined projection for maps of North and South America using a computer.
- *Bol'shoy Sovetskiy Atlas Mira* (*BSAM*, *Great Soviet World Atlas*): (1) Gall stereographic (arbitrary perspective) cylindrical projection modified to have standard parallels at 30°N and S. (2) Interrupted pseudocylindrical Eckert VI sinusoidal equal-area projection. (3) Arbitrary polyconic, made from sketches of the map graticule (Tolstova 1981a).
- F. V. Botley (1951): 'Oblique-gore' cylindrical or azimuthal equal-area projections. See also under Polyhedral projections.
- William Briesemeister (1953): Ellipse-bounded oblique Hammer–Aitoff projection centered at 45°N, 10°E, but with the major axis 1.75 rather than twice the length of the minor axis. The South Pole is shown in two places of the graticule, and parallels near the North Pole are nearly circular (Snyder 1993, pp. 239–40).
- L. M. Bugayevskiy: Ellipsoidal map projections made by generalizing Urmayev's method (Bugayevskiy 1986e). A Chebyshev projection for mapping polar regions (Bugayevskiy 1986d). Methods for making a set of Euler ellipsoidal projections; formulas for a particular variant of such projections (L. M. Bugayevskiy 1987b). Problems of automating computations, research into choosing projections and obtaining other elements of the mathematical basis (Vakhrameyeva *et al.* 1986).
- L. M. Bugayevskiy, V. N. Deputatova, and A. M. Portnov (1984): Zoning and economization of computations to transform map projections.
- L. M. Bugayevskiy and A. G. Ivanov (1971): Methods of approximate solutions of sets of Euler–Urmayev differential equations, their complete determination, and the preparation of various arbitrary projections (*R. Zh.* 1971, 9M26).
- L. M. Bugayevskiy and A. M. Portnov (1984): Individual images from space.
- L. M. Bugayevskiy and I. V. Slobodyanik (1985): Generalized Urmayev equidistant conic projection.
- L. M. Bugayevskiy and L. A. Vakhrameyeva (1987): Basic problems and directions for developing cartographic theory.
- L. M. Bugayevskiy, N. M. Volkov, and G. A. Ginzburg: Problems of research into the best possible map projections (*R. Zh.* 1977, 3M).
- Yu. L. Bugayevskiy: Methods of making map projections for designing anamorphous maps with transformation of space geometry (Yu. L. Bugayevskiy 1983; *R. Zh.* 1984, 9M15). Variscaled projections for anamorphous thematic maps (Bugayevskiy 1986). Varivalent projections for anamorphous maps (Yu. L. Bugayevskiy 1987a; *R. Zh.* 1987, 6M24DEP). Varivalent projection of the pseudocylindrical type for anamorphous maps (Yu. L. Bugayevskiy 1987b).
- Wellman Chamberlin: His trimetric projection is an arbitrary approximation of a three-point equidistant projection used by the National Geographic Society for maps of continents (Snyder 1993, p. 234).
- Édouard Collignon (1865): Equal-area pseudocylindrical projection for a world

map, with all meridians and parallels shown as straight lines, and a hemisphere shown as a square with poles as points.

- J. E. E. Craster (1929): Pseudocylindrical equal-area parabolic projection for which

$$x = R\lambda \sqrt{\frac{3}{\pi}} \left(2 \cos \frac{2\phi}{3} - 1\right), \quad y = R\sqrt{3\pi} \sin \frac{\phi}{3}$$

Craster (1938) proposed an oblique conformal conic projection for New Zealand.

- M. Daskalova: Equal-area conic projection with least-squares distortion (Daskalova 1964). Application of Airy criterion to conformal and equal-area conic projections (Daskalova 1982).
- J. N. De l'Isle (or De Lisle): Equidistant conic projection of the sphere, where the standard parallels were initially at equal distances from the central and outer parallels of the map (see also section 3.1.4).

$$x = \rho \sin \vartheta, \quad y = \rho_0 - \rho \cos \vartheta$$

where

$$\rho = R(G - \phi)$$
$$\vartheta = \alpha(\lambda - \lambda_0)$$
$$\rho_0 = R(G - \phi_0)$$
$$C = (\cos \phi_1)/\alpha + \phi_1$$
$$\alpha = (\cos \phi_1 - \cos \phi_2)/(\phi_2 - \phi_1)$$

where ϕ_0, λ_0 are latitude and longitude, respectively, for the origin of the rectangular coordinates; ϕ_1, ϕ_2 are the standard parallels and α is the cone constant.

- L. P. Denoyer: Semielliptical pseudocylindrical projection (Snyder and Voxland 1989, pp. 80–1, 222).
- B. D. Dent (1987): Polycentered oblique orthographic projection for a world map. Continental shapes on the world map are claimed to be close to the shapes on the globe, and are shown on the orthographic projection, computed separately for each of the five continents in its own orientation with its own central point. The resulting five maps of continents are arranged on a world map, but the matching of meridians and parallels is incomplete.
- V. N. Deputatova (1969a, 1969b): Perspective azimuthal projections with multiple transformations and development of their variants; generalized formulas for perspective projections.
- Athanasios Dermanis and Evangelos Livieratos (1984): Three-dimensional deformational analysis transforming the Earth's surface onto a telluroid.
- Athanasios Dermanis, Evangelos Livieratos, and I. Parashakis (1983): Use of strain criteria in addition to routine criteria for conformality, equivalency, etc., to assess projection advantages.
- A. I. Dinchenko (1938): Chebyshev projection for the USSR (*Tezarius* 1987).
- Double perspective projection of the sphere onto a cone and a plane, in which the globe is first projected from the center onto the cone (or onto two cones) having a common axis with it, and then this cone is projected onto the plane of one of the meridians with rays perpendicular to this plane (Konusova and Bertik 1985).

- Max Eckert (1906) presented six pseudocylindrical projections: three (2, 4, 6) equal area and three (1, 3, 5) arbitrary. Numbers 1 and 2 have straight lines for meridians, broken at the equator; numbers 3 and 4 have semiellipses for meridians; and 5 and 6 have sinusoidal meridians. Numbers 4 and 6 are described further in section 2.2.2.
- Elliptical or semielliptical projections have elliptical meridians, either semiellipses or smaller arcs. They may be equal area, like the Mollweide and Eckert IV, or arbitrary.
- M. M. Epshteyn (1964): Chebyshev-like projections.
- György Érdi-Krausz (1968; Snyder 1993, pp. 217–18): Combined sinusoidal (between 60°N and S latitude) and Mollweide (poleward) pseudocylindrical projection, equal area within each portion, but at two different area scales. Used in Hungarian atlases for world maps.
- Leonhard Euler: In addition to others described in section 7.7, a specific Euler projection is an equidistant conic projection of a sphere onto a conceptually secant cone, with the cone constant (and indirectly the standard parallels) determined so that the absolute values of differences between the lengths of arcs on the projection and on the sphere ($\rho\delta - R\lambda \cos \phi$) of the central parallel ϕ_0 and two outer parallels ϕ_n and ϕ_s are equal (Urmayev 1956):

$$\rho = \rho_0 - R(\phi - \phi_0), \quad \rho_0 = R \frac{\phi_n - \phi_s}{4} \cot \left(\frac{\phi_n - \phi_s}{4}\right) \cot \phi_0$$

$$\delta = \alpha\lambda, \quad \alpha = \frac{2}{\phi_n - \phi_s} \sin\left(\frac{\phi_n - \phi_s}{2}\right)\sin \phi_0$$

- C. B. Fawcett (1949): Equal-area projection which is a compound projection in either normal (polar) or oblique orientation, on which the inner hemisphere is a Lambert azimuthal equal-area projection, and the remainder (to 60°S) is shown interrupted as three and two unequal lobes, respectively, stressing continents.
- V. N. Gan'shin (1973): Research on projections for the International Map of the World at a scale of 1 : 1 000 000.
- V. N. Gan'shin and A. V. Lipin (1977): Ellipsoidal projection for air navigation.
- Bogusław Gdowski: On the theory of conformal, equal-area, and equidistant representation. Chebyshev projections (Gdowski 1973). Representation of surfaces providing minimum distortion (conformal and semiconformal projections) (Gdowski 1974a, 1974b). Optimal equal-area projections (Gdowski 1975).
- E. N. Gilbert: Minimizing distortion (Gilbert 1974). His two-world globe is a sphere on which the world is shown twice conformally using Lagrange transformations; this was orthographically projected by A. A. DeLucia (DeLucia and Snyder 1986) for a rounded globe appearance.
- G. A. Ginzburg and T. D. Salmanova (1957): Cylindrical and other projections using Urmayev's method for prescribed scale error at certain points.
- E. Grafarend and A. Niermann (1984): Finding that the best normal conformal, equal-area, or equidistant cylindrical projection satisfying the Airy or Airy–Kavrayskiy criteria is an equidistant cylindrical projection with standard parallels of $\pm 61.7°$ or $\pm 42°$, respectively.
- Wiktor Grygorenko (1975): Arbitrary polyconic projection with equally divided parallels, obtained by approximation of a map graticule sketch using Urmayev's method.

- Yasuchi Hagiwara (1984): Retroazimuthal projection for a world map. It shows the true azimuths of the center from any point on the map. Several variants of projections with the center at Mecca are considered.
- Masataka Hatano: Interrupted projection (Hatano 1966); elliptical pseudocylindrical equal-area projection (Hatano 1972).
- Philippe Hatt: Oblique azimuthal equidistant projection for the ellipsoid, using approximating series (Snyder 1993, pp. 161, 177).
- Vladislav Hojovec: Class of equidistant projections (Hojovec 1976; *R. Zh.* 1977, 2M60; see also section 2.2.2). Approximate solution of a variational type for geographical quadrangles, presented in 1976. Optimization of equal-area projections (Hojovec 1987).
- Vladislav Hojovec, Jaroslav Klášterka, and Hana Rendlová (1975): Variant of the Chebyshev projection to map Czechoslovakia.
- Holland projection is a double projection, for which the ellipsoid is transformed onto a sphere by the second Gauss method, and the sphere is transformed onto a plane with a stereographic projection for which the central point and central meridian of the projection coincide (Morozov 1979).
- Hölzel projection is an arbitrary pseudocylindrical projection with equally spaced parallels, an apparent modification of Eckert V (Snyder 1993, p. 308).
- I. B. Ivanov: Making new equal-area and other projections on the basis of vector analysis statements (*R. Zh.* 1962, 11B18).
- Henry James: Azimuthal perspective projection with negative transformation, ranging 113°30′ from the center (23°30′N, 15°E) but with the point of view 1.5 times the Earth radius from the center (Murchison 1857).
- Kei Kanazawa (1985): Formulas for cylindrical and azimuthal equal-area projections with optimal relationship of distortion for school maps.
- A. V. Kavrayskiy: Choice of arbitrary constants for the conformal conic projection (Kavrayskiy 1974). Quantitative assessment of orthodromic distortion on conformal projections (Kavrayskiy 1976). Development of algorithms for the relationship between geographical and plane rectangular coordinates using approximating functions (Kavrayskiy 1982).
- V. V. Kavrayskiy: Projection 1 is cylindrical, Mercator between $\pm 70°$ latitude and equirectangular poleward. Projections 2 and 4 are equidistant conic, minimizing error with standard parallels chosen by latitude range and least squares for the region mapped, respectively. Projection 3 is a conformal conic with special standard parallels. Projections 5, 6, and 7 are pseudocylindrical; 5 and 6 are equal area with sine-function and sinusoidal meridians, respectively; 7 has elliptical meridians with those at $\pm 120°$ circular (Snyder 1993). Some of these projections are described in previous chapters. Projection 8 is a conformal oval type; in a particular case it represents the entire Earth's surface inside the oval. Projection 9 is a compound equal-area pseudoconic projection on which all areas are reduced by a constant factor equal to the square of the basic scale of a map. The radius of parallels is $\rho = a + b(\phi_0 - \phi) + d(\phi_0 - \phi)^3$, where a, b, d are constants; d for latitudes less than 40°N has one value, and for latitudes greater than 40°N, another value. Projection 10 is an equal-area pseudoconic projection for world maps (Kavrayskiy 1958–60).
- S. F. Kobylyatskiy (1962): Equal-area cylindrical projection with Tissot projection properties.
- Antonin Koláčný (1964): Projections for atlas maps of Czechoslovakia.

- G. I. Konusova: Application of a series function of a complex variable for computing Chebyshev projections (Konusova 1964). Conformal projections with preset properties (Konusova 1970). Projections using local (point) and 'global' indexes of distortion for assessing their advantages (Konusova 1971). Projections for thematic maps of the USSR (Konusova 1979).
- A. N. Korkin: Theoretical proposal for equivalent projections and their variants (Tolstova 1981a).
- N. N. Kozlova (1972): Research into the convergence of series used for making stereographic projections.
- F. N. Krasovskiy (1922, 1925): (1) An equal-area conic projection and (2) and (3) approximately equidistant conic projections on which the distortion of the region between outer parallels is reduced by making scale error equal on outer and central parallels or by using least squares.
- Křovák projection is a double oblique conic conformal projection (Konusova and Bertik 1985).
- Krüger–Adams projection is a stereographic projection of an ellipsoid, which is defined by Gauss and Krüger as a special form of the Lagrange projection with $\alpha = 1$ and $k = 1$. Adams uses a double projection: the ellipsoid is conformally transformed onto a sphere (as proposed by Mollweide), and the sphere onto a tangent plane with formulas of the stereographic projection (Adams 1919, pp. 30–4; Krüger 1922, p. 7; Kavrayskiy 1934, p. 219).
- A. A. Kuznetsov (1984, 1987): General equations of conformal projections in a geocentric coordinate system.
- J. H. Lambert (1772) proposed seven projections: (1) conformal conic, (2) conformal world in a circle, later called a 'Lagrange' projection because of Lagrange's generalization in 1779, (3) transverse conformal cylindrical, now called a transverse Mercator, (4) cylindrical equal area with the equator the standard parallel, (5) transverse cylindrical equal area, (6) azimuthal equal area, and (7) equal-area conic with the pole as a point. Some of these have been discussed above.
- Lambert–Littrow projection is a transverse aspect of the 'Lagrange' projection developed by Lambert, with most of the world shown conformally. The meridians are hyperbolas and the parallels are ellipses, all confocal. It is also called the Littrow–Maurer equal-azimuth projection by Maurer (1935, S189).
- L. S. Ledovskaya (1969): Projection for maps of Eurasia made from a sketched approximation.
- L. P. Lee: Conformal projections for world maps (Lee 1976): The Earth is shown bounded by a rectangle, an ellipse, and various polygons; the world is also transformed conformally onto the faces of each Platonic polyhedron (tetrahedron to icosahedron). Conformal projection with oval isocols, determined with Miller's method, for a map of the Pacific Ocean (Lee 1974). Conformal projection with scales determined at given points (Lee 1975).
- I. G. Letoval'tsev (1963): Gnomonic projection for the ellipsoid. First the ellipsoid is transformed conformally onto a sphere, and then a gnomonic projection is used for the plane.
- Guozao Li: Double azimuthal projection (*R. Zh.* 1964, 10M23). Transverse conformal cylindrical projection with two standard meridians (Li 1981). A class of orthogonal projections obtained under the condition that $m = n^k$, where k is a constant parameter (it is pointed out that projections with $-1 \leq k \leq 1$ are important in cartographic practice) (Li 1987).

- Johan Lidman: Double-perspective azimuthal projection. First a hemisphere is projected from its center onto a cone tangent at 45°N latitude, and then from the cone surface onto the plane of the equator with perpendicular rays (Maurer 1935, pp. 19–20; English trans. pp. 29–30).
- T. Lipets and S. Napor: Conformal transformation of a plane onto a plane using Lagrange polynomials (*R. Zh.* 1969, 4.52.185).
- A. S. Lisichanskiy: Conformal projections with adaptable isocols which are derivatives of two initial conformal projections (Lisichanskiy 1962). Derivation of N. A. Urmayev's differential equations for transforming the Earth ellipsoid onto a plane (Meshcheryakov 1968). Formulas for oblique conformal projections of the ellipsoid (Lisichanskiy 1965).
- (A.-M.) Lorgna projection is the polar (normal) Lambert equal-area azimuthal projection (Schott 1882, p. 290).
- Lotus projection is compound. See under Bartholomew.
- Dyakhao Lyu and Guozao Li: Pseudoazimuthal projections for maps of China represented as a whole (*R. Zh.* 1964, 3M26).
- F. W. McBryde and P. D. Thomas (1949): Five equal-area pseudocylindrical projections. Number 1 has pointed poles and the formulas fit a sine function; number 2 is a modification of number 1, with poles one-quarter the length of the equator. Numbers 3–5 have poles one-third the length of the equator, and the meridians are sinusoids, quartic curves, and parabolas, respectively, derived from pointed-polar pseudocylindrical projections with similar curves for meridians. Number 4, the McBryde–Thomas flat-polar quartic projection, with the equator $\pi/\sqrt{2}$ times the length of the central meridian, may be calculated from the formulas

$$x = R\lambda[1 + 2\cos\theta/\cos(\theta/2)]/(3\sqrt{2} + 6)^{1/2}$$
$$y = 2\sqrt{3}R\sin(\theta/2)/(2 + \sqrt{2})^{1/2}$$

  where $\sin(\theta/2) + \sin\theta = (1 + \sqrt{2}/2)\sin\phi$, requiring iteration.
- R. R. Macdonald (1968): Compound projection of polyconic, equidistant conic, Bonne, and equirectangular projections with interruptions in the oceans.
- S. P. Malinin (1969): Non-coaxial perspective cylindrical projection, created by perspective projection from a fixed center onto a lateral cylindrical surface non-coaxial with the globe.
- A. A. Markov: Equidistant conic projection (Konusova and Bertik 1985). Research on the problems of representing spherical circles onto a plane (Vakhrameyeva and Bugayevskiy 1985).
- A. I. Martynenko: Algorithms and a general program for automatic computation and building elements of the mathematical basis of the map (Martynenko 1966; *R. Zh.* 1966, 7M20). The possibility of programming the process of creating the mathematical and geodetic bases of topographic maps (Martynenko 1965; *R. Zh.* 1965, 9M18).
- Hans Maurer: Two-point azimuthal and two-point equidistant projections are described in sections 7.8 and 7.9. On the two-point retroazimuthal projection (Maurer 1919b), the radio direction from a receiver anywhere on the map to either of two fixed stations on the equator may be measured as the angle between a straight line connecting the station and the receiver, and the straight vertical meridian through the receiver.
- Hans Maurer (1935) miscellaneous projections. These include a star projection, an

equal-area polyconic, a modification of van der Grinten 2, the Lambert–Littrow, globulars, and numerous others devised to complete categories in his Linnean classification.

- Franz Mayr (1964): Equal-area pseudocylindrical projection with the formulas

$$x = R\lambda\sqrt{\cos\phi}, \quad y = R\int_0^{\phi}\sqrt{\cos\phi}\, d\phi$$

- D. I. Mendeleyev (1907): Equidistant conic projection with pointed North Pole and a standard parallel ϕ_0:

$$\rho = R\left(\frac{\pi}{2} - \phi\right), \quad \delta = \lambda \cos\phi_0 \Big/ \left(\frac{\pi}{2} - \phi_0\right)$$

- G. A. Meshcheryakov: Research into the best possible Euler projection, with no distortion along the central rectilinear meridian (*R. Zh.* 1962, 2; see also section 7.7). The theory of perspective projections; azimuthal perspective projections transformed homographically (Meshcheryakov 1959; *R. Zh.* 1964, 2M21). Conformal transformation of surfaces onto a plane (Meshcheryakov 1963; *R. Zh.* 1964, 9M19). Most advantageous map projections (Meshcheryakov 1965a).
- G. A. Meshcheryakov and M. A. Topchilov (1970): New classes of map projections.
- A. A. Mikhaylov: Equidistant conic projection (Konusova and Bertik 1985). Research into the general theory of equidistant conic projections (Tolstova 1981a).
- O. M. Miller: Bipolar oblique conformal conic projection, cylindrical projection, and oblated stereographic projection are described in sections 6.3.8 and 7.3.4. Equal-area transformations of the sphere onto a plane (Miller 1965).
- John Milnor (1969): Minimizing distortion on projections.
- Multizone projections are of the same type for a set of zones, but with various parameters for each zone (e.g. the equidistant conic projection for a world map at the scale 1 : 2 500 000).
- Patrick Murdoch (1758): Numbers 1 and 3 are equidistant conic projections of the sphere, with true total area of the zone between the two outer parallels. There are two standard parallels, but Murdoch only determined formulas for overall error reduction. Number 2 is a perspective conic projection, projected from the center of the globe, and preserving the total area of the zone between outer parallels. These projections do not preserve local area scale.
- V. O. Murevskis (1975): Compound projection for charts of the world ocean (with two interruptions), a variant of an oblique Werner equal-area projection (*R. Zh.* 1975, 11M43).
- A. M. Nell: Arbitrary globular projection for one hemisphere averaging the equatorial stereographic and al-Bīrūnī (or Nicolosi) globular projection (Nell 1852). Equal-area pseudoconic combination of the Bonne and Lambert equal-area conic, with a limiting pseudocylindrical form (Nell 1890).
- Kh. Nischan (1962): Variants of transverse and oblique graticules of Lagrange projection for maps of the Pacific Ocean (*R. Zh.* 1963, 4M21; 1964, 10M25).
- Nordic projection is an oblique Hammer projection. See under Bartholomew.
- Ye. N. Novikova: Approximation method for research on ideal projections through the Lagrange multiplier method (*R. Zh.* 1984, 8M8; Novikova 1984).

Research on the best possible equidistant projection of variational type (for the class of projections orthogonal along meridians) (*R. Zh.* 1986, 7M25; Novikova 1986a). Solution of the problem of seeking ideal projections of a variational type (*R. Zh.* 1986, 7M26; Novikova 1986b).

- Orthodromic projection shows orthodromes (arcs of great circles) as straight lines (see gnomonic and two-point azimuthal projections).
- Edward Osada (1974): Class of generalized cylindrical projections; as a particular case formulas for the Gauss–Krüger projection are obtained, as well as a new conformal projection on which there is no distortion along two meridians.
- Jan Panasiuk (1973, 1974): Transverse pseudocylindrical projection and the oblique orthographic projection of an ellipsoid for maps of medium and large scales.
- Parabolic projections have meridians in the shape of parabolas. See Craster and Putniņš projections.
- Antoine Parent: Three perspective low-error polar azimuthal projections with $D = 1.594R$, $\sqrt{3}R$, and $2.105R$, respectively, presented in 1702 (Maurer 1935, p. 19; 1968 trans., pp. 27–8). They involve negative transformation and are similar to the La Hire projection.
- A. A. Pavlov: Application of Chebyshev polynomials to computation of map projections from sketches of the graticule (*R. Zh.* 1967, 11M34; Pavlov 1967). Applying polynomials to analytical transformation of projections (*R. Zh.* 1967, 11M34; Pavlov 1970). Optimal conformal projections for medium- and small-scale maps (Pavlov 1973).
- Lech Pernarowski (1970): Isoperimetric projections, preserving lengths along all meridians and parallels and having a non-orthogonal graticule (*R. Zh.* 1971, 5M31).
- August Petermann: Star-shaped compound projection, with the central part a polar azimuthal equidistant projection for the northern hemisphere, and eight points with straight sides on pseudoconic projections representing the southern hemisphere (Jäger 1865).
- M. Petrich and W. R. Tobler (1984): Generalization of the globular projection, with the world shown in a combination of the plate carrée and the standard (al-Bīrūnī or Nicolosi) globular projection, with interruption and semicircular lobes of increasing number and decreasing size ranging from two circular hemispheres to the full plate carrée.
- Petrovskiy projection is a trisighting projection for finding the radio direction to three fixed stations. The angles between three fixed stations on the map as measured from any other point equal the true angles measured between the corresponding orthodromes on the globe (Konusova and Bertik 1985).
- A. K. Philbrick (1953): Sinu–Mollweide projection is an oblique interrupted projection with the Mollweide north of 40°44′S latitude and the sinusoidal south of there on the base map. The graticule is rotated to place 55°N and 20°E at the center of the base map with the North Pole along the central vertical axis. The lower half is interrupted through oceans into three lobes and the northern half into two to emphasize continents.
- Polyhedral projections. The sphere is projected onto a polyhedron, using various projections. Proposed by Albrecht Dürer in 1538, and by numerous writers such as Irving Fisher (1943) and L. C. Adamo, Ledolph Baer, and J. P. Hosmer (1968) with gnomonic projection onto an icosahedron; L. P. Lee (1976) with a conformal

projection onto each of the Platonic polyhedra; A. J. Potter (1925) and Botley (1949) proposed a gnomonic tetrahedron; Stefania Gurba (1970) discussed a gnomonic dodecahedron (Snyder 1993). Karel Kuchař (1966–7) also studied gnomonic projection onto Platonic polyhedra (*R. Zh.* 1970, 1M46).

- Polyhedric projections are projections of the same type for a set of sheets, but with different parameters for each sheet (e.g. the Müffling projection used in Germany and the USSR, and others used in other parts of Europe).
- A. M. Portnov (1977): Choice of model for computing a Chebyshev projection.
- Claudius Ptolemy: His first projection is a prototype of the simple or equidistant conic projection of the globe with one standard parallel:

$$\rho = \rho_0 - R(\phi - \phi_0), \quad \rho_0 = R \cot \phi_0, \quad \theta = \lambda \sin \phi_0$$

  On Ptolemy's form, however, the straight meridians were broken at the equator. Ptolemy 2 is a prototype of the Bonne pseudoconic projection. Ptolemy 3 is a modified perspective projection (Snyder 1993, pp. 10–14).
- R. V. Putniņš (1934): Twelve pseudocylindrical projections. Numbered P_n and P'_n, if n is odd parallels are equally spaced and the projection is not equal area. If n is even, the projection is equal area. If number has a prime ($'$), poles are lines half the length of the equator. Without a prime, poles are points. Subscripts 1 and 2 indicate elliptical meridians, 3 and 4 parabolic meridians, and 5 and 6 hyperbolic meridians. P_4 is the same as Craster's parabolic, and P'_1 and P'_2 were presented by Wagner in 1932 (see Snyder 1993, pp. 205–7).
- Jiři Pyšek (1984): Methods for making equal-area, conformal and equidistant projections, and arbitrary conic projections from the solution of the equation $k_1(mn - 1) + k_2(m - n) = 0$, where k_1 and k_2 can be of any value.
- Regional projection (see under Bartholomew).
- Boris Rogev (1963): Conformal projections with preset values of scale at given points.
- Boris Rogev and L. Dimov (1970): Theory of conformal projections using the method of an integrating multiplier (*R. Zh.* 1972, 7M42).
- Henri Roussilhe (1922): Quasistereographic projection, a conformal projection of the ellipsoid. It can be considered a double conformal projection, with the ellipsoid transformed onto a sphere, preserving scale along the central meridian, and the sphere is transformed onto a plane by stereographic projection, with the central point and central meridian coinciding (Morozov 1979).
- Fernando Sansó (1972): Best possible conformal projection of a variational type.
- A. Z. Sazonov: Iterative formulas for computing the Gauss–Krüger projection for a wide zone (*R. Zh.* 1971, 8.52.68).
- Wilhelm Schjerning: Six pseudoconic projections, all modified azimuthal equidistant projections. Beginning with the polar aspect, he reduced the distances between some or all meridians, and then made oblique transformations. Number 3 shows the world bounded by two equal circles tangent at London. Number 4 is an oblique Werner. Number 6 has three petals on the Werner projection touching at the South Pole and interrupted to emphasize oceans (Snyder 1993, pp. 254–5).
- C. M. Schols: Equal-area pseudoconic projection (Konusova and Bertik 1985). See also section 7.3.1.
- Oskar Schreiber: Double conformal projection used for the Prussian Land Survey of 1876–1923, with the ellipsoid transformed onto a sphere by the Gauss second method, and the sphere transformed onto a plane with the transverse Mercator

projection of the sphere (Jordan and Eggert 1941; 1962 trans. pp. 241, 263, 266, 283).

- B. B. Serapinas (1983, 1984): Formulas for distortion and coordinates of conformal map projections of a triaxial ellipsoid.
- A. I. Shabanova: Variant of a conformal projection for maps of the USSR, close to a Chebyshev projection (Tolstova 1981a).
- Karl Siemon: Pseudocylindrical projections. (1) Loxodromes are plotted as straight lines and at correct length from a chosen central point (Siemon 1935). W. R. Tobler (1966) independently reintroduced this as the loximuthal projection. (2) Two equal-area projections with quartic curves for meridians. On one the equator is $\pi/\sqrt{2}$ or 2.22 times the length of the straight central meridian; it was independently reintroduced by O. S. Adams (1945, pp. 23, 46–50) as the quartic authalic projection. On the other the axes are affinely adjusted to a 2 : 1 ratio.
- Sinusoidal projections show meridians in the shape of sinusoids. The original has points for the poles. There are others, given other names, such as those by Eckert, Wagner, and V. V. Kavrayskiy, on which the poles are lines.
- J. P. Snyder: 'Magnifying-glass' azimuthal projections with an inner circle or rectangle shown (equal area, equidistant, conformal, etc.) at a larger scale than surrounding regions, which are at constant or tapered reduced scale (Snyder 1987a). See also sections 7.3.5, 7.5.1, and 7.13.1.
- Star projections are compound projections for world maps, where two projections are used, one for the central part, and the other for appendages, e.g. a polar azimuthal equidistant projection for the central part and a modified pseudoconic projection preserving scale along the central meridian but straight outer meridians for the interrupted southern hemisphere. Among others, G. Jäger, Petermann (see above), Berghaus (see section 6.3.3), and Maurer developed varieties beginning in 1865. (See Snyder 1993, pp. 138–40, 240.)
- V. D. Taich: Arbitrary polyconic projection, with the central meridian and equator equally divided straight lines, the outer meridian an elliptical arc, and the pole an arc of a circle (Tolstova 1981a).
- Tetrahedral projections (see under Bartholomew and Polyhedral projections).
- *The Times* projection is an arbitrary pseudocylindrical sinusoidal projection (Snyder 1993, pp. 213, 214).
- N. A. Tissot: Equal-area conic projection with the linear scale factor at both limiting parallels equal to the reciprocal of the least scale factor near the central parallel. V. V. Kavrayskiy (1934, pp. 104–6) found that this gives the least overall maximum angular deformation for an equal-area conic. Formulas for the ellipsoid are also given by V. V. Kavrayskiy (Konusova and Bertik 1985). Tissot perspective projections are outer-perspective azimuthal projections with negative transformation (Vakhrameyeva and Bugayevskiy 1985): (a) $R < D < 2R$; (b) $R < D < 3R$.
- W. R. Tobler (1973a): Equal-area hyperelliptical pseudocylindrical projection, with the outer meridians shaped intermediate between an ellipse and a rectangle. He also developed continuous cartograms (Tobler 1973b).
- T. I. Tolstova: Azimuthal, conic, and cylindrical projections applying the Airy criterion (Tolstova 1970). Theoretical aspects of map projection theory with elliptical parallels (Tolstova 1981a). Polyconic projections (Tolstova 1981b).
- M. A. Topchilov: Generalized class of conformal projections, represented by equation of elliptical type in the form of $\varepsilon = 0$, $n = km$, where $k > 0$ (Topchilov

1969). Class of orthogonal projections described by equations of a parabolic type (*R. Zh.* 1971, 4M28; Topchilov 1970).

- TsNIIGAiK projections. Number 1 is an arbitrary pseudoazimuthal; 2 is an arbitrary oblique cylindrical; 4 is an arbitrary transverse pseudoazimuthal with oval isocols (*Tezarius* 1987). Number 3 is a set of arbitrary polyconics (Tolstova 1981a).
- Ya. I. Tuchin (1973): Variant of the best possible Euler projection (*R. Zh.* 1974, 10M35).
- L. A. Vakhrameyeva: Perspective projections with multiple transformations (Vakhrameyeva 1961). Generalized formulas for conformal projections (Vakhrameyeva 1963). Conformal projections made with the help of harmonic polynomials (Vakhrameyeva 1969). Conformal projections obtained with the help of series (Vakhrameyeva 1971a). Conformal projections with asymmetric isocols (Vakhrameyeva 1971b).
- L. A. Vakhrameyeva and Yu. L. Bugayevskiy (1985): Variscaled projections for social and economic maps.
- A. J. van der Grinten (1904) presented two circular projections. The first is discussed in section 4.2, and shows the world in a circle, with parallels increasingly spaced away from the equator. His second, which he called apple shaped, shows the world in two circles intersecting so that the line connecting points of intersection is half the overall breadth of the two circles. Alois Bludau (Zöppritz and Bludau 1912, pp. 179–85) devised two modifications of the first with the same spacing of parallels along the central meridian, calling them 2 (with parallels intersecting meridians orthogonally) and 3 (with the parallels as parallel straight lines). He called van der Grinten's second projection number 4 (Snyder 1993, pp. 258–62).
- L. I. Vasilevskiy (1968): Varivalent projections for population maps.
- N. Ya. Vilenkin (1961): Conformal projections similar to Chebyshev projections.
- Vilmos Vincze (1968): Theory of generalized perspective projections made with an arbitrary position for the center of gnomonic, stereographic, and orthographic projections.
- N. M. Volkov: Map projections of imagery of celestial bodies taken from space (Volkov 1973; Bugayevskiy and Portnov 1984). Compound pseudocylindrical equal-area projections, with theory and methods for cartometric work (Volkov 1950).
- K.-H. Wagner (1949): His pseudocylindrical projections 1, 3, 4, and 6 (in 1949 order) appeared in 1932 with pole lines half the length of the equator. Number 1 is equal area with sinusoidal meridians, 3 is arbitrary with equidistant parallels, 4 is equal area with elliptical meridians, and 6 is arbitrary with equidistant parallels and elliptical meridians (see section 2.2.2). In 1949 he presented 2 and 5, which are modifications of 1 and 4, respectively, with gradual change in area scale. Number 7 is a modification of the Hammer–Aitoff, equal area but with the pole a curved line corresponding to $\pm 65°$ latitude. Number 8 is a modification of 7 to provide gradual area-scale changes. Number 9 is a modification like 7, but of the Aitoff projection and not equal area.
- Werner Werenskiold (1945): Independent reintroductions of three pseudocylindrical projections, except for scale. They are equal area with poles shown as straight lines half the length of the equator: (1) with parabolic meridians, the same as Putniņš P_4'; (2) with sinusoidal meridians, the same as Wagner I and Kavrayskiy

6; (3) with elliptical meridians, the same as Wagner IV.

- William William-Olsson (1968): Combined projection for world maps where part of the northern hemisphere from the pole to 20°N latitude is represented with a polar azimuthal equal-area projection; the remainder of the Earth's surface is represented in the form of four identical modified Bonne lobes with gaps in the oceans.
- Qi-he Yang (1965): Arbitrary conic projections utilizing numerical methods.
- A. E. Young (1920, pp. 22–3): Minimum error arbitrary conic projection almost identical with an equidistant conic.
- A. P. Yushchenko: Compound star-shaped five-pointed conformal projection (Tolstova 1981a).
- Yu. M. Yuzefovich: New Chebyshev projections (Yuzefovich 1971). Some new projections close to equal area (Yuzefovich 1972a); a generalized class of projections, where $\varepsilon = 0$, $m = n^k$, where k is a constant (when $k = 1$ we have conformal projections; when $k = -1$ we have the Euler projection; when $k = 0$ and $k \to \infty$ we have, respectively, projections equidistant along meridians and along parallels) (Yuzefovich 1972b).
- T. G. Zargaryan (1971): Conformal projections obtained with harmonic polynomials.
- L. K. Zatonskiy (1962): Arbitrary compound elliptical projection for world maps (*R. Zh.* 1963, 5M11).
- Ye-xun Zhong *et al.* (1965): Polyconic projection with unequally divided meridians and parallels (*R. Zh.* 1966, 5M23).
- Georgi Zlatyanov (1965): Arbitrary non-orthogonal circular projections for world maps.

References

All references are in English, Russian, German, French, Italian, or Spanish unless otherwise indicated.

Adamo, L. C., Baer, Ledolph, and Hosmer, J. P., 1968, Icosahedral–gnomonic projection and grid of the world ocean for wave studies. *Journal of Geophysical Research*, **73** (16), 5125–32.

Adams, B. R., 1984, Transverse cylindrical stereographic and conical stereographic projections, *American Cartographer*, **11** (1), 40–8.

Adams, O. S., 1919, *General theory of polyconic projections*, Washington, DC: US Coast and Geodetic Survey Special Publication 57.

—— 1925, *Elliptic functions applied to conformal world maps*, Washington, DC: US Coast and Geodetic Survey Special Publication 112.

—— 1927, *Tables for Albers projection*, Washington, DC: US Coast and Geodetic Survey Special Publication 130.

—— 1945, *General theory of equivalent projections*, Washington, DC: US Coast and Geodetic Survey Special Publication 236.

Airy, G. B., 1861, Explanation of a projection by balance of errors for maps applying to a very large extent of the Earth's surface and comparison of this projection with other projections, *London, Edinburgh, and Dublin Philosophical Magazine*, ser. 4, **22** (149), 409–21. See correction by Henry James and A. R. Clarke, 1862, On projections for maps applying to a very large extent of the Earth's surface, *London, Edinburgh, and Dublin Philosophical Magazine*, ser. 4, **23** (154), 306–12.

Albers, H. C., 1805, Beschreibung einer neuen Kegelprojektion, *Zach's Monatliche Correspondenz zur Beförderung der Erd- und Himmels-Kunde*, **12** (Nov.), 450–9.

Anderson, P. S., 1974, An oblique, poly-cylindrical, orthographic azimuthal equidistant cartographic projection, its purpose, construction and theory, *Cartography*, **8** (4), 182–6.

Andreyev, Marin, 1971, Poluchavaniye na varianti ortogonalni kartografski proyektsii (Bulgarian: Variants of an orthogonal cartographic projection), *Izvestiya na Tsentralnata Laboratoriya po Geodeziya*, **12**, 103–8.

—— 1972, Primeneniye giperbolicheskikh funktsiy pri poluchenii novykh kartograficheskikh proyektsiy, *Geodeziya i Aerofotos'emka*, (3), 95–100. Translated into English as A use of hyperbolic functions to obtain new map projections, *Geodesy and Aerophotography*, 1972, (3), 174–7.

—— 1974, Varianty tsilindricheskikh proyektsiy (Variants of cylindrical projections), *Geodeziya i Aerofotos'emka*, (4), 117–20.

—— 1975, Varianty azimutal'nykh proyektsiy, poluchennye putem obobshcheniya perspektivnykh azimutal'nykh proyektsiy s positivnym i negativnym izobrazhennem (Variants of azimuthal projections made by generalizing perspective azimuthal projections with positive and negative representations), *Geodeziya*,

Kartografiya i Zemeustroystvo, **15** (4), 27–9.
—— 1983, V'rkhu obshchata teorii ortogonal'nykh krugovykh proyektsiy (Bulgarian: Problems of the general theory of orthogonal circular projections), *Geodeziya, Kartografiya i Zemeustroystvo*, **23** (6), 27–9.
—— 1984, Ortogonal'nye krugovye proyektsiy c nebol'shimi iskazheniyami ploshchadey (Bulgarian: Orthogonal circular projections with slight area distortion), *Geodeziya, Kartografiya i Zemeustroystvo*, **24** (5), 21–5.
Avetyan, G. Kh., 1970, Nekotorye voprosy obshchey teorii ekvivalentnykh proyektsiy (Some problems of the general theory of equivalent projections), *Tezisy doklady*, 63, Novosibirsk Tekhn. Konferentsiya. Novosibirsk otd. Vses. Astronomo-Geodezich. o-va Novosib. in-t inzh. geodesii, aerofotos'emki i Kartografii.
—— 1972, O sisteme differentsial'nykh uravneniy kvazikonformnogo otobrazheniya ellipsoida vrashcheniya na ploskost' (About the system of differential equations for a quasiconformal representation of an ellipsoid of revolution onto a sphere), *Trudy*, TsNIIGAiK, (29).
Baetsle, P. L., 1970, Optimalisation d'une representation cartographique, *Bulletin trimestriel*, Société Belge de Photogrammetrie, 11–26.
Baker, J. G. P., 1986, The 'Dinomic' world map projection, *Cartographic Journal*, **23** (1), 66–7.
Balthasart, M., 1935, L'emploi de la projection cylindrique équivalente dans l'enseignement de la géographique, *Bulletin*, Société Belge d'Etudes Géographiques, **5** (2), 269–72.
Baranyi, János, 1968, The problems of the representation of the globe on a plane with special reference to the preservation of the forms of continents, in *Hungarian Cartographical Studies*, pp. 19–43, Budapest: Földmérési Intézet.
—— 1987, Konstruktion anschaulicher Erdabbildungen, *Kartographische Nachrichten*, **37** (1), 11–17.
Baranyi, János and Karsay, Ferenc, 1971, Alakhübb világtérkép-vetületek (Hungarian: World map projections of greater conformity), *Geodézia és Kartográfia*, **23** (2), 108–14.
—— 1972, World map projections with better shape-keeping properties, in *Hungarian Cartographical Studies 1972*, Budapest: Földmérési Intézet.
Bayeva, Ye. Yu., 1987, Issledovaniye polikonicheskikh proyektsiy s polyarnymi liniyami dlya kart mira. Variant proyektsii (Research on polyconic projections with polar lines for world maps. A projection variant), Moscow: MIIGAiK.
Behrmann, Walter, 1910, Die beste bekannte flächentreue Projektion der ganzen Erde, *Petermanns Mitteilungen*, **56–2** (3), 141–4.
Bespalov, N. A., 1980, *Metody resheniya zadach spheroidicheskoy geodezii* (*Methods of solving problems of spheroidal geodesy*), Moscow: Nedra.
Biernacki, Franciszek, 1975, Odwzorowania izoperymetryczne (Polish with English abstract: Isoperimetric projections), *Geodezja i Kartografia*, **24** (1), 57–61.
Boggs, S. W., 1929, A new equal-area projection for world maps, *Geographical Journal*, **73** (3), 241–5.
Boginskiy, V. M., 1967, Matematicheskiye sposoby polucheniya novykh kartograficheskikh proyektsiy melkomasstabnykh kart s ispol'zovaniyem EVM, *Geodeziya i Aerofotos'emka*, (4), 45–50. Translated into English as Mathematical methods of deriving new cartographic projections of small-scale maps using an electronic computer, *Geodesy and Aerophotography*, 1967, (4), 227–9.

—— 1969 (1972), *Sposob izyskaniya proizvol'nykh proyektsiy melkomasshtabnykh kart* (*A method for developing arbitrary projections of small-scale maps*), Moscow: Nedra.

—— 1985, Proyektsiya dlya sozdaniya originalov plastmassovykh globusov (Projections for making plastic globe prototypes), *Geodeziya i Kartografiya*, (11), 23–4.

Boginskiy, V. M. and Ginzburg, G. A., 1971, Izyskaniye yedinoy proyektsii dlya kart dvukh materikov s isopol'zovaniyem EVM (The search for a unified projection for the maps of two continents using a computer), *Geodeziya i Kartografiya*, (7), 54–60.

Botley, F. V., 1949, A tetrahedral gnomonic projection, *Geography*, **34**, (pt 3), 131–6.

—— 1951, A new approach to world distribution maps, *Geographical Journal*, **117** (pt 2), 215–17.

Braun, Carl, 1867, Ueber zwei neue geographische Entwurfsarten, *Wochenschrift für Astronomie, Meteorologie und Geographie*, **10** (33), 259–63; (34), 269–72; (35), 276–8.

Briesemeister, William, 1953, A new oblique equal-area projection, *Geographical Review*, **43** (2), 260–1.

Bugayevskiy, L. M., 1980, Vneshniye perspektivnye azimutalniye kartographicheskiye proyektsii ellipsoida s positivnym izobrazheniem (proyektsii kosmicheskikh fotosnimkov poverkhnostey nebesnykh tel), *Geodeziya i Aerofotos'emka*, (2), 85–92. Translated into English as External perspective azimuthal cartographic projections of an ellipsoid with positive image (projections of cosmic photographs of the surfaces of heavenly bodies), *Geodesy, Mapping and Photogrammetry*, 1978, **20** (3), 169–74.

—— 1982a, Kriterii otsenki pri vybore kartograficheskikh proyektsii (Criteria for the selection of cartographic projections), *Geodeziya i Aerofotos'emka*, (3), 92–6.

—— 1982b, Proyektsiya Gaussa–Kryugera dlya shirokoy polosy (The Gauss–Krüger projection for a wide belt), *Issledovaniye po Geodesiye Aerofotos'emke i Kartografiye*, 9–18.

—— 1986a, Opredeleniye promezhutochnykh tochek geodezicheskikh linii (Certain spacings of points on geodetic lines), *Geodeziya i Aerofotos'emka*, (3), 3–6.

—— 1986b, Polykonicheskiye ravnovelikiye proyektsii ellipsoida (Polyconic equal-area projections of the ellipsoid), *Geodeziya i Aerofotos'emka*, (4), 103–6.

—— 1986c, Kartograficheskiye proyektsii ellipsoida, poluchayemye na osnove obobshcheniya metoda N. A. Urmayeva (Cartographic projections of the ellipsoid calculated with the method of N. A. Urmayev), *Geodeziya i Aerofotos'emka*, (5), 121–8.

—— 1986d, Proyektsiya P. L. Chebysheva dlya kartografirovaniya polyarnykh oblastei (A projection of P. L. Chebyshev for mapping of polar regions), *Geodeziya i Aerofotos'emka*, (6), 102–6.

—— 1986e, Kartograficheskiye proyektsii ellipsoida, poluchayemye na osnove obobshcheniya metoda N. A. Urmayeva (Ellipsoidal cartographic projections made on the basis of generalizing N. A. Urmayev's method), *Geodeziya i Aerofotos'emka*, (5), 121–8.

—— 1987a, K voprosu poluchenii izometricheskikh koordinat i ravnougol'noy tsilindricheskoy proyektsii trekhosnogo ellipsoida (The problems of obtaining isometric coordinates and a conformal cylindrical projection of the triaxial

ellipsoid), *Geodeziya i Aerofotos'emka*, (4), 79–90.
—— 1987b, K voprosu o poluchenii proyektsiy Eulera (The problems of obtaining the Euler projection), *Geodeziya i Aerofotos'emka*, (6), 105–10.
—— 1989, Polykonicheskiye proyektsii ellipsoida, vrashcheniya s ortogonal'noy kartograficheskoy setkoy (Polyconic projections of the ellipsoid with orthogonal graticule), *Geodeziya i Aerofotos'emka*, (3), 140–3.
Bugayevskiy, L. M., Deputatova, V. N., and Portnov, A. M., 1984, Primeneniye metodov zonal'nosti i ekonomizatii vychisleniy v matematicheskoy kartografii (Application of zoning methods and economy of computations in mathematical cartography), *Geodeziya i Aerofotos'emka*, (6), 103–7.
Bugayevskiy, L. M. and Ivanov, A. G., 1971, O nekotorykh osnovakh preobrazovaniya i izykaniya kartograficheskikh proyektsiy (On some basic transformations and finding cartographic projections), *Trudy*, TsNIIGAiK, (189), 64–78.
Bugayevskiy, L. M. and Portnov, A. M., 1984, *Teoriya odinochnykh kosmicheskikh snimkov* (The theory of single space photographs), Moscow: Nedra.
Bugayevskiy, L. M. and Slobodyanik, I. V., 1985, Obobshchennaya ravnopromezhutochnaya vdol'meridianov konicheskaya proyektsiya N. A. Urmayeva (The generalized equidistant along meridians conic projection of N. A. Urmayev), *Geodeziya i Aerofotos'emka*, (5), 69–72.
Bugayevskiy, L. M. and Vakhrameyeva, L. A., 1987, Osnovnye zadachi i napravlennya razvitiya teorii kartograficheskikh proyektsiy (On basic problems and directions for developing cartographic theory), *Tezisy Doklady*, Moscow, Kartogr. v epokhu NTR: Teoriya, metody, prakt. Vses. Soveshch. po Kartogr., 19–20.
Bugayevskiy, Yu. L. 1983, K voprosu o sostavlenii kart s izmeneniyem metrik i prostranstva (The problem of compiling maps with metric and spatial changes), *Sbornik*, Nauch.-Tekhn. TsNIIGAiK Ser. Soversh, Tekhn. i. Tekhnol. i Peredov. opyt Vypolneniya Geod. i. Topogr. Rabot, (103), 54–5.
—— 1986, Peremenno-masshtabnye proyektsii dlya sozdaniya anamorfirovannykh tematicheskikh kart (Varying-scale projections for the production of anamorph-thematic maps), *Geodeziya i Aerofotos'emka*, (5), 139–44.
—— 1987a, Variavalentnye proyektsii dlya anamorfirovaniykh kart (Varivalent projections for anamorphous maps), *Trudy*, Konf. mol. Uchenykh MIIGAiK, Moscow, 25–26 March 1986; MIIGAiK.
—— 1987b, Variavalentnaya proyektsiya tipa psevdotsilindricheskoy dlya anamorfirovannykh kart (A varivalent projection of the pseudocylindrical type for anamorphic mapmaking), *Geodeziya i Aerofotos'emka*, (1), 100–7.
Collignon, Édouard, 1865, Recherches sur la répresentation plane de la surface du globe terrestre, *Journal*, École Polytechnique, **24**, 125–32.
Craster, J. E. E., 1929, Some equal-area projections of the sphere, *Geographical Journal*, **74**, (5), 471–4.
—— 1938, Oblique conical orthomorphic projection for New Zealand, *Geographical Journal*, **92**, (6), 537–8.
Daskalova, Mara, 1964, Normalna ekvivalentna konichna proyektsiya s nay-malka sredna kvadratia deformatsiya v dălzhina (Bulgarian: A normal equivalent conic projection with the least mean-square deformation of distance), *Sbornik Statey po Kartografii*, (6), 9–14.
—— 1982, Izvezhdaniye na konstantite pri komformnata i ravnopromezhdut'chna

konichna proyektsiya s prilozheniye na kriteriya na Eyri (Bulgarian: Deducing the constants of the conformal and equidistant conic projection with the application of Airy's criterion), *Godisnik–Vissiya Institut po Arkitektura i Stroitelstvo, 1981–2*, **29** (3), 41–52.

DeLucia, A. A. and Snyder, J. P., 1986, An innovative world map projection, *American Cartographer*, **13** (2), 165–7.

Dent, B. D., 1987, Continental shapes on world projections: the design of a polycentred oblique orthographic world projection, *Cartographic Journal*, **24** (2), 117–24.

Deputatova, V. N., 1969a, O razvitii perspektivnykh proyektsiy c mnogokratnymi izobrazheniyami, *Geodeziya i Aerofotos'emka*, (1), 121–31. Translated into English as Development of perspective projections with multiple representations, *Geodesy and Aerophotography*, 1969, (1), 58–62.

—— 1969b, Obobshchennye formuly perspektivnykh proyektsiy, *Geodeziya i Aerofotos'emka*, (2), 95–102. Translated into English as Generalized formulas of perspective projections, *Geodesy and Aerophotography*, (2), 117–21.

Dermanis, Athanasios and Livieratos, Evangelos, 1984, Deformation analysis of isoparametric telluroid mappings, *Bollettino di Geodesia e Scienze Affini*, **43** (4), 301–12.

Dermanis, Athanasios, Livieratos, Evangelos, and Parashakis, I., 1983, Applications of strain criteria in cartography, *Bulletin Géodésique*, **57**, (3), 215–25.

Dinchenko, A. I., 1938, Proyektsiya Chebysheva dlya Sovetskogo Soyuza (A Chebyshev projection for the Soviet Union), *Geodezist*, (10).

Driencourt, Ludovic and Laborde, Jean, 1932, *Traité des projections des cartes géographiques à l'usage des cartographes et des géodésiens*, Vol. 4, Paris: Hermann de cie.

Dyer, J. A. and Snyder, J. P., 1989, Minimum-error equal-area map projections, *American Cartographer*, **16** (1), 39–46.

Eckert, Max, 1906, Neue Entwürfe für Erdkarten, *Petermanns Mitteilungen*, **52** (5), 97–109.

Eisenlohr, Friedrich, 1870, Ueber Flächenabbildung, *Journal für Reine und Angewandte Mathematik* [Crelle's], **72** (2), 143–51.

—— 1875, Ueber Kartenprojection, *Zeitschrift der Gesellschaft für Erdkunde zu Berlin*, **10**, (59), 321–34.

Epshteyn, M. M., 1964, Izmenenie velichiny masshtaba v proyektsiyakh Chebysheva i blizkikh k nom, *Geodeziya i Aerofotos'emka*, (4), 79–84. Translated into English as Scale change in the Chebyshev and similar projections, *Geodesy and Aerophotography*, 1964, (4), 222–4.

Érdi-Krausz, György, 1968, Combined equal-area projection for world maps, in *Hungarian Cartographic Studies*, pp. 44–9, Budapest: Földmérési Intézet.

Fawcett, C. B., 1949, A new net for a world map, *Geographical Journal*, **114** (1–3), 68–70.

Fiala, Frantiček, 1952, *Kartografické zobrazováni*, Prague: Státní Pedagocké Naki. Translated from Czech into German as *Mathematische Kartographie*, Berlin: VEB Verlag Technik, 1957.

Fisher, Irving, 1943, A world map on a regular icosahedron by gnomonic projection, *Geographical Review*, **33** (4), 605–19.

Gan'shin, V. N., 1973, Issledovaniye proyektsiy mezhdunarodnoy karty masshtaba 1 : 1 000 000 (Research into projections of the international maps of the World at a scale of 1 : 1 000 000), *Geodeziya i Aerofotos'emka*, (6), 81–6.

Gan'shin, V. N. and Lipin, A. V., 1977, O vybore optimal'noy proyektsii poverkhnosti ellipsoida na sfery pri reshenii zadach vozdushnoy navigatsii, *Geodeziya i Aerofotos'emka*, (5), 20–5. Translated into English as Selecting the optimal projection of the surface of an ellipsoid onto a sphere when solving the problems of air navigation, *Geodesy, Mapping and Photogrammetry*, 1977, **19** (1), 13–16.

Gdowski, Bogusław, 1973, O pewnych własnościach odwzorowań Czebyszewa (Polish: Some properties of the Chebyshev projection), *Geodezja i Kartografia*, **22** (1), 69–74.

—— 1974a, Minimalizacja zniekształceń w klasie uogólnionych odwzorowań walcowych (Polish: Minimization of distortion in the general class of cylindrical projections), *Geodezja i Kartografia*, **23** (3), 205–11.

—— 1974b, Metody wariacyjne minimalizacji zniekształceń w odwzorowaniach powierzchni (Polish: Minimization of distortion of projections of the surface using the method of variations), *Geodezja i Kartografia*, **23** (4), 281–93.

—— 1975, Optymalne odwzorowania równopolowe (Polish: Optimal equivalent projections), *Geodezja i Kartografia*, **24** (3), 191–204.

Gilbert , E. N., 1974, Distortion in maps, *Review* (SIAM: Society for Industrial and Applied Mathematics), **16** (1), 47–62.

Ginzburg, G. A., Karpov, H. C., and Salmanova, T. D., 1955, Matematicheskaya kartografiya v SSSR (Mathematical cartography in the USSR), *Trudy*, TsNIIGAiK, (99) and (108).

Ginzburg, G. A. and Salmanova, T. D., 1957, Atlas dlya vybora kartograficheskikh proyektsiy (Atlas for the selection of cartographic projections), *Trudy*, TsNIIGAiK, (110).

—— 1962, Primeneniye v matematicheskoy kartografii metodov chislennogo analiza (Application of methods of numerical analysis in mathematical cartography), *Trudy*, TsNIIGAiK, (153), 6–80.

—— 1964, Posobiye po matematicheskoy kartografii (Manual of mathematical cartography), *Trudy*, TsNIIGAiK, (160).

Goode, J. P., 1925, The Homolosine projection–a new device for portraying the Earth's surface entire, *Annals of the Association of American Geographers*, **15** (3), 119–25.

Grafarend, E. and Niermann, A., 1984, Beste echte Zylinderabbildungen, *Kartographische Nachrichten*, **34** (3), 103–7.

Grave, D. A., 1896, *Ob osnovnykh zadachakh matematicheskoy teorii postroyeniya geograficheskikh kart* (*Basic problems of the mathematical theory of geographical map construction*), St Petersburg.

Grygorenko, Wiktor, 1975, Odwzorowanie Służby Topograficznej WP dla map świata (Polish: Polish Army Topographical Service projection for world maps), *Polski Przegląd Kartograficzny*, **7** (2), 65–71.

Gurba, Stefania, 1970, Ortodroma na globusie dwunastościennym (Polish: The orthodrome on a dodecahedron globe), *Polski Przegląd Kartograficzny*, **2** (4), 160–8.

Hagiwara, Yasuchi, 1984, A retro-azimuthal projection of the whole sphere, in *Technical Papers*, Vol. 1, pp. 840–8, Perth, Australia: International Cartographic Association, 12th Conference, 6–13 August 1984.

Hatano, Matasaka, 1966 (Japanese: A proposal of the new method for an interrupted map projection of the world), *Geographical Review of Japan. Chirigaku-*

hyoron, **39** (4), 229–39.
—— 1972, Consideration on the projection suitable for Asia–Pacific type world map and the construction of elliptical projection diagram, *Geographical Review of Japan*, **45** (9), 637–47. In English.
Henrici, Peter, 1986, Conformal maps, in his *Applied and computational complex analysis*, New York: John Wiley, Vol. 1, pp. 286–432. (See also Vol. 3, pp. 323–506, for Construction of conformal maps.)
Hinks, A. R., 1929, A retro-azimuthal equidistant projection of the whole sphere, *Geographical Journal*, **73** (3), 245–7.
Hojovec, Vladislav, 1976, Klass ravnopromezhutochnykh proyektsiy (A class of equidistant projections), *Sbornik*, (9), pp. 556–9. Moscow: VIII Mezhdunar. Kartogr. Konf. Tezisy Doklady Otkrytaya sessiya, 1976.
—— 1984, Některé možnosti získáni nových ekvivalentnich zobrazeni (Czech: Some possibilities of obtaining new equivalent projections), *Geodetický a Kartografický Obzor*, **30** (1), 2–5.
—— 1987, Optimalizace ekvivalentnich zobrazeni z hlediska zkresleni (Czech: Optimization of equivalent projections from the viewpoint of distortion), *Geodetický a Kartografický Obzor*, **33** (9), 258–65.
Hojovec, Vladislav, Klášterka, Jaroslav, and Rendlová, Hana, 1975, Užití Čebyševova a variač niho kritéria v konformnim zobrazeni (Czech: A variant of the Chebyshev projection criteria for conformal representation), *Geodetický a Kartografický Obzor*, **21** (1), 3–6.
Jäger, G., 1865, Der Nordpol, ein thiergeographisches Centrum, *Mittheilungen aus Justus Perthes' Geographischer Anstalt ... von Dr. A. Petermann*, Ergänzungsheft 16, pp. 67–70.
Jahnke, Eugen, Emde, Fritz, and Lösch, Friedrich, 1968, *Spetsial'nye funktsii; formuly, grafiki, tablitsiy* (Special functions; formulas, graphs, and tables), Moscow: Nauka. (Authors listed as Ianke, Emde, and Lesh in Russian edition.)
Jordan, Wilhelm and Eggert, Otto, 1939, 1941, *Handbuch der Vermessungskunde*, Vol. 3, Stuttgart: J. B. Metzlersche. 1st half, 1939; 2nd half, 1941. Translated into English by Martha W. Carta, 1962, as *Jordan's Handbook of Geodesy*, Vol. 3 (selected portions), US Army, Corps of Engineers, Army Map Service.
Kanazawa, Kei, 1985, Koti kogē koto semmon hakka hakujutsu kiē (Japanese: Projections for maps in school atlases), *Bulletin* (Kochi Technical College), (22), 27–36.
Kavrayskiy, A. V., 1974, O vybore proizvol'nykh postoyannykh ravnougol'noy konicheskoy proyektsii (On the choice of arbitrary constants for a conformal projection), *Geodeziya i Kartografiya* (4), 58–61.
—— 1976, O kolichestvennoy otsenke iskazheniya ortodromiy v ravnougol'nykh proyektsiyakh (The quantitative evaluation of distortion of orthodromes on conformal projections), *Geodeziya i Kartografiya*, (1), 61–3.
—— 1982, Preobrazovaniye pryamougol'nykh koordinat v geograficheskiye dlya nekotorykh proyektsiy ellipsoida (The transformation of rectangular coordinates to geographic for some projections of the ellipsoid), *Geodeziya i Kartografiya*, (7), 40–3.
Kavrayskiy, V. V., 1934, *Matematicheskaya kartografiya* (*Mathematical cartography*), Moscow–Leningrad.
—— 1958–60, *Izbrannye trudy: Izdaniye upravleniya nach-ka gilrograficheskoy sluzhby VMF* (*Selected works. Publishing house of Department of the Hydro-*

graphic Service Chief, Naval Institute), Nos 1, 2, 3.

Kobylyatskiy, S. F., 1962, Ravnovelikiye tsilindricheskaya proyektsii obladaynshchya svoystvani proyektsii Tisso (An equal-area cylindrical projection with the characteristics of Tissot's projection), *Uchenye Zapiski*, Institut, (2), 77–80.

Koláčný, Antonio, 1964, Matematický základ map Národního atlasu ČSSR (Czech: Mathematical basis of maps for the National Atlas of Czechoslovakia), *Geodetický a Kartografický Obzor*, **10** (7), 160–4.

Konusova, G. I., 1964, O primenenii ryadov funktsiy kompleksnogo peremennogo dlya vychisleniya proyektsiy Chebysheva (Use of series functions of a complex variable for the calculation of the Chebyshev projection), *Trudy*, NIIGAiK, (17), 95–102.

—— 1970, K voprosu o poluchenii konformnykh proyektsiy s ortogonal'noy zadannymi svoystvami (On the problem of making conformal projections with preset properties), *Tezisy Doklady*, NIIGAiK, Nauchno-Tekhn. Konferentsiya, Novosib. otb. Vses. Astronomo-Geodesich. o-va, p. 67.

—— 1971, K voprosu ob izyskanii kartograficheskikh proyektsiy (Problems of finding cartographic projections), *Geodeziya i Kartografiya*, (11), 61–5.

—— 1973, Ob izyskanii proyektsiy s ortogonal'noy kartograficheskoy setkoy, *Geodeziya i Aerofotos'emka*, (6), 93–8. Translated into English as Finding projections with an orthogonal cartographic grid, *Geodesy, Mapping and Photogrammetry*, 1973, **15** (4), 161–3.

—— 1975, K voprosu o nailuchskikh kartograficheskikh proyektsiyakh (On the problems of optimal cartographic projections), *Geodeziya i Aerofotos'emka*, (4), 105–10.

—— 1979, O proyektsiyakh tematicheskikh kart SSSR (On projections for thematic maps of the USSR), *Trudy*, NIIGAiK, Mezhvuz. sb. nauk. 5/45, pp. 3–11.

Konusova, G. I. and Bertik, I. A., 1985, Issledovaniye kartograficheskikh proyektsii s ravnootstoyashchimi parallelyami i ortogonal'noy setkoy (Research on cartographic projections with correctly located parallels and orthogonal graticule), *Trudy*, NIIGAiK, 7–17.

Kozlova, N. N., 1972, Otsenka tochnosti vychisleniya pryamougol'nykh koordinat konformnykh proyektsiy, poluchennykh s pomoshch'yu (Precise computation of rectangular coordinates for conformal projections, obtained with series), *Geodeziya i Kartografiya*, (5), 115–22.

Krasovskiy, F. N., 1922, *Novye kartograficheskiye proyektsii* (*New cartographic projections*), Vyshevo Geodezicheskovo Upravleniya. Reprinted in Zakatov, P. S. and Solov'ev, M. D., eds, 1956, *F. N. Krasovskiy: Izbrannye sochineniya*, Moscow.

—— 1925, Vychisleniye konicheskoy ravnopromezhutochnoy proyektsiy, nailuchshe prisposoblennoy dlya izobrazheniya dannoy strany, *Geodezist*, (6–7). Reprinted in Zakatov, P. S. and Solov'ev, M. D., (1956), *F. N. Krasovskiy: Izbrannye sochineniya*, Moscow.

Krüger, Louis, 1922, *Zur stereographischen Projektion*, Berlin: Preussisches Geodätisches Institut, Veröffentlichung, new ser., (80).

Kuchař, Karel, 1966–7, Einige Bemerkungen über polyedrische Globen, *Globusfreund*, (15–16), 133–5.

Kuznetsov, A. A., 1984, Obshchiye uravneniya ravnougol'nykh tsilindricheskikh proyektsiy v geotsentricheskoy sisteme koordinat (General equations of conformal cylindrical projections in the geocentric coordinate system), *Geodeziya i Kartografiya*, (8), 50–2

—— 1987, Obshchiye uravneniya ravnougol'nykh proyektsiy v geotsentricheskoy sisteme coordinat (General equations of conformal projections in the geocentric coordinate system), *Geodeziya i Kartografiya*, (10), 41–4.

Lambert, J. H., 1772, *Beiträge zum Gebrauche der Mathematik und deren Anwendung*, Part III, section 6: Anmerkungen und Zusätze sur Entwerfung der Land- und Himmelscharten, Berlin. Translated into English and introduced by W. R. Tobler as *Notes and comments on the composition of terrestrial and celestial maps*, Ann Arbor, MI: University of Michigan, 1972, Geographical Publication No. 8.

Ledovskaya, L. S., 1969, Novyy variant proyektsii dlya uchebnykh i drugikh massovykh kart Evrazii, *Geodeziya i Kartografiya*, (12), 61–6. Translated into English as *A new variant of projection for educational and other popular maps of the Eurasian continent*, Charlottesville, VA: Army Foreign Science and Technology Center, Report on FSTC-HT-23-881-71, 1971.

—— 1971, Obzor proyektsiy kart vsey zemnoy prverkhnosti i yeyekrupneyshikh chastey v sovremennykh sovetskikh i zarubezhnykhisdaniyakh (Map projection review of the whole Earth's surface and its largest parts in modern Soviet and foreign publications), *Trudy*, TsNIIGAiK, (189).

Lee, L. P., 1974, A conformal projection for the map of the Pacific, *New Zealand Geographer*, **30** (1), 75–7.

—— 1975, Conformal projection with specified scale at selected points, *Survey Review*, **23** (178), 187–8.

—— 1976, Conformal projections based on elliptic functions, *Cartographica*, Monograph 16, supplement 1 to *Canadian Cartographer*, **13**.

Letoval'tsev, I. G., 1963, Obobshcheniye gnomonicheskoy proyektsii dlya sferoida i osnovnye geodezicheskiye voprosy prolozheniya nazemnykh trass, *Geodeziya i Aerofotos'emka*, (5), 61–8. Translated into English as Generalization of the gnomonic projection for a spheroid and the principal geodetic problems involved in the alignment of surface routes, *Geodesy and Aerophotography*, 1963, (5), 271–4.

Li, Guozao, 1981 (Chinese: Cylindrical transverse conformal projection with two standard meridians), *Acta Geodetica et Cartographica Sinica*, **10** (4), 309–12.

—— 1987 (Chinese with English abstract: Orthogonal projections, obtained under the condition $m = n^k$), *Acta Geodetica et Cartographica Sinica*, **16** (2), 149–57.

Lisichanskiy, A. S., 1962, Proizvodnye ravnougol'nye proyektsii (Derivative conformal projections), *Nauchnye Zapiski*, (85), *Seriya Geodezicheskaya*, (9), L'vovskiy Politekhnichnyi Institut, 100–4.

—— 1965, Formuly kosykh konformnykh proyektsiy ellipsoida (Formulas for oblique conformal projection of the ellipsoid), *Geodeziya, Kartografiya i Aerofotos'emka*, (3), 105–11.

—— 1970a, Teoriya obobshchennogo klassa konformnykh azimutal'no–konicheskikh proyektsiy (A theory of the general class of conformal azimuthal–conic projections), *Geodeziya, Kartografiya i Aerofotos'emka*, (11), 83–7.

—— 1970b, Obobshchennyi klass ekvivalentnykh azimutal'no–konicheskikh proyektsiy (The general class of equivalent azimuthal–conic projections), *Geodeziya, Kartografiya i Aerofotos'emka*, (12), 99–103.

McBryde, F. W. and Thomas, P. D., 1949, *Equal-area projections for world statistical maps*, Washington, DC: US Coast and Geodetic Survey Special Publication 245.

Macdonald, R. R., 1968, An optimal continental projection, *Cartographic Journal*, **5** (1), 46–7.

Maling, D. H., 1960, A review of some Russian map projections, *Empire Survey Review*, **15** (115), 203–15; (116), 255–66; (117), 294–303.

—— 1989, *Measurements from Maps – Principles and Methods of Cartometry*, Oxford: Pergamon press.

—— 1992, *Coordinate Systems and Map Projections*, Oxford: Pergamon Press, 2nd edn.

Malinin, S. P., 1969, Nesoosnye perspektivno-tsilindricheskiye proyektsii (Non-coaxial perspective cylindrical projections), *Geodeziya i Kartografiya*, (1), 58–68.

Martynenko, A. I. 1965, O vozmozhnosti programmirovaniya protsessa sozlaniya matematicheskoy i geodezicheskoy osnovy topograficheskoy karty (On the possibility of programming the process of creating mathematical and geodetical bases of topographic maps), *Geodeziya i Kartografiya*, (2), 50–5.

—— 1966, Avtomaticheskoy sposob sozdaniya matematicheskoy osnovy kart (An automatic method of creating the mathematical bases of maps), *Geodeziya i Kartografiya*, (1), 57–67.

Maurer, Hans, 1914, Die Definitionen in der Kartenentwurfslehre im Anschlusse an die Begriffe zenital, azimutal und gegenazimutal, *Petermanns Mitteilungen*, **60–2** (Aug.), 61–7; (Sept.), 116–21.

—— 1919a, „Doppelbüschelstrahlige, orthodromische" statt „doppelazimutale, gnomonische" Kartenentwürfe. Doppel-mittabstandstreue Kartogramme. (Bemerkungen zu den Aufsätzen von W. Immler und H. Thorade. Ann. d. Hydr. usw 1919, S. 22 und 35.), *Annalen der Hydrographie und Maritimen Meteorologie*, **47** (3–4), 75–8.

—— 1919b, Das winkeltreue gegenazimutale Kartennetz nach Littrow (Weirs Azimutdiagramm), *Annalen der Hydrographie und Maritimen Meteorologie*, **47**, (1–2), 14–22.

—— 1935, Ebene Kugelbilder: Ein Linnésches System der Kartenentwürfe, *Petermanns Mitteilungen*, Ergänzungsheft 221. Translated into English by Peter Ludwig as *Plane globe projection – a Linnean system of map projection*, with editing and foreword by William Warntz, Harvard Papers in Theoretical Geography, Geography and the Properties of Surfaces series, paper 22, 1968.

Mayr, Franz, 1964, Flachentreue Plattkarten. Eine bisher vernachlassigte Gruppe unechter Zylinderprojektionen, *International Yearbook of Cartography*, **4**, 13–19.

Mendeleyev, D. I., 1907, K poznanyu Rossi (Toward knowing Russia), *Izvestiya*, Akademiya Nauk (St Petersburg), ser. 6, **1**, 141–57.

Meshcheryakov, G. A., 1959, K teorii perspektivnykh proyektsiy (The theory of perspective projections), *Trudy*, Inz.-Stroitel'n. Fakulteta Chelyabinskogo Politekhn. Inst., **2**, 54–6.

—— 1963, O konformnom otobrazhenii poverkhnostey na ploskost' (On conformal representations of surfaces onto a plane), *Trudy*, NIIGAiK, **17** (1), 75–9.

—— 1965a, K probleme naivygodneyshikh kartograficheskikh proyektsiy, *Geodeziya i Aerofotos'emka*, (4), 115–27. Translated into English as The problem of choosing the most advantageous projections, *Geodesy and Aerophotography*, 1965, (4), 263–8.

—— 1965b, Ob osnovykh uravneniyakh matematicheskoy kartografii (Basic equations of mathematical cartography), *Trudy*, NIIGAiK, **18** (2), 65–75.

—— 1968, *Teoreticheskiye osnovy matematicheskoy kartografii (Theoretical bases of mathematical cartography)*, Moscow: Nedra.

Meshcheryakov, G. A. and Topchilov, M. A., 1970, O novykh klassakh kartograficheskikh proyektsiy, *Geodeziya i Aerofotos'emka*, (2) 98–106. Translated into English as New classes of cartographic projections, *Geodesy and Aerophotography*, 1970, (2), 119–23.

Miller, O. M., 1941, A conformal map projection for the Americas, *Geographical Review*, **31**(1), 100–4.

—— 1942, Notes on cylindrical projections, *Geographical Review*, **32** (3), 424–30.

—— 1953, A new conformal projection for Europe and Asia (*sic*; should read Africa), *Geographical Review*, **43** (3), 405–9.

—— 1965, Some equivalent map projection transformations in the plane, *Survey Review*, **18** (136), 73–7.

Milnor, John, 1969, A problem in cartography, *American Mathematical Monthly*, **76** (10), 1101–12.

Morozov, V. P., 1979, *Sferoidicheskaya geodeziya* (*Spheroidal geodesy*), Moscow: Nedra.

Murchison, R. I., 1857, New geometrical projection of two-thirds of a sphere, in his Address to the Royal Geographical Society of London, *Journal*, **27**, cxli–iii.

Murdoch, Patrick, 1758, On the best form of geographical maps, Royal Society, *Philosophical Transactions*, **50** (2) 553–62. Reprinted in abridged *Philosophical Transactions*, London: C. and R. Baldwin, 1809, Vol. 11, pp. 215–18.

Murevskis, V. O., 1975, O novoy ravnovelikoy proyektsii dlya kart mirovogo okeana (New equal-area projections for maps of the world ocean), *Nauka*, Kolsbaniya urovnya Mirovogo okeana i vopr. mor. geomorfol., 136–42.

Nell, A. M., 1852, *Vorschlag zu einer neuen charten Projektion*, Mainz: Inauguralschrift ..., F. Kupferberg.

—— 1890, Äquivalente Kartenprojektionen, *Petermanns Mitteilungen*, **36** (4) 93–8.

Nischan, H. (Nishan, Khorsta), 1962, O primenenii konformnykh proyektsiy dlya kart Tikhogo okeana (On applications of conformal projections for maps of the Pacific Ocean), *Trudy*, MIIGAiK (49), 105–16.

Novikova, Ye. N., 1984, K voprosu otyskaniya ideal'nykh proyektsiy s ispol'zovaniyem metoda mnozhiteley Lagranzha (Method for research of ideal projections through the use of the Lagrange multiplier method), *Krivorozh. gornorud. in-t. Krivoy Rog.* (Rukopis' dep. v UkrNIINTI, 17 Mar. 1984, no. 544Uk-84 Dep.).

—— 1986a, Nailuchshaya ravnopromezhutochnaya vdol' meridianov proyektsii variatsionnogo tipa (The best possible equidistant-along-meridians projection of the variational type), *Krivorozh. gornorud. in-t. Krivoy Rog.* (Rukopis' dep. v UkrNIINTI, 2 Jan. 1986, no. 49-Uk).

—— 1986b, Resheniye problemy otyskaniya ideal'nykh proyektsii variatsonnogo tipa (Solution of the problem of researching ideal projections of a variational type), *Krivorozh. gornorud. in-t. Krivoy Rog.* (Rukopis' dep. v UkrNIINTI, 12 Feb. 1986, no. 547-Uk.)

Osada, Edward, 1974, O odwzorowaniu równopolowym i skalach długości (Polish: Equal-area projection and linear scale), *Geodezja i Kartografia*, **23** (3), 199–204.

Panasiuk, Jan, 1973, Siatki kartograficzne w pewnym poprzecznym pseudowalcowym i równoodległościowym odwzorowaniu powierzchni elipsoidy obrotowej spłaszczonej na płaszczynzną (Polish: Cartographic graticules for a set of transverse pseudocylindrical and equidistant projections of the surface of an ellipsoid of revolution flattened onto a plane), *Geodezja i Kartografia*, **22** (2), 129–38.

—— 1974, Siatki kartograficzne w ortograficznym rzucie okośnym powierzchni elipsoidy obrotowej spłaszczonej na płaszczyznę (Polish: Cartographic graticules for the oblique orthographic projection of the surface of an ellipsoid of revolution), *Geodezja i Kartografia*, **23** (1), 53–60.

Pavlov, A. A., 1967, Primeneniye polinomov Chebysheva dlya rascheta kartograficheskoy proyektsii po eskizu setki (Application of Chebyshev polynomials for cartographic projection computation from grid sketches), *Nauka*, Leningrad: Novye probl. i metody kartogr., 32–9.

—— 1970, Matematicheskaya kartografiya (Mathematical cartography), *Kartografiya 1967–1969*, (4) (Itog. Nauki. VIN171 AN SSSR), 32–42.

—— 1973, O primenenii ravnougol'nykh proyektsiy, prisposobylyayemykh k ochertaniyam izobrazhayemoy territorii (On applications of conformal projections adapted to outlining representations of territories), *Vestnik*, Leningradskiy Universitet, *Geologiya–Geografiya*, **6** (1), 125–8.

—— 1974, *Prakticheskoye posobiye po matematicheskoy kartografii* (*Practical manual on mathematical cartography*), Leningrad: LGU.

Pernarowski, Lech, 1970, Odwzorawania izoparametryczne (Isoparametric projections; should read isoperimetric), *Geodezja i Kartografia*, **19** (4) 279–303.

Petrich, M. and Tobler, W. R., 1984, The globular projection generalized, *American Cartographer*, **11** (2), 101–5.

Philbrick, A. K., 1953, An oblique equal area map for world distributions, *Annals of the Association of American Geographers*, **43** (3), 201–15.

Portnov, A. M., 1977, O vybore modeli kontura dlya rascheta proyektsiy P. L. Chebysheva (About the choice of models for the calculation of projections of P. L. Chebyshev), *Geodeziya i Kartografiya*, (5), 62–3.

Potter, A. J., 1925, The tetrahedral gnomonic projection, *Geographical Teacher*, **13** (pt. 1, No. 71) 52–6.

Putniņš, R. V., 1934, Jaunas projekci jas pasaules kartēm (Latvian: New projections for world maps), *Geografiski raksti, Folia Geographica 3 and 4*, 180–209. Extensive French résumé.

Pyšek, Jiři, 1984, Obecné řešení kuželových zobrazení (Czech with English abstract: General use of conic projections), *Geodetický a Kartografický Obzor*, **30** (9), 211–20.

Reilly, W. I., 1973, A conformal mapping projection with minimum scale error, (part 1), *Survey Review*, **22** (168), 57–71.

Reilly, W. I. and Bibby, H. M., 1975–6, A conformal mapping projection with minimum scale error (parts 2 and 3), *Survey Review*, 1975, **23** (176), 79–87; 1976, **23** (181), 302–15.

Robinson, A. H., 1974, A new map projection: Its development and characteristics, *International Yearbook of Cartography*, **14**, 145–55.

Rogev, Boris, 1963, Konformni kartni proyektsii, mashtabăt na koito priema zadadeni stoynosti ve dadeni tochki (Bulgarian: Conformal map projections with prescribed scale values at given points), *Izvestiya*, Bălgarska Akademiya na Naukite, Tsentralna Laboratoriya po Geodeziya, **4**, 101–4.

Rogev, Boris and Dimov, L., 1970, Poluch eniye konformnoy kartograficheskoy proyektsii po metody integriruyushchego mnozhitelya (Bulgarian: Obtaining conformal cartographic projections by the method of an integrating multiplier), *Godishnik na Visshiya Minno-Geolozhki Institut 1966–1967*, sv. 3, 13, 159–78.

Roussilhe, Henri, 1922, Emploi des coordonnées rectangulaires stéréographiques pour le calcul de la triangulation dans un rayon de 560 kilomètres autour de l'origine, Section of Geodesy, International Union of Geodesy and Geophysics, May.

Sansó, Fernando, 1972, Carta conforme con minime deformazioni areali (Italian with English abstract), *Accademia Nazionale dei Lincei, Rendiconti, Classe di Scienze, Fisiche, Matematiche e Naturali*, **52** (2) 197–205.

Schmid, Erwin, 1974, *World maps on the August Epicycloidal Conformal projection*, Washington, DC: National Oceanic and Atmospheric Administration, NOAA Technical Report NOS 63.

Schott, C. A., 1882, A comparison of the relative value of the polyconic projection used on the Coast and Geodetic Survey, with some other projections, Washington, DC: US Coast and Geodetic Survey, *Report of the Superintendent*, June 1880, App. 15, pp. 287–96.

Serapinas, B. B., 1983, Otsenka iskazheniy v kartograficheskikh proyektsiyakh trekhosnogo ellipsoida (Distortion estimation in cartographic projections of the triaxial ellipsoid), *Geodeziya i Kartografiya*, (8), 55–6.

—— 1984, O poluchenii ravnougol'nykh kartograficheskikh proyektsiy trekhosnogo ellipsoida (About the calculation of conformal cartographic projections for the triaxial ellipsoid), *Geodeziya i Kartografiya*, (8), 48–50.

Siemon, Karl, 1935, Wegtreue Ortskurskarten, *Mitteilungen des Reichamts für Landesaufnahme*, **11** (2), 88–95.

—— 1937, Flächenproportionales Umgraden von Kartenentwürfen, *Mitteilungen des Reichamts für Landesaufnahme*, **13** (2), 88–102.

Snyder, J. P., 1977, A comparison of pseudocylindrical map projections, *American Cartographer*, **4** (1), 59–81.

—— 1984, Minimum-error map projections bounded by polygons, *Cartographic Journal*, **21** (2), 112–20. See corrections in 1985, **22** (1), 73.

—— 1985, *Computer-assisted map projection research*, Washington, DC: US Geological Survey Bulletin 1629.

—— 1987a, 'Magnifying-glass' azimuthal map projections, *American Cartographer*, **14** (1), 61–8.

—— 1987b, *Map projections – a working manual*, Washington, DC: US Geological Survey Professional Paper 1395.

—— 1988, New equal-area map projections for non-circular regions, *American Cartographer*, **15** (4), 341–55.

—— 1993, *Flattening the Earth – Two thousand years of map projections*, Chicago: University of Chicago Press.

Snyder, J. P. and Steward, Harry, 1988, eds, *Bibliography of map projections*, Washington, DC: Geological Survey Bulletin 1856.

Snyder, J. P. and Voxland, P. M., 1989, *An album of map projections*, Washington, DC: US Geological Survey Professional Paper 1453. (Graticules for 90 projections, with brief description and formulas.)

Sprinsky, W. H. and Snyder, J. P., 1986, The Miller Oblated Stereographic Projection for Africa, Europe, Asia and Australasia, *American Cartographer*, **13** (3),

253–61.

Starostin, F. A., Vakhrameyeva, L. A., and Bugayevskiy, L. M., 1981, Obobshchennaya klassifikatsiya kartograficheskikh proyektsiy po vidu izobrazheniya meridianov i paralleley (Generalized classification of cartographic projections with regard to the representation of meridians and parallels), *Geodeziya i Aerofotos'emka*, (6), 111–16.

Suvorov, A. K. and Bugayevskiy, Yu. L., 1987, Preobrazovaniye dlin liniy kartograficheskikh modeley v sootvetstvii s funktsional'nymi edinitsami (The transformation of line lengths on cartographic models in accordance with numerical functions), *Izvestiya*, Akademiya Nauk SSSR, serya geograficheskaya (4), 105–12.

Tezarius informatsionno-poiskovoy po soderzhaniyu i proyektsiyam kartograficheskikh istochnikov (*Thesaurus for information retrieval on contents and projections of cartographic resources*), 1987, Moscow: TsNIIGAiK.

Thomas, P. D., 1952, *Conformal projections in geodesy and cartography*, Washington, DC; US Coast and Geodetic Survey Special Publication 251.

—— 1970, *Spheroidal geodesics, reference systems, and local geometry*, Washington, DC: US Naval Oceanographic Office SP-138.

Tobler, W. R., 1966, Notes on two projections, *Cartographic Journal*, **3** (2), 87–9.

—— 1973a, The hyperelliptical and other new pseudo cylindrical equal area map projections, *Journal of Geophysical Research*, **78** (11), 1753–9.

—— 1973b, A continuous transformation useful for districting, *Annals of the New York Academy of Sciences*, **219**, 215–20.

Tolstova, T. I., 1970, Obobshchennye formuly konicheskiye proyektsii (Generalized formulas for conic projections), *Trudy*, MIIGAiK, (57), 94–116.

—— 1981a, K teorii kartograficheskikh proyektsii (Toward the theory of cartographic projections), *Doklady*, Akademiya Nauk SSSR, **259** (2), 316–20.

—— 1981b, Obobshchennye polikonicheskiye proyektsii (Generalized polyconic projections), *Doklady*, Akademiya Nauk SSSR, **259** (1), 61–5.

Topchilov, M. A., 1969, Ob odnom obobshchenii klass konformmykh proyektsiy, *Geodeziya i Aerofotos'emka*, (2), 103–8. Translated into English as A generalization of a class of conformal projections. *Geodesy and Aerophotography*, 1969, (2), 121–4.

—— 1970, Sistema Eylera–Urmayeva-apparat izyskaniya novykh kartograficheskikh proyektsiy (The system of the Euler–Urmayev apparatus in seeking new cartographic projections), *Tezisy Doklady*, Nauchno-tekhn. Konferentsiya Novosibirsk std. Vses. astronomo-geodezich. ostrova NIIGAiK, 71–2.

Tuchin, Ya. I., 1973, Odin iz variantov nailuchshey Eyleyrovoy proyektsii (One of the variants for the best Euler projection), *Trudy*, NIIGAiK, (30), 65–7.

Tyuflin, Yu. S. and Abalakin, V. K., 1979, Systemy koordinat i elementy vrashcheniya planet i ikh sputnikov (Coordinate systems and elements of revolution of planets and their satellites), *Geodeziya i Kartografiya*, (12), 34–41.

Urmayev, M. S., 1981, *Orbital'nye metody kosmicheskoy geodezii* (Orbital methods for space geodesy), Moscow: Nedra.

Urmayev, N. A., 1941, *Matematicheskaya kartografiya* (Mathematical cartography), Moscow: Voyennoinzhenernaya Akademiya.

—— 1947, *Metody izkaniya novykh kartograficheskikh proyektsiy* (*Method of establishment of new cartographic projections*), Moscow: VTU GSh VSSSR.

—— 1950, Izyskaniye nekotorykh novykh tsilindricheskikh, azimutal'nykh i psevdotsilindricheskikh proyektsiy (The search for some new cylindrical, azimuthal, and pseudocylindrical projections), *Sbornik*, Nauchno-Tekhnicheskiye i Proizvodstvennye Statey GUGK, (29).

—— 1953, Issledovaniya po matematicheskoy kartografii (Studies in mathematical cartography), *Trudy*, TsNIIGAiK, (98).

—— 1956, Teoriya gomograficheskogo preobrazovaniya i yeye primeneniye k matematicheskoy kartografii i sostavleniyu kart (Theory of homographic transformation and its application to mathematical cartography and map design), *Trudy*, TsNIIGAiK, (113).

—— 1962, Osnovy matematicheskoy kartografii (The fundamentals of mathematical cartography), *Trudy*, TsNIIGAiK, (144).

Vakhrameyeva, L. A., 1961, O metode polucheniya perspektivnykh proyektsiy s mnogokratnymi izobrazheniyami, predlozhennom Prof. M. D. Solov'evym (Obtaining a perspective projection with multiple aspects as suggested by Prof. M. D. Solov'ev), *Trudy*, MIIGAiK, (47), 117–26.

—— 1963, Obobshchennye formuly konformnykh proyektsiy, *Geodeziya i Aerofotos'emka*, (1), 109–15. Translated into English as Generalized formulas for conformal projections, *Geodesy and Aerophotography*, 1963, (1), 48–51.

—— 1969, Konformnye proyektsii, poluchennye s pomoshch'yu garmonicheskikh polinomov, *Geodeziya i Aerofotos'emka*, (1), 111–16. Translated into English as Conformal projections obtained with the aid of harmonic polynomials, *Geodesy and Aerophotography*, 1969, (1), 53–5.

—— 1971a, Konformnye proyektsii, poluchayemye s pomoshch'yu ryadov, i ikh primeneniye, *Geodeziya i Aerofotos'emka*, (6), 91–4. Translated into English as Conformal projections obtained from series and their applications, *Geodesy and Aerophotography*, 1971, (6), 338–40.

—— 1971b, Konformnye proyektsii s asimmetrichnymi izokolami, *Geodeziya i Aerofotos'emka*, (1), 133–8. Translated into English as Conformal projections with asymmetric distribution of distortion, *Geodesy and Aerophotography*, 1971, (1), 52–4.

Vakhrameyeva, L. A., Bugayevskiy, L. M. and Kazakova, Z. L., 1986, *Matematicheskaya kartografiya* (*Mathematical cartography*), Moscow: Nedra.

Vakhrameyeva, L. A. and Bugayevskiy, Yu. L., 1985, Issledovaniya peremennomasshtabnykh proyektsii dlya sotsial'no-ekonomicheskikh kart (Analysis of variable-scale projections for socio-economic maps), *Geodeziya i Aerofotos'emka*, (3), 95–9.

van der Grinten, A. J., 1904, Darstellung der ganzen Erdoberfläche auf einer kreisformigen Projektionsebene, *Petermanns Mitteilungen*, **50** (7), 155–9.

Vasilevskiy, L. I., 1968, Variavalentnye proyektsii i ikh primeneniye dlya kart naseleniya i drugikh special'nykh geograficheskikh kart (Many-valued projections and their applications for settlement and other special maps), *Materialy dvtorogo Mezhduvedomstvennogo Soveshchaniya po Geograficheskogo Naseleniya*, **3**, 1967, 22–5.

Vilenkin, N. Ya., 1961, O konformnykh proyektsiyakh, blizkikh k chebyshevskoy proyektsii (On conformal projections similar to the Chebyshev projection), *Geodeziya i Aerofotos'emka*, (5), 79–89.

Vincze, Vilmos, 1968, Einheitliche Ableitung und allgemeingültige Grundgleichungen der reelen Projektionen, *Acta Geodaetica et Montanistica*, **3** (1–2), 69–103.

Vitkovskiy, V. V., 1907, *Kartografiya; teoriya kartograficheskikh proyektsiy*

(Cartography; the theory of cartographic projections), St Petersburg: Erlikh.

Volkov, N. M., 1950, *Princhipy i metody kartometrii* (*Principles and methods of cartometry*), Moscow–Leningrad: Akademiya Nauk SSSR.

—— 1973, Kartograficheskaya proyektsiya fotosnimkov nebesnykh tel'poluyayennykh iz kosmosa, *Geodeziya i Aerofotos'emka*, (2), 91–9. Translated into English as Cartographic projection of photographs of celestial bodies taken from outer space, *Geodesy, Mapping and Photogrammetry*, 1973, **15** (2), 87–91.

Wagner, Karlheinz, 1949, *Kartographische Netzentwürfe*, Leipzig: Bibliographisches Institut.

Werenskiold, Werner, 1945, A class of equal area map projections, *Avhandlinger*, (11). Norske Videnskaps-Akademi, Oslo, Matematisk-naturvidenskapelig Klasse 1944.

Wiechel, H., 1879, Rationelle Gradnetzprojectionen, *Civilingenieur*, new series, **25**, cols 401–22.

William-Olsson, William, 1968, A new equal area projection of the world, *Acta Geographica (Helsinki)*, **20**, 389–93.

Yang, Qi-he, 1965 (Chinese: The use of numerical method for establishing optional conical projections), *Acta Geodetica et Cartographica Sinica*, **8** (4), 295–318.

Young, A. E., 1920, *Some investigations in the theory of map projections*, London: Royal Geographical Society, Technical Series, No. 1.

—— 1930, Conformal map projections, *Geographical Journal*, **76** (4), 348–51.

Yuzefovich, Yu. M., 1971, Raspostraneniye teoremy Chebysheva–Grave na odin novyy klass kartograficheskikh proyektsiy, *Geodeziya i Aerofotos'emka*, (3), 97–102. Translated into English as Extension of the Chebyshev–Grave theorem to a new class of cartographic projections, *Geodesy and Aerophotogrammetry*, 1971, (3), 155–7.

—— 1972a, O nekotorykh novykh proyektsiyakh, blizkikh k ekvivalentnym (On some new projections close to equivalent), *Sbornik*, (86), 245–53, Nauch. tr. Belorus. S.-Kh. Akad.

—— 1972b, O vychislenii odnogo integrala pri issledovanii kartograficheskikh proyektsiy (On the computation of a single integral by researching cartographic projections), *Sbornik*, Nauch. tr. Belorus. S.-Kh. Akad., (86), 260–7.

Zargaryan, T. G., 1971, Nekotorye issledovaniya konformnykh proyektsiy, poluchennykh s pomoshch'yu garmonicheskikh polinomov (Some research into conformal projections obtained by means of harmonic polynomials), *Trudy*, MIIGAiK, (58), 132–45.

Zatonskiy, L. K., 1962, Proizvol'naya sostavnaya ellipticheskaya proyektsiya dlya kart mira (An arbitrary composite elliptical projection for a world map), *Trudy*, **55**, 173–4, Akademiya Nauk SSSR, Institut Okeanologii.

Zhong, Ye-xun, *et al.*, 1965 (Chinese: Projecting a polyconic projection with unequally divided meridians and parallels and methods for its analytical computation), *Acta Geodetica et Cartographica Sinica*, **8** (3), 218–36.

Zlatyanov, Georgi, 1965, Proizvolna neortogonalna krăgova proyektsiya za drebnomashchabni svetovni. Karti (Bulgarian: An arbitrary non-orthogonal circular projection for small-scale world maps), *Sbornik Statey po Kartografii*, (7), 7–11.

Zöppritz, K. J. and Bludau, Alois, 1912, Die Projectionslehre, in their *Leitfaden der Kartenentwurfslehre*, Leipzig: Teubner, 3rd edn.

Appendix 1

Map systems (1)	Map groups by methods of perception, estimation, and analysis of cartographic information (2)	Limiting values of distortion for which effect can still be ignored (3)	Approximate hierarchy of requirements for map projections (4)	Approximate map scales and types of projection distortion (5)
Scientific–technical and technical maps	Analysis of cartographic information mainly with the help of a computer, and to a lesser extent with manual cartometry	When reductions from an analysis of cartographic information are introduced into values measured	Type of distortion desired; type of distortion distribution; ease of introducing reductions; minimum distortion of linear scale, areas, curvature of geodetic and other position lines (with a corresponding priority of one or more requirements for type of distortion)	1 : 1 000 000 and larger: conformal projections; 1 : 2 000 000–1 : 10 000 000: equal-area, conformal, and arbitrary projections. Local linear scale factors should be constant (or equal to unity) along the lines of one direction, scale should vary mainly along orthogonal directions
	Analysis of cartographic information mainly on the basis of cartometry of higher accuracy and to a lesser extent with the help of a computer	Distortion of linear scale and area up to $\pm$(0.2–0.4%), of angles up to $\frac{1}{4}^\circ$–$\frac{1}{2}^\circ$	Type of distortion desired; minimum distortion of linear scale, angles, areas, curvature of position lines; desired type of distortion distribution; permissive distortion ranges (setting the priority for requirements as to types of distortion); ease of introducing reduction into the values measured	1 : 1 000 000 and larger: conformal projections; 1 : 2 000 000–1 : 10 000 000: equal-area, conformal, and arbitrary projections. Projections should be among the so-called best projections or be close to them. Outer isocols of conformal projections and those close to them should approach the outline of the region to be mapped

Map systems (1)	Map groups by methods of perception, estimation, and analysis of cartographic information (2)	Limiting values of distortion for which effect can still be ignored (3)	Approximate hierarchy of requirements for map projections (4)	Approximate map scales and types of projection distortion (5)
Maps for general use	Analysis and utilization of cartographic information by rough measurements and by estimation of dimensions, shape, relative position, and significance of the region	Distortion of linear scale and area up to ±(2–3)%, of angles up to 2°–3°	Types of distortion; minimum distortion of linear scale angles, areas, curvature of position lines; type of distribution and permissive distortion ranges (setting the priority of requirements for distortion types); minimum distortion; symmetry of map graticule; type of map usage (in combination or separately); availability of regional overlap	1 : 1 000 000 and larger: conformal projection; 1 : 2 000 000–1 : 10 000 000 equal-area, conformal, and arbitrary projections. Projection should be among the so-called best projections or be close to them. Outer isocols of conformal projections and those close to them should approach the outline of the region to be mapped
	Cartographic information is determined and estimated mainly visually, sometimes by very rough measurements. Mainly wall maps, some maps in atlases and in textbooks, and illustrative maps in various publications	Distortion of linear scale and area up to ±(6–8)%, of angles up to 6°–8° (no visual distortion is noted)	Curvature of graticule; its equidistance and orthogonality; types of pole representation; graticule symmetry, dimensions, and shape of the central meridian and the equator; visual perception of projection (sphericity effect, regional relationships, shapes (stereographicity), geographical positions of regions); availability of regional overlap. Type of distortion; minimum distortion of linear scale, angles, areas, permissive distortion ranges	1 : 2 000 000–1 : 10 000 000: equal-area, arbitrary, and conformal projections; scales rarely smaller than 1 : 10 000 000: arbitrary, equal-area, and conformal projections

Map systems (1)	Map groups by methods of perception, estimation, and analysis of cartographic information (2)	Limiting values of distortion for which effect can still be ignored (3)	Approximate hierarchy of requirements for map projections (4)	Approximate map scales and types of projection distortion (5)
	Cartographic information is perceived and estimated only visually; no measurements are taken (includes many wall maps, educational and school maps, a number of illustrative maps in various publications	Distortion of linear scale and area up to ±(10–12)%, of angles up to 10°–12° (some visual distortion is noted)	Requirements for visual perception of projection (sphericity effect, regional relationships, their areas and shapes); presence of regional overlap; type of pole representation; map graticule symmetry; shape of the central meridian and the equator; curvature of the graticule, equidistance and orthogonality; minimum distortion, character of its distribution	Smaller than 1 : 10 000 000: arbitrary, equal-area, and conformal projections. Scales rarely 1 : 2 000 000–1 : 10 000 000: equal-area, arbitrary, and conformal projections
	Maps in reference and educational publications, in technical manuals and textbooks	Distortion of linear scale and area up to ±(10–12)% of angles up to 10°–12° (some visual distortion is noted)	Simplicity of map graticule (orthogonality, curvature of meridians and parallels, their equidistance, appearance of the central meridian and the equator)	Smaller than 1 : 10 000 000 arbitrary, equal-area, and conformal projections. Scales rarely 1 : 2 000 000–1 : 10 000 000 equal-area, arbitrary, and conformal projections

Appendix 2

Radius of curvature: meridian M, in meters; first vertical N, in meters. GRS 80 ellipsoid

ϕ	M	$\Delta/30$	N	$\Delta/30$
0°00′	6 335 439	0.16	6 378 137	0.05
30′	6 335 444	0.48	6 378 139	0.16
1°00′	6 335 459	0.81	6 378 144	0.27
30′	6 335 483	1.13	6 378 152	0.38
2°00′	6 335 517	1.45	6 378 163	0.49
30′	6 335 560	1.77	6 378 178	0.60
3°00′	6 335 614	2.09	6 378 195	0.70
30′	6 335 676	2.42	6 378 217	0.81
4°00′	6 335 749	2.74	6 378 241	0.92
30′	6 335 831	3.05	6 378 268	1.03
5°00′	6 335 923	3.37	6 378 299	1.13
30′	6 336 024	3.69	6 378 333	1.24
6°00′	6 336 134	4.01	6 378 370	1.34
30′	6 336 255	4.32	6 378 411	1.45
7°00′	6 336 384	4.63	6 378 454	1.56
30′	6 336 523	4.95	6 378 501	1.66
8°00′	6 336 672	5.26	6 378 551	1.76
30′	6 336 829	5.57	6 378 603	1.87
9°00′	6 336 996	5.87	6 378 660	1.97
30′	6 337 173	6.18	6 378 719	2.07
10°00′	6 337 358	6.48	6 378 781	2.18
30′	6 337 553	6.79	6 378 846	2.28
11°00′	6 337 756	7.09	6 378 914	2.38
30′	6 337 969	7.38	6 378 986	2.48
12°00′	6 338 190	7.68	6 379 060	2.58
30′	6 338 421	7.97	6 379 137	2.67
13°00′	6 338 660	8.26	6 379 218	2.77
30′	6 338 908	8.55	6 379 301	2.87
14°00′	6 339 164	8.84	6 379 387	2.96
30′	6 339 430	9.12	6 379 476	3.06
15°00′	6 339 703	9.40	6 379 568	3.15
30′	6 339 985	9.68	6 379 662	3.25
16°00′	6 340 276	9.96	6 379 760	3.34
30′	6 340 574	10.23	6 379 860	3.43
17°00′	6 340 881	10.50	6 379 963	3.52
30′	6 341 196	10.76	6 380 068	3.61
18°00′	6 341 519	11.03	6 380 177	3.70
30′	6 341 850	11.28	6 380 288	3.78
19°00′	6 342 188	11.54	6 380 401	3.87
30′	6 342 535	11.79	6 380 517	3.95
20°00′	6 342 888	12.04	6 380 636	4.04
30′	6 343 250	12.29	6 380 757	4.12
21°00′	6 343 618	12.53	6 380 881	4.20
30′	6 343 994	12.77	6 381 007	4.28
22°00′	6 344 377	13.00	6 381 135	4.36
30′	6 344 767	13.23	6 381 266	4.44
23°00′	6 345 164	13.46	6 381 399	4.51
30′	6 345 568	13.68	6 381 534	4.59
24°00′	6 345 978	13.90	6 381 672	4.66
30′	6 346 395	14.11	6 381 812	4.73
25°00′	6 346 819	14.32	6 381 953	4.80
30′	6 347 249	14.53	6 382 097	4.87
26°00′	6 347 684	14.73	6 382 244	4.94
30′	6 348 126	14.93	6 382 392	5.00

ϕ	M	$\Delta/30$	N	$\Delta/30$
27°00′	6 348 574	15.12	6 382 542	5.07
30′	6 349 028	15.31	6 382 694	5.13
28°00′	6 349 487	15.49	6 382 848	5.19
30′	6 349 951	15.67	6 383 003	5.25
29°00′	6 350 422	15.84	6 383 161	5.31
30′	6 350 897	16.01	6 383 320	5.36
30°00′	6 351 377	16.17	6 383 481	5.42
30′	6 351 862	16.33	6 383 643	5.47
31°00′	6 352 352	16.49	6 383 808	5.52
30′	6 352 847	16.64	6 383 973	5.57
32°00′	6 353 346	16.78	6 384 141	5.62
30′	6 353 850	16.92	6 384 309	5.67
33°00′	6 354 357	17.06	6 384 479	5.71
30′	6 354 869	17.19	6 384 651	5.76
34°00′	6 355 385	17.31	6 384 823	5.80
30′	6 355 904	17.43	6 384 997	5.84
35°00′	6 356 427	17.54	6 385 172	5.87
30′	6 356 953	17.65	6 385 348	5.91
36°00′	6 357 482	17.75	6 385 526	5.94
30′	6 358 015	17.85	6 385 704	5.98
37°00′	6 358 551	17.94	6 385 883	6.01
30′	6 359 089	18.03	6 386 063	6.03
38°00′	6 359 630	18.11	6 386 244	6.06
30′	6 360 173	18.19	6 386 426	6.09
39°00′	6 360 719	18.26	6 386 609	6.11
30′	6 361 266	18.32	6 386 792	6.13
40°00′	6 361 816	18.38	6 386 976	6.15
30′	6 362 367	18.43	6 387 161	6.17
41°00′	6 362 920	18.48	6 387 346	6.18
30′	6 363 475	18.52	6 387 531	6.20
42°00′	6 364 030	18.56	6 387 717	6.21
30′	6 364 587	18.59	6 387 903	6.22
43°00′	6 365 145	18.62	6 388 090	6.23
30′	6 365 703	18.64	6 388 277	6.23
44°00′	6 366 263	18.65	6 388 464	6.24
30′	6 366 822	18.66	6 388 651	6.24
45°00′	6 367 382	18.66	6 388 838	6.24
30′	6 367 942	18.66	6 389 026	6.24
46°00′	6 368 501	18.65	6 389 213	6.24
30′	6 369 061	18.64	6 389 400	6.23
47°00′	6 369 620	18.62	6 389 587	6.22
30′	6 370 178	18.59	6 389 774	6.22
48°00′	6 370 736	18.56	6 389 960	6.20
30′	6 371 293	18.52	6 390 146	6.19
49°00′	6 371 849	18.48	6 390 332	6.18
30′	6 372 403	18.43	6 390 517	6.16
50°00′	6 372 956	18.38	6 390 702	6.14
30′	6 373 507	18.32	6 390 886	6.12
51°00′	6 374 057	18.25	6 391 070	6.10
30′	6 374 604	18.18	6 391 253	6.08
52°00′	6 375 150	18.10	6 391 435	6.05
30′	6 375 693	18.02	6 391 617	6.02
53°00′	6 376 234	17.94	6 391 797	5.99
30′	6 376 772	17.84	6 391 977	5.96
54°00′	6 377 307	17.74	6 392 156	5.93
30′	6 377 839	17.64	6 392 334	5.89

ϕ	M	$\Delta/30$	N	$\Delta/30$
55°00′	6 378 368	17.53	6 392 511	5.86
30′	6 378 894	17.42	6 392 686	5.82
56°00′	6 379 417	17.30	6 392 861	5.78
30′	6 379 936	17.17	6 393 034	5.74
57°00′	6 380 451	17.04	6 393 206	5.69
30′	6 380 962	16.90	6 393 377	5.65
58°00′	6 381 469	16.76	6 393 546	5.60
30′	6 381 972	16.62	6 393 714	5.55
59°00′	6 382 471	16.47	6 393 881	5.50
30′	6 382 965	16.31	6 394 046	5.45
60°00′	6 383 454	16.15	6 394 209	5.39
30′	6 383 938	15.98	6 394 371	5.34
61°00′	6 384 418	15.81	6 394 531	5.28
30′	6 384 892	15.63	6 394 689	5.22
62°00′	6 385 361	15.45	6 394 846	5.16
30′	6 385 825	15.27	6 395 001	5.10
63°00′	6 386 283	15.08	6 395 154	5.03
30′	6 386 735	14.88	6 395 305	4.97
64°00′	6 387 181	14.68	6 395 454	4.90
30′	6 387 622	14.48	6 395 601	4.83
65°00′	6 388 056	14.27	6 395 745	4.76
30′	6 388 484	14.05	6 395 888	4.69
66°00′	6 388 906	13.84	6 396 029	4.62
30′	6 389 321	13.61	6 396 168	4.54
67°00′	6 389 729	13.39	6 396 304	4.47
30′	6 390 131	13.16	6 396 438	4.39
68°00′	6 390 526	12.92	6 396 570	4.31
30′	6 390 913	12.69	6 396 699	4.23
69°00′	6 391 294	12.44	6 396 826	4.15
30′	6 391 667	12.20	6 396 950	4.07
70°00′	6 392 033	11.95	6 397 073	3.99
30′	6 392 392	11.69	6 397 192	3.90
71°00′	6 392 742	11.44	6 397 309	3.81
30′	6 393 086	11.18	6 397 424	3.73
72°00′	6 393 421	10.91	6 397 535	3.64
30′	6 393 748	10.64	6 397645	3.55
73°00′	6 394 067	10.37	6 397 751	3.46
30′	6 394 379	10.10	6 397 855	3.37
74°00′	6 394 682	9.82	6 397 956	3.28
30′	6 394 976	9.54	6 398 054	3.18
75°00′	6 395 262	9.26	6 398 150	3.09
30′	6 395 540	8.97	6 398 242	2.99
76°00′	6 395 809	8.68	6 398 332	2.89
30′	6 396 070	8.39	6 398 419	2.80
77°00′	6 396 321	8.10	6 398 503	2.70
30′	6 396 564	7.80	6 398 584	2.60
78°00′	6 396 798	7.50	6 398 662	2.50
30′	6 397 023	7.20	6 398 737	2.40
79°00′	6 397 239	6.89	6 398 809	2.30
30′	6 397 446	6.59	6 398 878	2.20
80°00′	6 397 643	6.28	6 398 943	2.09
30′	6 397 832	5.97	6 399 006	1.99
81°00′	6 398 011	5.66	6 399 066	1.89
30′	6 398 180	5.34	6 399 123	1.78
82°00′	6 398 341	5.03	6 399 176	1.68
30′	6 398 492	4.71	6 399 226	1.57

ϕ	M	$\Delta/30$	N	$\Delta/30$
83°00′	6 398 633		6 399 273	
		4.39		1.46
30′	6 398 765		6 399 317	
		4.07		1.36
84°00′	6 398 887		6 399 358	
		3.75		1.25
30′	6 398 999		6 399 396	
		3.43		1.14
85°00′	6 399 102		6 399 430	
		3.11		1.04
30′	6 399 195		6 399 461	
		2.78		0.93
86°00′	6 399 279		6 399 489	
		2.46		0.82
30′	6 399 353		6 399 513	
		2.13		0.71
87°00′	6 399 416		6 399 535	
		1.80		0.60
30′	6 399 471		6 399 553	
		1.48		0.49
88°00′	6 399 515		6 399 567	
		1.15		0.38
30′	6 399 549		6 399 579	
		0.82		0.27
89°00′	6 399 574		6 399 587	
		0.49		0.16
30′	6 399 589		6 399 592	
		0.16		0.05
90°00′	6 399 594		6 399 594	

$M = a(1 - e^2)/(1 - e^2 \sin^2 \phi)^{3/2}$
$N = a/(1 - e^2 \sin^2 \phi)^{1/2}$
$a = 6\,378\,137$ meters, $e^2 = 0.006\,694\,380\,0$

Appendix 3

Mean radius of curvature R_0, in meters; radii of parallels r, in meters. GRS 80 ellipsoid

ϕ	R_0	$\Delta/30$	r
0°00′	6 356 752	0.108	6 378 137
30′	6 356 756	0.324	6 377 896
1°00′	6 356 765	0.540	6 377 172
30′	6 356 781	0.756	6 375 966
2°00′	6 356 804	0.971	6 374 278
30′	6 356 833	1.186	6 372 107
3°00′	6 356 869	1.401	6 369 454
30′	6 356 911	1.616	6 366 320
4°00′	6 356 959	1.830	6 362 704
30′	6 357 014	2.043	6 358 606
5°00′	6 357 076	2.256	6 354 028
30′	6 357 143	2.468	6 348 969
6°00′	6 357 217	2.680	6 343 429
30′	6 357 298	2.890	6 337 409
7°00′	6 357 384	3.100	6 330 910
30′	6 357 477	3.309	6 323 932
8°00′	6 357 577	3.517	6 316 475
30′	6 357 682	3.723	6 308 540
9°00′	6 357 794	3.929	6 300 128
30′	6 357 912	4.134	6 291 238
10°00′	6 358 036	4.337	6 281 873
30′	6 358 166	4.539	6 272 032
11°00′	6 358 302	4.739	6 261 716
30′	6 358 444	4.939	6 250 926
12°00′	6 358 592	5.136	6 239 662
30′	6 358 746	5.333	6 227 926
13°00′	6 358 906	5.527	6 215 719
30′	6 359 072	5.720	6 203 040
14°00′	6 359 244	5.911	6 189 892
30′	6 359 421	6.101	6 176 274
15°00′	6 359 604	6.288	6 162 189
30′	6 359 793	6.474	6 147 637
16°00′	6 359 987	6.658	6 132 619
30′	6 360 187	6.840	6 117 135
17°00′	6 360 392	7.019	6 101 189
30′	6 360 603	7.197	6 084 779
18°00′	6 360 818	7.373	6 067 909
30′	6 361 040	7.546	6 050 578
19°00′	6 361 266	7.717	6 032 788
30′	6 361 498	7.886	6 014 540
20°00′	6 361 734	8.052	5 995 836
30′	6 361 976	8.216	5 976 678
21°00′	6 362 222	8.377	5 957 065
30′	6 362 474	8.536	5 937 001
22°00′	6 362 730	8.693	5 916 485
30′	6 362 990	8.847	5 895 521
23°00′	6 363 256	8.998	5 874 109
30′	6 363 526	9.146	5 852 250
24°00′	6 363 800	9.292	5 829 947
30′	6 364 079	9.435	5 807 201
25°00′	6 364 362	9.575	5 784 014
30′	6 364 649	9.712	5 760 387
26°00′	6 364 941	9.846	5 736 323
30′	6 365 236	9.978	5 711 822
27°00′	6 365 535	10.106	5 686 886
30′	6 365 838	10.231	5 661 518
28°00′	6 366 145	10.353	5 635 720
30′	6 366 456	10.472	5 609 493
29°00′	6 366 770	10.588	5 582 838
30′	6 367 088	10.701	5 555 759
30°00′	6 367 409	10.810	5 528 257
30′	6 367 733	10.917	5 500 333
31°00′	6 368 061	11.019	5 471 991
30′	6 368 391	11.119	5 443 232
32°00′	6 368 725	11.215	5 414 058
30′	6 369 061	11.308	5 384 472
33°00′	6 369 400	11.398	5 354 475
30′	6 369 742	11.484	5 324 070
34°00′	6 370 087	11.566	5 293 258
30′	6 370 434	11.645	5 262 043
35°00′	6 370 783	11.721	5 230 427
30′	6 371 135	11.793	5 198 411
36°00′	6 371 489	11.861	5 165 999
30′	6 371 844	11.926	5 133 192
37°00′	6 372 202	11.987	5 099 993
30′	6 372 562	12.045	5 066 405
38°00′	6 372 923	12.098	5 032 429
30′	6 373 286	12.149	4 998 069
39°00′	6 373 651	12.195	4 963 327
30′	6 374 016	12.238	4 928 206
40°00′	6 374 384	12.277	4 892 708
30′	6 374 752	12.313	4 856 835
41°00′	6 375 121	12.344	4 820 591
30′	6 375 492	12.372	4 783 978
42°00′	6 375 863	12.396	4 746 999
30′	6 376 235	12.417	4 709 656
43°00′	6 376 607	12.433	4 671 953
30′	6 376 980	12.446	4 633 892
44°00′	6 377 354	12.455	4 595 476
30′	6 377 727	12.461	4 556 708
45°00′	6 378 101	12.462	4 517 591
30′	6 378 475	12.460	4 478 127
46°00′	6 378 849	12.454	4 438 320
30′	6 379 222	12.444	4 398 173
47°00′	6 379 596	12.430	4 357 688
30′	6 379 968	12.412	4 316 868
48°00′	6 380 341	12.391	4 275 718
30′	6 380 713	12.366	4 234 239
49°00′	6 381 084	12.337	4 192 435
30′	6 381 454	12.304	4 150 309
50°00′	6 381 823	12.268	4 107 864
30′	6 382 191	12.228	4 065 104
51°00′	6 382 558	12.184	4 022 031
30′	6 382 923	12.136	3 978 649
52°00′	6 383 287	12.085	3 934 960
30′	6 383 650	12.030	3 890 970
53°00′	6 384 011	11.971	3 846 680
30′	6 384 370	11.909	3 802 094

ϕ	R_0	$\Delta/30$	r
54°00′	6 384 727	11.843	3 757 215
30′	6 385 082	11.773	3 712 047
55°00′	6 385 436	11.700	3 666 594
30′	6 385 787	11.623	3 620 857
56°00′	6 386 135	11.543	3 574 842
30′	6 386 482	11.459	3 528 552
57°00′	6 386 825	11.371	3 481 990
30′	6 387 167	11.280	3 435 159
58°00′	6 387 505	11.186	3 388 063
30′	6 387 841	11.088	3 340 707
59°00′	6 388 173	10.987	3 293 092
30′	6 388 503	10.882	3 245 224
60°00′	6 388 829	10.774	3 197 105
30′	6 389 152	10.663	3 148 739
61°00′	6 389 472	10.548	3 100 130
30′	6 389 789	10.430	3 051 282
62°00′	6 390 102	10.309	3 002 198
30′	6 390 411	10.185	2 952 883
63°00′	6 390 717	10.058	2 903 339
30′	6 391 018	9.927	2 853 571
64°00′	6 391 316	9.794	2 803 582
30′	6 391 610	9.657	2 753 377
65°00′	6 391 900	9.517	2 702 959
30′	6 392 185	9.375	2 652 332
66°00′	6 392 466	9.230	2 601 499
30′	6 392 743	9.081	2 550 466
67°00′	6 393 016	8.930	2 499 235
30′	6 393 284	8.777	2 447 811
68°00′	6 393 547	8.620	2 396 197
30′	6 393 806	8.461	2 344 398
69°00′	6 394 059	8.299	2 292 417
30′	6 394 308	8.135	2 240 259
70°00′	6 394 552	7.968	2 187 928
30′	6 394 791	7.799	2 135 427
71°00′	6 395 025	7.627	2 082 760
30′	6 395 254	7.453	2 029 932

ϕ	R_0	$\Delta/30$	r
72°00′	6 395 478	7.276	1 976 947
30′	6 395 696	7.098	1 923 809
73°00′	6 395 909	6.917	1 870 521
30′	6 396 116	6.734	1 817 089
74°00′	6 396 318	6.549	1 763 516
30′	6 396 515	6.362	1 709 806
75°00′	6 396 706	6.172	1 655 963
30′	6 396 891	5.981	1 601 992
76°00′	6 397 070	5.789	1 547 897
30′	6 397 244	5.594	1 493 681
77°00′	6 397 412	5.398	1 439 350
30′	6 397 574	5.200	1 384 907
78°00′	6 397 730	5.000	1 330 357
30′	6 397 880	4.799	1 275 703
79°00′	6 398 024	4.596	1 220 950
30′	6 398 162	4.392	1 166 103
80°00′	6 398 293	4.186	1 111 165
30′	6 398 419	3.979	1 056 141
81°00′	6 398 538	3.771	1 001 034
30′	6 398 651	3.562	945 851
82°00′	6 398 758	3.352	890 593
30′	6 398 859	3.141	835 267
83°00′	6 398 953	2.928	779 875
30′	6 399 041	2.715	724 423
84°00′	6 399 122	2.501	668 915
30′	6 399 197	2.286	613 355
85°00′	6 399 266	2.070	557 747
30′	6 399 328	1.854	502 096
86°00′	6 399 384	1.637	446 406
30′	6 399 433	1.420	390 681
87°00′	6 399 476	1.202	334 926
30′	6 399 512	0.984	279 145
88°00′	6 399 541	0.766	223 342
30′	6 399 564	0.547	167 521
89°00′	6 399 581	0.328	111 688
30′	6 399 590	0.109	55 846
90°00′	6 399 594		0

$R_0 = (MN)^{1/2}$
$r = N \cos \phi$
See Appendix 2 for M and N.

Appendix 4

Arc length s(M) of the meridian from the equator to the parallel of latitude ϕ; arc length of meridians for 30′ of latitude; arc length of parallels for 30′ of longitude; values of N cot ϕ. GRS 80 ellipsoid

Lat. ϕ	Arc length of meridian from the equator to the parallel of latitude ϕ (m)	Arc length of meridian for 30′ of latitude (m)	Arc length of parallel for 30′ of longitude (m)	$N \cot \phi$
0°00′	0	55 287	55 660	∞
30′	55 287	55 287	55 658	730 862.295
1°00′	110 574	55 287	55 651	365 403.597
30′	165 862	55 288	55 641	243 571.784
2°00′	221 149	55 288	55 626	182 646.691
30′	276 437	55 288	55 607	146 084.285
3°00′	331 726	55 289	55 584	121 703.220
30′	387 015	55 289	55 557	104 282.919
4°00′	442 304	55 290	55 525	91 213.094
30′	497 595	55 291	55 489	81 043.584
5°00′	552 885	55 291	55 449	72 904.293
30′	608 177	55 292	55 405	66 241.522
6°00′	663 470	55 293	55 357	60 686.139
30′	718 764	55 294	55 304	55 982.591
7°00′	774 058	55 295	55 248	51 948.340
30′	829 354	55 297	55 187	48 449.523
8°00′	884 652	55 298	55 122	45 385.745
30′	939 950	55 299	55 052	42 680.232
9°00′	995 250	55 301	54 979	40 273.271
30′	1 050 552	55 302	54 901	38 117.720
10°00′	1 105 855	55 304	54 820	36 175.864
30′	1 161 160	55 306	54 734	34 417.174
11°00′	1 216 466	55 307	54 644	32 816.670
30′	1 271 774	55 309	54 550	31 353.717
12°00′	1 327 084	55 311	54 451	30 011.118
30′	1 382 397	55 313	54 349	28 774.429
13°00′	1 437 711	55 315	54 242	27 631.427
30′	1 493 027	55 317	54 132	26 571.700
14°00′	1 548 346	55 320	54 017	25 586.323
30′	1 603 666	55 322	53 898	24 667.603
15°00′	1 658 990	55 324	53 775	23 808.870
30′	1 714 315	55 327	53 648	23 004.319
16°00′	1 769 643	55 329	53 517	22 248.866
30′	1 824 974	55 332	53 382	21 538.046
17°00′	1 880 307	55 335	53 243	20 867.918
30′	1 935 643	55 337	53 100	20 234.992
18°00′	1 990 982	55 340	52 952	19 636.165
30′	2 046 324	55 343	52 801	19 068.669
19°00′	2 101 668	55 346	52 646	18 530.030
30′	2 157 016	55 349	52 487	18 018.025
20°00′	2 212 366	55 352	52 324	17 530.653
30′	2 267 720	55 355	52 156	17 066.110
21°00′	2 323 077	55 359	51 985	16 622.762
30′	2 378 437	55 362	51 810	16 199.129
22°00′	2 433 800	55 365	51 631	15 793.863
30′	2 489 167	55 369	51 448	15 405.738

Lat. ϕ	Arc length of meridian from the equator to the parallel of latitude ϕ (m)	Arc length of meridian for 30′ of latitude (m)	Arc length of parallel for 30′ of longitude (m)	$N \cot \phi$
23°00′	2 544 538	55 372	51 261	15 033.634
30′	2 599 911	55 376	51 071	14 676.524
24°00′	2 655 289	55 379	50 876	14 333.469
30′	2 710 670	55 383	50 677	14 003.607
25°00′	2 766 054	55 386	50 475	13 686.143
30′	2 821 442	55 390	50 269	13 380.346
26°00′	2 876 835	55 394	50 059	13 085.538
30′	2 932 230	55 398	49 845	12 801.097
27°00′	2 987 630	55 402	49 627	12 526.443
30′	3 043 034	55 406	49 406	12 261.041
28°00′	3 098 442	55 410	49 181	12 004.390
30′	3 153 853	55 414	48 952	11 756.030
29°00′	3 209 269	55 418	48 719	11 515.527
30′	3 264 689	55 422	48 483	11 282.480
30°00′	3 320 113	55 426	48 243	11 056.513
30′	3 375 542	55 430	47 999	10 837.276
31°00′	3 430 974	55 435	47 752	10 624.440
30′	3 486 411	55 439	47 501	10 417.698
32°00′	3 541 852	55 443	47 247	10 216.761
30′	3 597 298	55 448	46 988	10 021.358
33°00′	3 652 748	55 452	46 727	9 831.236
30′	3 708 202	55 457	46 461	9 646.155
34°00′	3 763 661	55 461	46 192	9 465.890
30′	3 819 125	55 466	45 920	9 290.228
35°00′	3 874 593	55 470	45 644	9 118.971
30′	3 930 065	55 475	45 365	8 951.928
36°00′	3 985 543	55 480	45 082	8 788.922
30′	4 041 024	55 484	44 796	8 629.784
37°00′	4 096 511	55 489	44 506	8 474.353
30′	4 152 002	55 494	44 213	8 322.480
38°00′	4 207 498	55 498	43 916	8 174.020
30′	4 262 999	55 503	43 616	8 028.838
39°00′	4 318 504	55 508	43 313	7 886.805
30′	4 374 014	55 513	43 007	7 747.799
40°00′	4 429 529	55 517	42 697	7 611.702
30′	4 485 049	55 522	42 384	7 478.404
41°00′	4 540 573	55 527	42 068	7 347.801
30′	4 596 103	55 532	41 748	7 219.791
42°00′	4 651 637	55 537	41 425	7 094.279
30′	4 707 176	55 542	41 100	6 971.173
43°00′	4 762 720	55 546	40 770	6 850.388
30′	4 818 269	55 551	40 438	6 731.839
44°00′	4 873 822	55 556	40 103	6 615.448
30′	4 929 381	55 561	39 765	6 501.139
45°00′	4 984 944	55 566	39 423	6 388.838
30′	5 040 513	55 571	39 079	6 278.478
46°00′	5 096 086	55 576	38 732	6 169.991
30′	5 151 664	55 581	38 381	6 063.314
47°00′	5 207 247	55 585	38 028	5 958.386
30′	5 262 835	55 590	37 672	5 855.149

Lat. ϕ	Arc length of meridian from the equator to the parallel of latitude ϕ (m)	Arc length of meridian for 30′ of latitude (m)	Arc length of parallel for 30′ of longitude (m)	$N \cot \phi$
48°00′	5 318 428	55 595	37 313	5 753.546
30′	5 374 025	55 600	36 951	5 653.524
49°00′	5 429 628	55 605	36 586	5 555.031
30′	5 485 235	55 610	36 218	5 458.017
50°00′	5 540 847	55 615	35 848	5 362.436
30′	5 596 464	55 619	35 475	5 268.240
51°00′	5 652 086	55 624	35 099	5 175.386
30′	5 707 712	55 629	34 720	5 083.832
52°00′	5 763 344	55 634	34 339	4 993.537
30′	5 818 980	55 638	33 955	4 904.460
53°00′	5 874 620	55 643	33 569	4 816.565
30′	5 930 266	55 648	33 180	4 729.814
54°00′	5 985 916	55 653	32 788	4 644.173
30′	6 041 571	55 657	32 394	4 559.607
55°00′	6 097 230	55 662	31 997	4 476.084
30′	6 152 894	55 666	31 598	4 393.572
56°00′	6 208 563	55 671	31 196	4 312.039
30′	6 264 236	55 675	30 792	4 231.457
57°00′	6 319 914	55 680	30 386	4 151.797
30′	6 375 596	55 684	29 977	4 073.030
58°00′	6 431 283	55 689	29 566	3 995.131
30′	6 486 974	55 693	29 153	3 918.073
59°00′	6 542 669	55 698	28 738	3 841.831
30′	6 598 369	55 702	28 320	3 766.381
60°00′	6 654 073	55 706	27 900	3 691.698
30′	6 709 781	55 710	27 478	3 617.761
61°00′	6 765 494	55 715	27 054	3 544.546
30′	6 821 210	55 719	26 627	3 472.033
62°00′	6 876 931	55 723	26 199	3 400.200
30′	6 932 656	55 727	25 769	3 329.027
63°00′	6 988 385	55 731	25 336	3 258.493
30′	7 044 117	55 735	24 902	3 188.581
64°00′	7 099 854	55 739	24 466	3 119.271
30′	7 155 595	55 743	24 028	3 050.545
65°00′	7 211 339	55 746	23 588	2 982.385
30′	7 267 087	55 750	23 146	2 914.774
66°00′	7 322 839	55 754	22 702	2 847.696
30′	7 378 595	55 757	22 257	2 781.133
67°00′	7 434 354	55 761	21 810	2 715.070
30′	7 490 116	55 764	21 361	2 649.491
68°00′	7 545 883	55 768	20 911	2 584.382
30′	7 601 652	55 771	20 459	2 519.727
69°00′	7 657 425	55 775	20 005	2 455.511
30′	7 713 201	55 778	19 550	2 391.722
70°00′	7 768 981	55 781	19 093	2 328.344
30′	7 824 763	55 784	18 635	2 265.365
71°00′	7 880 549	55 787	18 176	2 202.770
30′	7 936 338	55 790	17 715	2 140.548
72°00′	7 992 129	55 793	17 252	2 078.685
30′	8 047 924	55 796	16 788	2 017.170

Lat. ϕ	Arc length of meridian from the equator to the parallel of latitude ϕ (m)	Arc length of meridian for 30′ of latitude (m)	Arc length of parallel for 30′ of longitude (m)	$N \cot \phi$
73°00′	8 103 721	55 799	16 323	1 955.989
30′	8 159 521	55 801	15 857	1 895.131
74°00′	8 215 324	55 804	15 390	1 834.584
30′	8 271 130	55 807	14 921	1 774.337
75°00′	8 326 938	55 809	14 451	1 714.379
30′	8 382 748	55 812	13 980	1 654.698
76°00′	8 438 561	55 814	13 508	1 595.283
30′	8 494 376	55 816	13 035	1 536.124
77°00′	8 550 193	55 818	12 561	1 477.211
30′	8 606 013	55 821	12 086	1 418.532
78°00′	8 661 834	55 823	11 610	1 360.078
30′	8 717 658	55 825	11 133	1 301.838
79°00′	8 773 483	55 826	10 655	1 243.802
30′	8 829 311	55 828	10 176	1 185.962
80°00′	8 885 140	55 830	9 697	1 128.306
30′	8 940 971	55 832	9 217	1 070.826
81°00′	8 996 803	55 833	8 736	1 013.512
30′	9 052 637	55 835	8 254	956.355
82°00′	9 108 472	55 836	7 772	899.346
30′	9 164 309	55 837	7 289	842.474
83°00′	9 220 147	55 839	6 806	785.732
30′	9 275 986	55 840	6 322	729.110
84°00′	9 331 827	55 841	5 837	672.600
30′	9 387 668	55 842	5 353	616.192
85°00′	9 443 510	55 843	4 867	559.878
30′	9 499 353	55 844	4 382	503.648
86°00′	9 555 197	55 844	3 896	447.496
30′	9 611 042	55 845	3 409	391.411
87°00′	9 666 887	55 845	2 923	335.385
30′	9 722 733	55 846	2 436	279.411
88°00′	9 778 579	55 846	1 949	223.478
30′	9 834 425	55 847	1 462	167.579
89°00′	9 890 272	55 847	975	111.705
30′	9 946 119	55 847	487	55.848
90°00′	10 001 966	55 847	0	0

Appendix 5

Isometric latitudes: values of ln *U. Zone areas: area S of zone from equator to latitude per radian of longitude. GRS 80 ellipsoid*

ϕ	ln U	S (in 10 km^2)
0°00′	0.000 000 0	0.0
30′	0.008 668 3	35 262.5
1°00′	0.017 337 3	70 522.3
30′	0.026 007 7	105 776.9
2°00′	0.034 680 0	141 023.7
30′	0.043 355 1	176 260.0
3°00′	0.052 033 5	211 483.3
30′	0.060 715 9	246 690.9
4°00′	0.069 403 0	281 880.2
30′	0.078 095 4	317 048.6
5°00′	0.086 794 0	352 193.5
30′	0.095 499 2	387 312.3
6°00′	0.104 211 9	422 402.3
30′	0.112 932 7	457 461.1
7°00′	0.121 662 3	492 485.9
30′	0.130 401 3	527 474.3
8°00′	0.139 150 5	562 423.5
30′	0.147 910 6	597 330.9
9°00′	0.156 682 3	632 194.1
30′	0.165 466 3	667 010.4
10°00′	0.174 263 3	701 777.2
30′	0.183 074 0	736 491.9
11°00′	0.191 899 1	771 152.0
30′	0.200 739 5	805 754.9
12°00′	0.209 595 7	840 297.9
30′	0.218 468 6	874 778.6
13°00′	0.227 358 9	909 194.3
30′	0.236 267 4	943 542.5
14°00′	0.245 194 7	977 820.7
30′	0.254 141 8	1 012 026.2
15°00′	0.263 109 4	1 046 156.5
30′	0.272 098 1	1 080 209.2
16°00′	0.281 109 0	1 114 181.5
30′	0.290 142 7	1 148 071.1
17°00′	0.299 200 1	1 181 875.3
30′	0.308 281 9	1 215 591.6
18°00′	0.317 389 1	1 249 217.6
30′	0.326 522 5	1 282 750.6
19°00′	0.335 682 9	1 316 188.3
30′	0.344 871 2	1 349 528.0
20°00′	0.354 088 3	1 382 767.3
30′	0.363 335 0	1 415 903.6
21°00′	0.372 612 4	1 448 934.6
30′	0.381 921 2	1 481 857.7
22°00′	0.391 262 4	1 514 670.4
30′	0.400 637 1	1 547 370.2
23°00′	0.410 046 0	1 579 954.8
30′	0.419 490 3	1 612 421.7
24°00′	0.428 970 9	1 644 768.3
30′	0.438 488 8	1 676 992.3
25°00′	0.448 045 0	1 709 091.2
30′	0.457 640 7	1 741 062.7
26°00′	0.467 276 8	1 772 904.2
30′	0.476 954 5	1 804 613.5
27°00′	0.486 674 8	1 836 188.0
30′	0.496 438 9	1 867 625.4
28°00′	0.506 247 9	1 898 923.3
30′	0.516 103 0	1 930 079.3
29°00′	0.526 005 5	1 961 091.1
30′	0.535 956 4	1 991 956.3
30°00′	0.545 957 1	2 022 672.6
30′	0.556 008 8	2 053 237.6
31°00′	0.566 112 8	2 083 649.1
30′	0.576 270 5	2 113 904.5
32°00′	0.586 483 2	2 144 001.8
30′	0.596 752 2	2 173 938.5
33°00′	0.607 079 1	2 203 712.5
30′	0.617 465 2	2 233 321.3
34°00′	0.627 912 1	2 262 762.8
30′	0.638 421 1	2 292 034.6
35°00′	0.648 994 0	2 321 134.6
30′	0.659 632 2	2 350 060.5
36°00′	0.670 337 6	2 378 810.1
30′	0.681 111 6	2 407 381.2
37°00′	0.691 956 0	2 435 771.6
30′	0.702 872 5	2 463 979.0
38°00′	0.713 863 0	2 492 001.4
30′	0.724 929 3	2 519 836.5
39°00′	0.736 073 4	2 547 482.2
30′	0.747 297 2	2 574 936.4
40°00′	0.758 602 6	2 602 196.9
30′	0.769 991 8	2 629 261.7
41°00′	0.781 466 8	2 656 128.6
30′	0.793 029 9	2 682 795.5
42°00′	0.804 683 3	2 709 260.5
30′	0.816 429 3	2 735 521.3
43°00′	0.828 270 4	2 761 576.1
30′	0.840 208 8	2 787 422.7
44°00′	0.852 247 3	2 813 059.1
30′	0.864 388 3	2 838 483.4
45°00′	0.876 634 7	2 863 693.5
30′	0.888 989 0	2 888 687.5
46°00′	0.901 454 4	2 913 463.5
30′	0.914 033 6	2 938 019.4
47°00′	0.926 729 8	2 962 353.4
30′	0.939 546 1	2 986 463.5
48°00′	0.952 485 8	3 010 347.9
30′	0.965 552 3	3 034 004.7
49°00′	0.978 749 1	3 057 432.1
30′	0.992 079 8	3 080 628.1
50°00′	1.005 548 3	3 103 591.0
30′	1.019 158 3	3 126 318.9
51°00′	1.032 913 9	3 148 810.1
30′	1.046 819 4	3 171 062.7
52°00′	1.060 879 1	3 193 075.0
30′	1.075 097 6	3 214 845.4
53°00′	1.089 479 5	3 236 372.0
30′	1.104 029 7	3 257 653.1

ϕ	ln U	S (in 10 km^2)
54°00′	1.118 753 4	3 278 687.0
30′	1.133 655 9	3 299 472.2
55°00′	1.148 742 6	3 320 006.9
30′	1.164 019 4	3 340 289.5
56°00′	1.179 492 3	3 360 318.3
30′	1.195 167 5	3 380 091.9
57°00′	1.211 051 6	3 399 608.6
30′	1.227 151 5	3 418 866.8
58°00′	1.243 474 3	3 437 865.0
30′	1.260 027 6	3 456 601.7
59°00′	1.276 819 2	3 475 075.4
30′	1.293 857 3	3 493 284.7
60°00′	1.311 150 7	3 511 228.0
30′	1.328 708 3	3 528 903.9
61°00′	1.346 539 8	3 546 311.0
30′	1.364 655 1	3 563 447.9
62°00′	1.383 064 9	3 580 313.2
30′	1.401 780 2	3 596 905.6
63°00′	1.420 812 9	3 613 223.7
30′	1.440 175 3	3 629 266.3
64°00′	1.459 880 6	3 645 031.9
30′	1.479 942 5	3 660 519.4
65°00′	1.500 375 9	3 675 727.6
30′	1.521 196 2	3 690 655.0
66°00′	1.542 420 1	3 705 300.7
30′	1.564 065 0	3 719 663.3
67°00′	1.586 149 8	3 733 741.7
30′	1.608 694 3	3 747 534.8
68°00′	1.631 719 8	3 761 041.5
30′	1.655 249 1	3 774 260.6
69°00′	1.679 306 5	3 787 191.1
30′	1.703 918 2	3 799 832.0
70°00′	1.729 112 1	3 812 182.2
30′	1.754 918 4	3 824 240.6
71°00′	1.781 369 7	3 836 006.4
30′	1.808 501 0	3 847 478.6
72°00′	1.836 350 4	3 858 656.2
30′	1.864 959 0	3 869 538.3
73°00′	1.894 371 7	3 880 124.1
30′	1.924 637 2	3 890 412.6
74°00′	1.955 808 8	3 900 403.0
30′	1.987 944 9	3 910 094.6
75°00′	2.021 109 6	3 919 486.5
30′	2.055 373 6	3 928 577.9
76°00′	2.090 814 8	3 937 368.2
30′	2.127 519 8	3 945 856.5
77°00′	2.165 585 2	3 954 042.3
30′	2.205 118 4	3 961 924.8
78°00′	2.246 240 6	3 969 503.3
30′	2.289 088 0	3 976 777.4
79°00′	2.333 815 1	3 983 746.3
30′	2.380 598 2	3 990 409.4
80°00′	2.429 639 0	3 996 766.4
30′	2 481 170 7	4 002 816.5
81°00′	2.535 464 0	4 008 559.4
30′	2.592 836 2	4 013 994.6
82°00′	2.653 662 3	4 019 121.5
30′	2.718 390 1	4 023 939.9
83°00′	2.787 559 9	4 028 449.2
30′	2.861 832 6	4 032 649.2
84°00′	2.942 027 8	4 036 539.4
30′	3.029 179 4	4 040 119.6
85°00′	3.124 617 6	4 043 389.5
30′	3.230 093 9	4 046 348.8
86°00′	3 347 980 5	4 048 997.2
30′	3.481 603 3	4 051 334.5
87°00′	3.635 833 2	4 053 360.6
30′	3.818 221 8	4 055 075.3
88°00′	4.041 420 1	4 056 478.4
30′	4.329 144 8	4 057 569.8
89°00′	4.734 640 4	4 058 349.5
30′	5.427 805 9	4 058 817.3
90°00′	∞	4 058 973.2

Appendix 6

A. Dimensions of the Earth ellipsoid

Ellipsoid	a (m) b (m)	Flattening $f = (a - b)/a$	Countries using the reference ellipsoid
GRS 80	6 378 137.0 6 356 752.3	1/298.26	Satellite determined, internationally adopted
Krasovskiy (1940)	6 378 245.0 6 356 863.0	1/298.30	Former socialist states, Antarctic continent, China
Bessel (1841)	6 377 397.2 6 356 079.0	1/299.15	Europe and Asia
Hayford (1909) (Int. 1924)	6 378 388.0 6 356 911.9	1/297.00	Europe, Asia, South America, Antarctic continent
Clarke (1880)	6 378 249.1 6 356 514.9	1/293.46	Africa, Barbados, Israel, Jordan, Iran, Jamaica
Clarke (1866)	6 378 206.4 6 356 583.8	1/294.98	North and Central America
Airy (1830)	6 377 491 6 356 185	1/299.32	Great Britain, Ireland
Everest (1830)	6 377 276.3 6 356 075.4	1/300.80	India, Pakistan, Nepal, Sri Lanka
	e^2	e'^2	
GRS 80	0.006 694 4	0.006 739 5	—
Krasovskiy	0.006 693 4	0.006 738 5	—
Bessel	0.006 674 3	0.006 719 2	—
Hayford	0.006 722 7	0.006 768 2	—
Clarke (1866)	0.006 768 7	0.006 814 8	—

$e^2 = (a^2 - b^2)/a^2$, $e'^2 = (a^2 - b^2)/b^2 = e^2/(1 - e^2)$
a = semimajor axis, b = semiminor axis
e = first eccentricity, e' = second eccentricity

B. Radius R of a sphere representing the Earth ellipsoid. GRS 80 ellipsoid

Type of transformation	Latitude on sphere	Sphere radius R (m)
Conformal	ϕ'	6 378 137
Equal area	ϕ''	6 371 007
Equidistant along meridian	ϕ'''	6 367 449

Mars is assumed to be an ellipsoid of revolution with semiaxes $a = 3394.5$, $b = 3376.4$ km, Venus and the Moon as spheres of radii 6051 and 1738 km, respectively. Conformal radius is taken as a, but different values do not affect conformality.

Appendix 7

Mathematical constants

π		3.141 592 653 589 793 238
π^2		9.869 604 401 089 358 619
$\sqrt{\pi}$		1.772 453 850 905 516 027
$1/\pi$		0.318 309 886 183 790 671 5
e		2.718 281 828 459 045 235
M (modulus of common logs)		0.434 294 481 903 251 827 6
Circumference of circle		
degrees	360	
angular minutes	21 600	
angular seconds	1 296 000	
Radius of circle		
$\rho^\circ = \frac{360}{2\pi}$ degrees		57.295 779 513 082 320 88
$\rho' = \frac{360}{2\pi} \times 60$, angular minutes		3 437.746 770 784 939 253
$\rho'' = \frac{360}{2\pi} \times 60 \times 60$, angular seconds		206 264.806 247 096 355 2
arc $1^\circ = 1/\rho^\circ$ radians		0.017 453 292 519 943 295 77
arc $1' = 1/\rho'$ radians		0.000 290 888 208 665 721 60
arc $1'' = 1/\rho''$ radians		0.000 004 848 136 811 095 36

Appendix 8

Map projection types in different languages

Russian	English	French	German*
Azimutal'naya proyektsiya	Azimuthal projection	Projection azimutale	Azimutalabbildung
Konicheskaya proyektsiya	Conic(al) projection	Projection conique	Kegelabbildung
Polikonicheskaya proyektsiya	Polyconic(al) projection	Projection polyconique	Abbildung polykonische
Mnogogrannaya proyektsiya	Polyhedric projection	Projection polyedrique	Polyederabbildung
Proyektsiya Gaussa–Kryugera	Gauss–Krüger projection	Projection de Gauss–Krüger	Gauss–Krüger-Abbildung
Tsilindricheskaya proyektsiya	Cylindrical projection	Projection cylindrique	Zylinderabbildung
Proyektsiya Merkatora	Mercator projection	Projection de Mercator	Mercatorabbildung
Proizvol'naya proyektsiya	Arbitrary projection	Projection aphylactique	Abbildung vermittelnde
Normal'naya proyektsiya	Normal aspect of a map projection	Projection directe	Abbildung normalachsig
Poperechnaya proyektsiya	Transverse aspect of a map projection	Projection transverse	Abbildung querachsig
Kosaya proyektsiya	Oblique aspect of a map projection	Projection oblique	Abbildung schiefachsig
Ravnougol'naya proyektsiya	Conformal projection	Projection conforme	Abbildung konforme
Ravnovelikaya proyektsiya	Equal-area (equivalent) projection	Projection equivalente	Abbildung flächentreue
Ravnopromezhutochnaya proyektsiya	Equidistant projection	Projection equidistante	Abbildung abstandstreue
Ravnopromezhutochnaya po parallelyam proyektsiya	—	Projection d'echelle constante le long des paralleles	Abbildung abweitungstreue
Ravnopromezhutochnaya azimutal'naya proyektsiya	Azimuthal equidistant projection	Projection azimutale equidistante	Abbildung mittabstandstreue

* Note: The terms *Projektion* and *Entwurf* are also frequently used in German in place of *Abbildung*.

Index

See Chapter 11 and references on pages 279–294 for names not listed here.